Advances in Intelligent Systems and Computing

Volume 229

Series Editor

Janusz Kacprzyk, Warsaw, Poland

For further volumes:
http://www.springer.com/series/11156

Bogumił Kamiński · Grzegorz Koloch
Editors

Advances in Social Simulation

Proceedings of the 9th Conference of the European Social Simulation Association

Editors
Bogumił Kamiński
Division of Decision Analysis and Support
Institute of Econometrics
Warsaw School of Economics
Warsaw
Poland

Grzegorz Koloch
Division of Decision Analysis and Support
Institute of Econometrics
Warsaw School of Economics
Warsaw
Poland

ISSN 2194-5357 ISSN 2194-5365 (electronic)
ISBN 978-3-642-39828-5 ISBN 978-3-642-39829-2 (eBook)
DOI 10.1007/978-3-642-39829-2
Springer Heidelberg New York Dordrecht London

Library of Congress Control Number: 2013944366

Printed on acid-free paper

Springer is part of Springer Science+Business Media (www.springer.com)

Preface

This book is the conference proceedings of ESSA 2013, the 9th Conference of the European Social Simulation Association. ESSA conferences constitute annual events which serve as an international platform for the exchange of ideas and discussion of cutting-edge research in the field of social simulations, both from the theoretical as well as the applied perspective. This year's conference attracted 120 presentation submissions out of which 33 were selected for publication in the proceedings. We would like to express our thanks to members of the conference Scientific Committee for their excellent expertise in the process of reviewing and selection of the articles for the proceedings.

This book consists of four parts covering a wide scope of current advances of theoretical and applied social simulation research.

The first part, entitled *Methods for the development of simulation models,* consists of 8 articles, which present the theoretical grounding for the validation of social networks, the empirical validation of agent based models, testing models robustness, analyzing their quality, modeling periodic normalized behavior and the specification of models based on argumentative data.

The second part collects 14 articles, which involve *Applications of agent-based modeling* in a wide variety of disciplines mainly focusing on sociology, micro- and macroeconomics, economics of public goods and game theory. In this section one can find studies on the effects of tax cuts on the valuation of firms, the higher order theory of mind, the analysis of cooperation in public goods game, minority games, adaptive learning under incomplete information and fictitious play, the interaction of social conformity and orientation, simulating innovation, moral guilt, the estimation of production rate limits, macroeconomic, bottom-up dynamics, shadow economy, innovation diffusion, journal publication and disease transmission.

Part three provides insights into the modeling of *Adaptive behavior, social interactions and global environmental change* and consists of 6 articles. The works presented concern such issues as evolving societies, diffusion dynamics, adaptive economics behavior, simulation of opinion dynamics and drought management.

The last part is constituted of 7 papers organized around the problem of *Using qualitative data to inform behavioral rules.* Articles collected here explore such topics as grounded simulation, model generation based on qualitative data, narratives of drought, context- and scope-sensitive analysis of behavior, water governance, ethnographic decision tree modeling and biomass use modeling.

The scope of articles presented in the present proceedings encompasses a wide range of issues ranging from methodological studies to explicitly applied research projects. This proves that, while the research agenda of social simulation has grown extensively, it is not satiated, but successfully aims at exploring new venues. The accepted papers show that cooperation between social scientists, computer scientists and scholars and practitioners from other disciplines using computational models and methodologies can be successfully applied to advance social understanding. We are convinced that this volume will serve as a useful compendium which presents in a nutshell the most recent advances at the frontiers of social simulation research.

June 2013

Bogumił Kamiński
Grzegorz Koloch

Contents

Towards Validating Social Network Simulations

Syed Muhammad Ali Abbas[1], Shah Jamal Alam[2], and Bruce Edmonds[1]

[1] Centre for Policy Modelling, Manchester Metropolitan University
ali@cfpm.org, bruce@edmonds.name
[2] School of Geosciences, University of Edinburgh
sj.alam@ed.ac.uk

Abstract. We consider the problem of finding suitable measures to validate simulated networks as outcome of (agent-based) social simulations. A number of techniques from computer science and social sciences are reviewed in this paper, which tries to compare and 'fit' various simulated networks to the available data by using network measures. We look at several social network analysis measures but then turn our focus to techniques that not only consider the position of the nodes but also their characteristics and their tendency to cluster with other nodes in the network – subgroup identification. We discuss how static and dynamic nature of networks may be compared. We conclude by urging a more comprehensive, transparent and rigorous approach to comparing simulation-generated networks against the available data.

Keywords: networks, validation, measures, ABM, underdetermination.

1 Introduction

Agent-based models (ABM) provide a method of representing local interactions using the interactions between software agents [16]. Agents' interactions can lead to the formation of ties with other agents and thus the generation of simulated networks. Often, the social processes governing the agents' interactions influence the evolution of such networks. In such cases, identifying the generative mechanisms is non-trivial, but ABM can help do this, leading to an understanding of emerging network structures [14, 30]. Moreover, where the agents form several types of relations, multiple overlapping networks result.

Validating agent-based simulations is hard. Their relative complexity means that it is easier to 'fit' simple sets of empirical data so that a simple comparison of one or two measures with empirical data is not sufficient to ensure that a simulation is an adequate representation of what is being modeled. Different agent-based simulations might well end up producing the same outputs if only compared in a few ways. The more ways, in which a simulation is measured, especially if these are of very different aspects, the greater the chance that any differences in simulation composition will be revealed in a significant difference in the simulation outputs.

Many agent-based simulations directly generate explicit social networks, e.g., [1, 2, 4, 10, 14, 19, 20, 24, 35, 38]. There are many other simulations where relatively stable

B. Kamiński and G.Koloch (eds.), *Advances in Social Simulation*,
Advances in Intelligent Systems and Computing 229,
DOI: 10.1007/978-3-642-39829-2_1, © Springer-Verlag Berlin Heidelberg 2014

patterns of interaction emerge, from which networks might be inferred. In either case, this raises the possibility of attempting to validate these simulation-generated networks against observed real-world networks. Such network validation would be another way in which to check simulations, and hence increase our confidence in them. This paper reviews some of such approaches that may be useful for social simulators.

2 The Central Problem

Networks have a high degree of freedom. There are $2^{\frac{n(n-1)}{2}}$ ways of connecting n nodes with undirected edges, even when multiple edges, undirected edges and self-loops are not allowed. Thus, there are many more ways of joining 25 nodes into such a network than atoms in the universe (which is around 10^{80}, according to [34]). A single measure projects the set of possible networks onto a single dimension, thus it is inevitable that a great many different networks will be projected onto small ranges of the measure. Thus, a lack of divergence of networks with respect to any particular measure will not be sufficient to distinguish between many distinct networks.

Of course, one rarely wants to prove that one has exactly the same network but rather a similar one (in some sense); however, this does not change the *scaling* of the problem. While the tactic of increasing the number of measures used for comparison may be effective for relatively small networks, it is unlikely to be feasible for 'large' networks. However, since many permutations of a network exist that are structurally equivalent [39], this can be used to reduce the number of networks that need to be considered. One such attempt has been to study the 'universality classes' of the structural mcasures/statistics from the network science community but such methods are rarely addressed in the analysis of agent-based simulated networks [23].

As with any simulation output, not all of its aspects can be considered relevant or significant with respect to what is being modeled. Some identifiable aspects of the observed network might be due to the underlying social mechanisms but others merely due to happenstance (i.e., factors irrelevant to the purpose of the model). Such a scenario is discussed in [1, 2, 3], where simulated networks from an agent-based model of friendship networks are compared to a real data of Facebook™ users in a college campus [1 , 2, 3] . Moreover, not all aspects of the network will have been intended to be reproduced by the modeler. Thus, when developing agent-based models that generate (social) networks, it is important to specify:

1. Which class of networks one might expect to observe if one could "re-run" reality under the same basic conditions as assumed in the model,
2. What properties of the simulated networks, one would expect to observe given how the model has been constructed.

The first comes down to what assumptions one is making about the target social system, which (one hopes) is informed by a multitude of empirical sources. The second depends upon the first but also one's purpose for developing the model and may be constrained by the available resources.

3 Social Network Analysis Measures

Using social network analysis (SNA) measures to evaluate graphs that are output from simulations is common. However, a single measure (or a small set of measures) is unlikely to establish the validity of a simulated network. When a particular measure captures some aspect that is crucial to the kind of process being investigated, the implicit assumption is that the closer the value of the measure on the synthetic graph is to that of the reference graph, the more similar the two graphs will be in that respect. How well this works depends upon what we mean by the 'closeness' of a measure when comparing networks. For instance, [21] reports on that centrality measures degrade gracefully with the introduction of random changes in the network, however, this result might depend on the topology of the network [27]. Below, we briefly preview some of the most popular approaches [32, 37] used by agent-based modelers. For a more detailed review on networks in agent-based simulations, we refer to [5, 6], and for formal definitions and extensive discussions of network measures [31, 37].

Node Degree Distribution: is the number of edges each node has in a graph. For a directed graph, this includes incoming, outgoing and total edges; while for undirected graphs, it means just the total edges of each node has. In order to determine the node degree distribution of a network, one can measure its deviance from another distribution using such measures, such as the Least Square Error (LSE) or the Kolmogorov-Smirnov statistic. When a theoretical distribution is assumed, against which the distribution of a simulated network is compared, maximum likelihood estimation (MLE) techniques are useful. The degree distribution gives a good idea of the prevalence of different kinds of node but can be inadequate for validation purposes (e.g., [35]).

Assortativity Mixing: assortative mixing means that nodes with similar degrees tend to be connected with each other, calculated by working out the correlation of the degrees of all connected node pairs in the graph. It ranges from -1 to 1: a positive value indicating that nodes tend to connect with others with similar degrees and a negative value otherwise [7]. Usually, the averaged value of all node pairs is calculated for the whole graph for comparison – known as the global clustering coefficient. A value approaching 1 would result from a graph where there are uniformly densely interconnected nodes and other areas that are uniformly but sparsely connected. A negative value might result from a graph where nodes with a high degree are distributed away from each other. This measure is useful when the local uniformity of degree distribution (or otherwise) is being investigated.

Cluster Coefficient: is the ratio of the number of edges that exist between a node's immediate neighbors and the maximum number of edges that could exist between them. If all of one's friends know one another this would result in a cluster coefficient of 1; in the other extreme, where none knew each other, would result in a value of 0. This measure is important, for instance, if one had hypothesized that, the process of making new friends is driven by the "friend of a friend". The local cluster coefficients can be displayed as a distribution or a simple average. This measure might be useful when the empirical evidence being compared against was in the form of "ego nets" [39].

Geodesic Distance: The general degree of node separation in a graph is usually measured through three metrics: average path length, network radius and network diameter. A geodesic distance is the shortest path between any two nodes (in "hops"). The average path length is the average of all-geodesic distances on the graph [2]. The diameter is the largest geodesic distance within the graph. For each node, the eccentricity is the geodesic distance to the node furthest from it; the radius is then the minimum eccentricity in the graph. Geodesic distances are important in the presence of "flood-fill" gossip mechanisms, where messages are passed on to all of a node's neighbors. The radius of a graph is a lower bound for the "time" (network jumps) for a message to reach all other nodes, the average path length the average "time" for nodes to get the message, and the diameter giving an upper bound.

Similarity Measures: If a set of nodes is the same in two different graphs, we can calculate the hamming distance between their adjacency matrices to determine as how many edges are different in each of them [26, 37]. For example, in [1], similarity between snapshots of the simulated and observed networks was compared in which the number of agents in the simulated network matched the number of real individuals in the observed network. This gives a direct count of how many links one may need to change in order to obtain structurally equivalent graphs. A similar approach is correlating the columns (or rows) in corresponding adjacency matrices (e.g., using the Pearson Correlation Coefficient) and then using these numbers to measure the difference. A major drawback of these approaches is that they require corresponding nodes to exist in both networks.

Eigenvalues and Eigenvector: The eigenvector [31] approach is an effort to find the most central actors (i.e., those with the smallest distance from others), in terms of the "global" or the "overall" structure of the network. The idea is to pay lesser attention to "local" patterns. A node's importance is determined by the sum of the degree of its neighboring nodes.

Feature Extraction: Many network analysis measures require extensive computational power (e.g., [31]), making their use infeasible for very large graphs. In [9], Bagrow et al. have introduced a technique to extract a rather small feature representation matrix from a large graph. For such a graph, a geodesic distance based matrix, B-Matrix, is calculated where the nth row contains the degree distribution separated by n 'hops'. For instance, the first row would contain the usual degree distribution of all nodes; while the last row with the highest n, would contain the network diameter. The distance between the nodes is calculated using the Breadth-First Search (BFS) algorithm. The dimension of this matrix is *the total number of nodes* x *network diameter*. For structurally equivalent graphs, the calculated B-Matrices would exactly be the same. This technique can be used to extract featured matrices from large graphs and then compare them.

4 Subgroup Identification

When validating simulated networks generated from agent-based simulations, merely looking at the average graph properties may be inadequate. Different generative mechanisms may lead to varying subgraph characteristics in the simulated network even if their global characteristics may be similar. A comparison of global and subgraph network characteristics is necessary when a generative mechanism is introduced into agent-based simulations. Global properties of a given dynamic network may give clues about the robustness of the underlying processes, whilst local properties may show the variability that may occur for different settings for the same processes. Subgraph analysis may also be suitable for networks with changing ties and nodes.

The ratio of the number of particular subgroups in the 'real' or simulated network (compared to their random network counterpart in terms of nodes and edges) is independent of the network's size and thus is a good candidate for analyzing dynamical networks when both the edge and node sets changes over time. The social network analysis community has long known network substructures as 'fragments', especially, 'triads' for testing hypotheses such as balance, transitivity, and power [37, 15]. Milo et al. [28], inspired from sociological foundations, introduced the concept of local structures as statistically significant 'motifs', to study 'building blocks' of complex networks. The process includes counting the number of occurrences for each subgraph that appears in the network. This is followed by testing for statistical significance for each subgraph by comparison with randomized networks. Faust [17] compared networks of different sizes and domains w.r.t. their subgraph and lower-order properties, i.e., the nodal degree distributions and dyad census.

Graph clusters, in a social context, are called *communities* we briefly look at three approaches to their definition and detection.

Community Detection [31]. This is a technique to identify subgraphs where nodes are closely linked, i.e., when there are more edges between the nodes of the subgraph than external links to other nodes. There are many algorithms available to identify such clusters, see e.g., [8, 9]. To identify how closely connected these communities are, the concept of modularity was introduced [31], which is the fraction of the edges that fall within the groups (or communities) minus the expected fraction under the assumption that the edges were randomly distributed. It ranges from 0 to 1, the higher the value the more cohesive the community. An average may be used for comparing networks.

For graphs where some attributes of nodes are known (e.g., demographic data), several techniques may be used to match the semantic structure of the graphs. These can be used to identify which parameters may be more important (e.g., [1, 2]). Two are affinity and the silo index, now discussed.

Affinity [29]. Affinity measures the ratio of the fraction of edges between attribute-sharing nodes, relative to what would be expected if attributes were randomly distributed. It ranges from 0 to infinity. Values greater than 1 indicate a positive correlation whereas values between 0 and 1 have a negative correlation. For an attribute, say residential block as dormitory, one could first calculate the fraction of edges having the same residence location. It is calculated as follows:

$$S_a = \frac{|\{(i,j) \in E : s.t. a_i = a_j\}|}{|E|}$$

where, a_i represents the value *a* for a node *i*. In other words, we are identifying the total number of matched nodes with the same attribute values for an attribute *a*. *E* represents the total number of edges. Next, we calculate E_A which represents the expected value when attributes are randomly assigned.

$$E_a = \frac{\sum_{i=0}^{k} T_i \ (T_i - 1)}{|U|(|U| - 1)}$$

where, T_i represents the number of nodes with each of the possible *k* attribute values and U is the sum of all T_i nodes, i.e., $U = \sum_{i=0}^{k} T_i$. The ratio of the two is known as *affinity*: $A_a = S_a/E_a$. This measure is then used to discover "attribute level communities", that is, subgraphs with high affinity. Either the affinity of a whole network could be compared with another, or this can be used to identify communities in simulated and observed networks.

Silo Index. This index identifies the proportion of edges between nodes with the same attribute value in a network. If all the nodes that have a value *Y* for an attribute only have links to other such nodes, and not to nodes with any other values for that attribute, it implies that a strong community exists, which is disconnected from the rest. Such a set would have a maximum value of this index. In short, this index helps us identify how cohesive inter-attribute edges are. It ranges from -1 to 1, for the extreme cases (no in-group edges to only in-group edges respectively) and can be written as $(I - E)/(I + E)$, where *I* is the number of internal edges and *E* the number of external edges. It is quite similar to the E-I index [23] but with an opposite sign.

5 Dynamic Networks

Most, if not all, agent-based models of social networks assume a fixed-sized population of agents, which makes it easier to compare networks across time as only the ties change during the simulation, i.e., the '*n*' denoting the size of nodes in the networks remains fixed. The analysis and validation of networks becomes more difficult when the underlying network changes with respect to the individuals, which, in addition to the dynamics of interactions, affects the edge set. Depending upon the phenomenon being simulated agents may leave or join the networks over a course of a simulation. Consequently, it is possible that the shape and size of the network at any time t_i could be radically different from that at any later time t_j, ($i< j$). This is also almost inevitable for many real social networks that are observed at different times.

Techniques such as Quadratic Assignment Procedure (QAP) have been used in comparing longitudinal networks. QAP is nonparametric and thus unlike the *t*-test, requires no *a priori* assumption about the distribution of the observed data [22].

However, it requires that the networks must be of the same size. As discussed in Section 4, when the node set does not change, one may use distance measures to compare features (a set of network characteristics) across different time steps. For example, Legendi and Gulyas [24] have proposed to study cumulative measures of network characteristics by applying time-dependent processes of changes in agents' ties in a network.

To compare networks of varying size that are generated from agent-based models, Alam et al. [4] proposed the use of the Kolmogorov-Smirnov (KS) statistic as a general schema to compare pairwise simulated networks at different time steps by comparing distributions of node-statistics from the graphs (e.g., degree distribution). As a nonparametric test, the KS test does not assume that an underlying normal distribution and may be applied on entire distributions of network analysis measures. McCulloh and Carley [26] adopted a technique from statistical control theory called cumulative sum control chart (*CUSUM*) to detect changes in longitudinal networks. As they show, this statistical technique can be applied to a variety of network measures to compare simulated and real networks. Alt et al. [7] used *CUSUM* to detect changes in agent-based simulated networks by tracking changes in the *betweenness* and *closeness* centrality measures. McCulloh et al. [27] also used spectral analysis in conjunction with the *CUSUM* statistic to detect periodicity as well as significant changes in longitudinal networks.

Concerning the handling of dynamic networks, a way forward is to combine nonparametric tests as above [4] with the kinds of subgraph analysis and change detection methods discussed in the previous section. An open question is whether motifs identified in a dynamic social network can, in fact, be coherently interpreted or not [19]. Another possibility may follow from work by Asur et al. [8], who presented an event-driven framework for analyzing networks where new members join and leave communities at different time steps.

6 Testing against Exponential Random Graph Models

One set of approaches where the target class and similarity measures are well defined is under the label "Exponential Random Graph Models" (ERGM). Here the probability of any two nodes being directly connected is (indirectly) given a probability, then one 'fits' this class to some data and constraints, resulting in the most likely network (or a set of networks) from this class. That is, given an adjacency matrix, $\boldsymbol{X}$, indicating the presence of ties between nodes, $X_{ij} = 0/1$ for no tie/tie, then an objective function, $f_i(\beta, \boldsymbol{x})$, indicates the preferred 'directions' of change (where i is the node and $\boldsymbol{x}$ the network), which gives a probability that X_{ij} changes to $1 - X_{ij}$ $(for\ i \neq j)$ as:

$$p_{ij}(\beta, \boldsymbol{x}) = \frac{\exp\left(f_i(\beta, \boldsymbol{x}^{\pm ij})\right)}{\sum_{h=1}^{n} \exp\left(f_i(\beta, \boldsymbol{x}^{\pm ih})\right)}$$

where $\boldsymbol{x}^{\pm ij}$ denotes the network where X_{ij} is changed to $1 - X_{ij}$ and p_{ii} is the probability of not changing anything [36].

Different assumptions, such as transitivity of edges can be built in by choosing different objective functions. The likelihood of this class of models in explaining a set of data can be calculated and computer algorithms are used to find the best fit to the data (given a particular kind of objective function) including statistics as to how likely/unlikely that best model was (e.g., [13]).

For example, if one has panels of survey data about actor properties; choose some assumptions about the nature of the relationships between actors and an algorithm will come up with a dynamic network model for these.

This process is analogous to regressing an equation (with variables and parameters) to a set of data – one determines the form of the equation and then finds the best fit to the data. By comparing the fit of different models one gains information about the data, for example, to what extent a linear model + random noise explains the data, or whether including a certain variable explains the data better than without. In the case of ERGM, the above machinery allows one to find which class of networks, resulting from a given objective function, best fits a set of data. In other words, it gives a "surprise free" view of what a network that gave the data might look like.

From the point of view of validating a simulation, the available empirical data could be used to infer the best kind of ERGM model against which the actual network produced by a simulation could be judged by calculating the probability of the output network being generated by that particular ERGM model. This would be a kind of a "null model" comparison giving an indication of the extent to which the simulation deviated from the "surprise free" ERGM base-line model. This scheme has a number of advantages: (1) a well-defined basis on which the comparison was made, (2) the ability to use non-network data (e.g., waves of panel data) and (3) the ability to use assumptions, such as transitivity, to be made explicit. However, it still boils down to a single measure each time against a particular class of ERGM models, and it does depend on the extent to which the network is *in fact* "surprise free".

7 Towards a Scheme for Validating Networks

In this section, we give an example of how some of the aforementioned network measures were used to compare simulated results against real life data and to explore the parameter space of an agent-based model. Elsewhere [1, 2, 3], we used the affinity measure to identify the compatibility of nodes based on their attributes. For the corresponding agent-based model, we developed an algorithm that enabled agents to find other agents with similar attributes. In particular, it identified the significance of each attribute in order to determine the compatibility among the agents. The higher the affinity measure, the higher the importance of an attribute. Hence, this insight helped us determine parameter space for our local processes for our model.

After collecting simulation results, we then compared degree distribution (both distribution type and its fitted parameter values), clustering coefficient, average community modularity (along with the identified number of communities) and the calculated Silo indices of the attribute space. We observed time series of the clustering coefficient and the standard deviation in number of links over the course of each simulation run. This gave us a good insight about how each interaction strategy e.g., 'random' meeting, 'friend-of-a-friend', 'party' affect the above global measures of the simulated social network. Based on these measures and exploration techniques for parameter space, we identified one set of the interaction model as the best candidate to explain the underlying reference dataset.

To initiate the debate on the need for standards to validate simulated networks, we suggest to begin with calculating topological measures such as degree distribution, community structure, clustering coefficient, geodesic measures such as diameter, radius and average path lengths and then compare them with reference dataset using the techniques we described earlier. For social networks with nodal attributes, either categorical or not, we propose to first calculate affinity measures for each attribute, and then calculate, for each attribute, Silo indices for the whole attribute space. Finally, perform statistical tests, if possible nonparametric; to test how best they fit the data. These set of measures develop confidence in validating our models and could be deemed as a standard set of measures for the social network comparison.

8 Concluding Discussion

There does not seem to be a "golden bullet" for comparing simulation-generated networks and empirically derived networks. This paper provides a brief review of some of the approaches that are in contemporary use. These approaches, which are by no means exhaustive, could play some part in determining the extent to which a generated graph matches a target graph. Clearly, the more complex a model is (and agent-based simulations tend to be at the more complex end of the modeling spectrum), the more independent validation it needs, suggesting that a multiple approach is desirable. The kind of network validation attempted should depend upon the model's underlying assumptions and goals behind the simulation – namely, what is and is not, deemed significant about the reference and simulated networks. Ideally, these should be documented in any description of the model validation so a reader is in a position to judge the claimed goodness of the simulation network "fit". There does not seem to be an established norm for either describing such underlying assumptions nor for measuring the extent to which a particular set of generated networks match their targets.

A majority of agent-based social simulation models uses stereotypic networks (e.g., the Watts-Strogatz network), over which agents interact. A smaller set of models aim at simulating social networks through more descriptive and contextualized rules (e.g., [4]) but these often lack the data to sufficiently validate the resulting networks. The recent rise of online social networks offers the possibility to acquire the necessary data for the network validation. For instance, Abbas [1, 2, 3] used the friendship networks for a college campus in the United States to help validate an agent-based model of friendship choice. Here networks generated from different set of rules from the agent-based model were compared using several of the network measures above.

Clearly, we feel that more attention is needed concerning the validation of networks resulting from simulation models. This lack may be due to the disconnection between the different scientific communities looking at the emergence of networks. We hope that the social simulation workshops and conferences such as ESSA and MABS could play a role in bringing these groups together and thus help establish: (a) the importance of this problem (b) the development of approaches to better validate generated networks against empirical data and ultimately, (c) some norms and standards of what constitutes an adequate attempt to do this when publishing research where simulations produce networks as an important part of their purpose.

References

1. Abbas, S.: An agent-based model of the development of friendship links within Facebook. In: Seventh European Social Simulation Association Conference (2011)
2. Abbas, S.: Homophily, Popularity and Randomness: Modelling Growth of Online Social Network. In: Proc. of the Twelfth International Joint Conference on Autonomous Agents and Multi-agent Systems, St. Paul, USA (2013)
3. Abbas, S.: An agent-based model of the development of friendship links within Facebook. J. of Computational and Mathematical Organization Theory, 1–21 (2013)
4. Alam, S.J., Edmonds, B., Meyer, R.: Identifying Structural Changes in Networks Generated from Agent-based Social Simulation Models. In: Ghose, A., Governatori, G., Sadananda, R. (eds.) PRIMA 2007. LNCS (LNAI), vol. 5044, pp. 298–307. Springer, Heidelberg (2009)
5. Alam, S.J., Geller, A.: Networks in agent-based social simulation. In: Heppenstall, A.J., et al. (eds.) Agent-Based Models of Geographical Systems, pp. 199–216. Springer (2012)
6. Amblard, F., Quattrociocchi, W.: Social Networks and Spatial Distribution. In: Edmonds, B., Meyer, R. (eds.) Simulating Social Complexity: Understanding Complex Systems 2013, pp. 401–430. Springer (2013)
7. Alt, J.K., Lieberman, S.: Representing Dynamic Social Networks in Discrete Event Social Simulation. In: Proc. 2010 Winter Simulation Conference, pp. 1478–1489 (2010)
8. Asur, S., Parthasarathy, S., Ucar, D.: An event-based framework for characterizing the evolutionary behavior of interaction graphs. In: Proc. of the Thirteenth ACM SIGKDD International Conference on Knowledge Discovery and Data Mining, pp. 913–921 (2007)
9. Bagrow, J.P., Bollt, E.M., Skufca, J.D.: Portraits of complex networks. Europhysics Letters 81, 68004 (2008)
10. Blondel, V., Guillaume, J.: Fast unfolding of communities in large networks. Journal of Statistical Mechanics: Theory and Experiment 10008 (2008)
11. Borgatti, S.P., Carley, K.M., Krackhardt, D.: On the Robustness of Centrality Measures under Conditions of Imperfect Data. Social Networks 28, 124–136 (2006)
12. Bornholdt, S.: Statistical Mechanics of Community Detection. Physical Review E 74, 016110 (2008)
13. Burk, W.J., Steglich, C.E.G., Snijders, T.A.B.: Beyond dyadic interdependence: Actor-oriented models for co-evolving social networks and individual behaviors. International Journal of Behavioral Development 31 (2007)

14. Carley, K.M.: Dynamic Network Analysis. In: Dynamic Social Network Modeling and Analysis: Workshop Summary and Papers, Committee on Human Factors, pp. 133–145. National Research Council (2003)
15. deNooy, W., Mrvar, A., Batagelj, V.: Exploratory Social Network Analysis with Pajek, 2nd edn. Cambridge University Press, Cambridge (2012)
16. Epstein, J., Axtell, R.: Growing Artificial Societies: Social Science from the Bottom Up. The MIT Press, Boston (1996)
17. Faust, K.: Comparing Social Networks: Size, Density, and Local Structure. Metodološkizveski 3(2), 185–216 (2006)
18. Frantz, T.L., Carley, K.M.: Relating Network Topology to the Robustness of Centrality Measures. CASOS technical report CMU-ISRI-05-117 (2005), http://www.casos.cs.cmu.edu/publications/papers/CMU-ISRI-05-117.pdf
19. Hales, D., Arteconi, S.: Motifs in Evolving Cooperative Networks Look Like Protein Structure Networks. Networks and Heterogeneous Media 3(2), 239–249 (2008)
20. Hamill, L., Gilbert, G.: A Simple but More Realistic Agent-based Model of a Social Network. In: Fifth European Social Simulation Association Conference, Brescia, Italy (2008)
21. Kermack, W.O., McKendrick, A.G.: Contributions to the mathematical theory of epidemics. part I. Proceedings of the Royal Society A 115, 700–721 (1927)
22. Krackhardt, D.: QAP Partialling as a Test of Spuriousness. Social Networks 9, 171–186 (1987)
23. Krackhardt, D., Stern, R.N.: Informal Networks and Organizational Crises: An Experimental Simulation. Social Psychology 51(2), 123–140 (2011)
24. Legendi, R.O., Gulyas, L.: Effects of Time-Dependent Edge Dynamics on Properties of Cumulative Networks. In: Emergent Properties in Natural and Artificial Complex Systems, Vienna, Austria, pp. 29–36 (2011)
25. Macindoe, O., Richards, W.: Graph Comparison Using Fine Structure Analysis. In: IEEE Second International Conference on Social Computing, pp. 193–200 (2010)
26. McCulloh, I., Carely, K.M.: Detecting Change in Longitudinal Social Networks. Journal of Social Structure 12 (2009)
27. McCulloh, I., Norvell, J.A., Carley, K.M.: Spectral Analysis of Social Networks to Identify Periodicity. Journal of Mathematical Sociology 36(2), 80–96 (2012)
28. Milo, R., Shen-Orr, S., Itzkovitz, S., et al.: Network Motifs: Simple Building Blocks of Complex Networks. Science 298, 824–827 (2002)
29. Mislove, A., Marcon, M., Gummadi, K.P., et al.: Measurement and analysis of online social networks. In: Proc. of the Seventh ACM SIGCOMM Conference on Internet Measurement (IMC 2007), pp. 29–42. ACM, New York (2007)
30. Moss, S., Edmonds, B.: Sociology and Simulation: Statistical and Qualitative Cross-Validation. American Journal of Sociology 110(4), 1095–1131 (2005)
31. Newman, M.E.J.: Networks: An Introduction. Oxford University Press, USA (2010)
32. Newman, M.E.J.: Networks: Assortative mixing in networks. Physics Review Letters 89, 208701 (2002)
33. Newman, M.E.J., Girvan, M.: Finding and evaluating community structure in networks. Physics Review E 69(2), 026113 (2004)
34. Observable Universe. Wikipedia (n.d.), http://en.wikipedia.org/wiki/Observable_universe (retrieved on January 28, 2013)

35. Papadopoulos, F., Kitsak, M., Serrano, M.A., et al.: Popularity versus similarity in growing networks. Nature 2012, 537–540 (2012)
36. Snijders, T.A.B.: Longitudinal Methods of Network Analysis. In: Meyers, B. (ed.) Encyclopedia of Complexity and System Science, pp. 5998–6013. Springer (2008)
37. Wasserman, S., Faust, K.: Social Network Analysis: Methods and Applications. Cambridge University Press (1994)
38. Watts, D.J., Dodds, P.S., Newman, M.E.J.: Identity and search in social networks. Science 296, 1302–1305 (2002)
39. Wellman, B., Potter, S.: The Elements of Personal Communities. In: Wellman, B. (ed.) Networks in the Global Village. Westview Press, Boulder Co. (1999)

On the Quality of a Social Simulation Model: A Lifecycle Framework*

Claudio Cioffi-Revilla

Center for Social Complexity and Dept. Computational Social Science
Krasnow Institute for Advanced Study, George Mason University
Fairfax, Virginia 22030 (Washington DC) USA
ccioffi@gmu.edu

Abstract. Computational social science grows from several research traditions with roots in The Enlightenment and earlier origins in Aristotle's comparative analysis of social systems. Extant standards of scientific quality and excellence have been inherited through the history and philosophy of science in terms of basic principles, such as formalization, testing, replication, and dissemination. More specifically, the properties of Truth, Beauty, and Justice proposed by C.A. Lave and J.G. March for mathematical social science are equally valid criteria for assessing quality in social simulation models. Helpful as such classic standards of quality may be, social computing adds new scientific features (complex systems, object-oriented simulations, network models, emergent dynamics) that require development as additional standards for judging quality. Social simulation models in particular (e.g., agent-based modeling) contribute further specific requirements for assessing quality. This paper proposes and discusses a set of dimensions for discerning quality in social simulations, especially agent-based models, beyond the traditional standards of verification and validation.

Keywords: Quality standards, evaluation criteria, social simulations, agent-based models, comparative analysis, computational methodology, Simon's paradigm.

1 Introduction: Motivation and Background

The field of social simulation in general, and agent-based modeling in particular, has begun to generate methodological proposals for assessing and promoting quality across diverse and related areas (Gilbert and Troitzsch, 2005; Taber and

* Prepared for the 9th Conference of the European Social Simulation Association, Warsaw School of Economics, Warsaw, Poland, September 16–20,2013. This paper was inspired by the First Workshop on Quality Commons, Maison de la Recherche, Paris, 28–29 January, 2010. Thanks to Petra Ahrweiler, Edmund Chattoe-Brown, Bruce Edmonds, Corinna Elsenbroich, Nigel Gilbert, David Hales, Dirk Helbing, Andreij Nowak, and Paul Ormerod for stimulating discussions at the Center for Mathematics and Analysis in the Social Sciences, University of Paris Sorbonne.

B. Kamiński and G.Koloch (eds.), *Advances in Social Simulation*,
Advances in Intelligent Systems and Computing 229,
DOI: 10.1007/978-3-642-39829-2_2,

Timpone, 1996).[1] For instance, proposals exist in the area of communicating social simulation models (Cioffi and Rouleau, 2010; Grimm et al., 2005), comparing models (Rouchier et al., 2008; Cioffi, 2011; Cioffi and Gotts, 2003), and assessing complex projects that involve large interdisciplinary teams (Cioffi, 2010). Consensus on quality standards in social simulation has not yet emerged, but a promising discussion by practitioners is already under way.

The properties of "Truth," "Beauty," and "Justice" have been proposed by Charles A. Lave and James G. March (1993) and are widely used for discerning quality in *social science mathematical models.* The three terms "Truth," "Beauty," and "Justice" (or "TBJ," for short) are labels for quality dimensions referring to fundamentally good—i.e., normatively desirable—features of social science modeling. Accordingly, the TBJ terms must be interpreted as labels, and not literally (Lave and March, 1993: Chapter 3).

Truth refers to the empirical explanatory content of a model—its contribution to improving causal understanding of social phenomena—in the sense of developing positive theory. For example, truth is normally judged by internal and external validation procedures, corresponding to axiomatic coherence and empirical veracity, respectively (Kaplan, 1964; Sargeant, 2004). Truthfulness is the main classical criterion for evaluating empirical science (Hempel, 1965; Cover and Curd, 1998; Meeker, 2002), whether the model is statistical, mathematical, or computational. "Truth" must be a constituent feature in a social science model, such that without it a model has no overall quality contribution.

Beauty refers to the esthetic quality of a model, to its elegance in terms of properties such as parsimony, formal elegance, syntactical structure, and similar stylistic features. Beauty is about art and form. For example, the mathematical beauty of some equations falls within this criterion, including features such as the style of a well-annotated system of equations where notation is clear, well-defined, and elegant. Unlike truth, beauty is not necessarily a constituent attribute, but is certainly a desirable scientific quality.

Justice refers to the extent to which a model contributes to a better world—to improvement in the quality of life, the betterment of the human condition, or the mitigation of unfairness. Justice is a normative criterion, unlike the other two that are positive and esthetic. For example, a model may improve our understanding of human conflict, inequality, refugee flows, or miscommunication, thereby helping to mitigate or improve social relations and well-being through conflict resolution, poverty reduction, humanitarian assistance, or improved cross-cultural communication, respectively. Policy analysis can be supported by modeling.

These Lave-March criteria of truth, beauty, and justice are useful for evaluating the quality of social simulation models. For example, in the classic Schelling (1971) model of segregation all three criteria are well-recognized. This is a

[1] This paper focuses on social simulations, so the broader field of computational social science (e.g., social data algorithms or socioinformatics, complexity models, social networks, social GIS, and related areas of social computing) lies beyond the scope of this paper. Quality research in those other areas is subject to its own standards.

fundamental reason why Schelling's model is so highly appreciated. Other examples that satisfy the Lave-March TBJ criteria might also include the Sugarscape model (Epstein and Axtell, 1996), the Iruba model (Doran, 2005), and Pick-a-Number (Hoffmann, 2002, 2005).

However, a further challenge exists because social simulations have features that render truth, beauty, and justice insufficient as criteria for assessing quality. This is because social simulation models are instantiated or rendered in code (a computer program in some language), so one can easily imagine a social simulation that would be of high quality in terms of truth, beauty, and justice, but fail in overall quality because simulation models pose additional challenges beyond other social science models (i.e., beyond the features of statistical or mathematical models).

As illustrated in Figure 1 (UML class diagram), social simulations have properties that are shared with all models in science generally and social science in particular, based on inheritance as a specialized class, in addition to having other features of their own. For example, the specific programming language of an agent-based model (Java, C++, or other) would be a defining feature.

The inheritance relation between social science models and social simulations readily suggests the specific features that distinguish the latter from the former, as illustrated in Table 1.

Additional criteria for social simulations—i.e., criteria beyond classical standards for social social models—should allow us to judge quality in terms of "The Good, The Bad, and The Ugly."

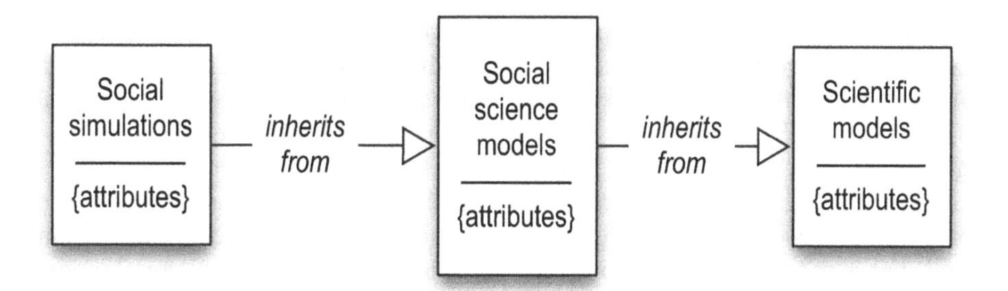

Fig. 1. UML class diagram illustrating the hierarchy of scientific models (left), social science models (center), and social simulations (right), each having increasingly specific standards for judging quality (left to right). *Source:* Prepared by the author.

Table 1. Quality Criteria for Evaluating Models in Domains of Science

Models in ...	Truth	Beauty	Justice	Additional
Science	Yes	Yes	No	No
Social science	"	"	Yes	"
Social simulation	"	"	"	Yes

Source: Prepared by the author.

Common practices such as verification and validation are well-known quality control procedures for assessing the quality of scientific models in general (Cover and Curd, 1998). However, verification and validation are insufficient criteria for assessing the quality of social science models, specifically for social simulations. An important implication is that current emphasis on model verification and validation is warranted (Cioffi, 2010; Sargent, 2004), but *verification and validation are insufficient by themselves for judging the quality of a social simulation model* (agent-based or other).

Therefore, a key methodological question concerning quality is: Which additional criteria—i.e., beyond Truth, Beauty, and Justice—could or should be used to assess the quality of a social simulation model? The next section addresses this question by proposing a set of dimensions for evaluating the quality of a given social simulation model.

2 Dimensions of Quality in Social Simulation Models

The quality of any complex artifact—whether a social simulation model or the International Space Station—is a multifaceted property, not a single dimension. Dimensions of quality can be used for evaluation as well as for a checklist of desirable attributes for building and developing a social simulation model. Arguably, there are two levels of quality assessment for computational social simulations, corresponding to the concepts of *a model* and *modeling*, respectively.

First, from a model's perspective, any set of quality dimensions for evaluating a social simulation must be based on its specific attributes or uniquely constituent features as a computational artifact in the sense of Simon (1996). Moreover, whether the overall quality of a given model should be an additive or a multiplicative function of individual qualitative features is less important than the idea that overall quality depends on a set of dimensions or desirable features beyond the Lave-March criteria, not on some single preeminent feature (e.g., simulation environment or programming language).

Second, from a modeling perspective, quality assessment should cover the broader modeling or model-building process as such, beyond the social simulation model that is produced in a narrow sense. This is because a computational model in final (i.e., committed) instantiated code is the result of a sequence of earlier modeling stages that precede the model itself, such as the critical stage of model design prior to implementation. Quality in design affects quality in the product of implementation, even when implementation *per se* is carried out in a proper manner (i.e., competently, with effectiveness and efficiency).

The following framework for quality assessment combines both perspectives by focusing on the classical methodological stages of social simulation model development:

1. Formulation
2. Implementation
3. Verification
4. Validation

5. Analysis
6. Dissemination

Such a Lifecycle Framework provides a viable checklist of quality dimensions to consider, based on the preceding methodological principles for social simulation. Note that verification and validation constitute only two contexts for assessing quality and, as shown below, some of the others involve quite a number of aspects regarding quality evaluation.

1. **Formulation.** Quality can be assessed starting from the formulation of a research problem that a given social simulation is supposed to solve. A first set of quality assessments regards research questions: Is the research question or class of research questions clearly formulated? Is the focal or referent empirical system well defined? Beyond clarity, is the research question original and significant? Originality should be supported by complete and reasoned surveys of prior extant literature to assess scientific progress. Every computational simulation model is designed to address a research question, so clarity, originality, and significance are critical. Motivation is a related aspect of problem formulation. Is the model properly motivated in terms of relevant extant literature? Or, is the simulation model the very first of its kind? If so, are there prior statistical or mathematical models in the same domain? Regrettably, incomplete, poorly argued, or totally missing literature reviews are rather common in social simulation and computational social science.
2. **Implementation.** Rendering an abstracted model in code involves numerous aspects with quality-related implications, starting with aspects of instantiation selection. Does the code instantiate relevant social theory? Is the underlying social theory instantiated using a proper program or programming language? Code quality brings up other aspects that may be collectively referred to as "The Grimson-Guttag Standards" (Guttag, 2013): Is the code well-written? Is the style safe/defensive? Is it properly commented? Can it be understood with clarity one year after it was written? In addition, what type of implementation strategy is used? I.e., is the model written in native code or using a toolkit? If toolkit (Nikolay and Madey, 2009), which one, why, and how good is the application? Is the choice of code (native or toolkit) well-justified, given the research questions? In terms of "nuts and bolts," quality questions include such things as what is the quality of the random number generator (RNG)? Think Mersenne Twister (Luke, 2011), MT19937, or other PRNG. Which types of data structures are used, given the semantics? Are driven threshold dynamics used? If so, how are the firing functions specified? In terms of algorithmic efficiency, What is the implementation difficulty of the problem(s) being addressed by the model? How efficient is the code in terms of implementing the main design ideas? In terms of computational efficiency, how efficient is the code in terms of using computational resources? This differs from algorithm efficiency. From the perspective of architectural design, is the code structured in a proper and elegant manner commensurate to the research question? In terms of object ontology, does the model instantiate the object-based ontology of the focal

system for the chosen level of abstraction? These quality-related questions precede verification and validation.

3. **Verification.** Which passive and active tests were conducted to verify that the model is behaving in the way it is intended to behave? Social scientists also call this internal validity. Verification tests include but are not limited to the following: Code walkthrough, debugging, unit testing, profiling, and other common procedures used in software development (Sergeant, 2004). What where the results of such verification tests? Quality assessment should cover investigation of which verification procedures were in fact used, since results can range widely depending on the extent of verification methods employed. Unfortunately, most social simulations are reported without much (or any) information regarding verification procedures, as if "results speak for themselves."
4. **Validation.** Similarly, validation of a social simulation, what social scientist call external validation (or establishing external validity), consists of a suite of tests, not a single procedure. Such tests are important for assessing quality in a social simulation. Which tests (histograms, RMSE for assessing goodness of fit, time series, spatial analysis, network structures, and other forms of real vs. artificial pattern matching tests) were conducted to validate the model? What were the results? Validation tests are often the focus of reporting results, at the expense of all other phases in the lifecycle of a social simulation model.
5. **Analysis.** The preceding aspects provide a basis for establishing overall confidence in a given model. What is the level of confidence in the model's results, given the combined set of verification and validation tests? If networks are present and significant in the focal system, does the model exploit theory and research in social network analysis (Wasserman and Faust, 2005)? Does the model facilitate analysis of complexity in the system of nonlinear interactions and emergent properties? Which features of complexity (emergence, phase transitions, power-laws or other heavy-tailed distributions, criticality, long-range dynamics, near-decomposability, serial-parallel systems, or other structural features) are relevant to the particular model? If spatial features are significant, does the simulation employ appropriate spatial metrics and statistical tools for spatial data? What is the overall analytical plan in terms of simulation runs and how is it justified? How does computational analysis advance fundamental or applied understanding of social systems? In terms of overall effectiveness, does the model render what is necessary for answering the initial research question or class of research questions? This differs from efficiency. In terms of the simulation's computational facilities, does the model possess the necessary functionality for conducting extensive computational analysis to answer the research questions or go even beyond? How powerful is the model in terms of enabling critical or insightful experiments? For example, in terms of parameter exploration (evolutionary computation) and record-keeping. What is the quality of the physical infrastructure that renders the most effective simulation experience?

6. **Dissemination.** Finally, the quality of a social simulation should be assessed in terms of its "life-beyond-the-lab." For instance, in terms of pedagogical value: Does the model teach well? I.e., does it teach efficiently and effectively? In terms of communicative clarity and transparency: Are useful flowcharts and UML diagrams of various kinds (class, sequence, state, use case) provided for understanding the model? Are they drawn with graphic precision and proper style (Ambler, 2005)?In terms of replicability, what is the model's replication potential or feasibility? How is reproducibility facilitated? Aspects related to a model's graphics are also significant for assessing quality, not just as "eye candy." In terms of GUI functionality, is the user interface of high quality according to the main users? Is the GUI fundational for answering the research questions? More specifically, in terms of visualization analytics: Is visualization implemented according to high standards (Thomas and Cook, 2005)? This does not concern only visual quality (Tufte, 1990), but analytics for drawing valid inferences as well (Cleveland, 1993; Few, 2006; Rosenberg and Grafton, 2010). In terms of "long-term care:" What is the quality of the model in terms of curatorial sustainability? How well is the model supported in terms of being easily available or accessible from a long-term perspective? In which venue (Google Code, Sourceforge, OpenABM, Harvard-MIT Data Center/Dataverse, or documentation archives such as the Social Science Research Network SSRN) is the model code and supplementary documentation made available? Finally, some social simulations are intended as policy analysis tools. Is the model properly accredited for use as a policy analysis tool, given the organizational mission and operational needs of the policy unit? Does the model add value to the overall quality of policy analysis? Does it provide new actionable information (new insights, plausible explanations, projections, margins of error, estimates, Bayesian updates) that may be useful to decision-makers?

3 Discussion

Verification and validation are obviously essential dimensions of quality in social simulation models; but they are just two among other dimensions of interest in assessing quality across the full spectrum of model development stages. Thus, and contrary to common belief, verification and validation may be viewed as necessary but insufficient conditions of a high-quality social simulation. The many dimensions of quality in social simulations extend far beyond and much deeper than the Lave-March TBJ criteria. Some of these criteria may be viewed as utilitarian (e.g., based on resources), while others are non-utilitarian (based on style).

Quality has dimensions because it is a latent concept, rather than a single directly measurable property. Therefore, proxies (i.e., measurable dimensions or attributes) are needed. The quality dimensions proposed in the preceding section provide a viable framework while social simulation develops as a field, not as a permanent set of fixed criteria. Each stage in the Lifecycle Framework contains numerous dimensions for quality assessment because social simulations are complex artifacts, in the sense of Simon.

Interestingly, Osgood's first dimension in cognitive EPA-space is Good-Bad (evaluation). This is why quality evaluation (good-bad-ugly) is essential (Osgood, May, and Miron, 1975). The proposed criteria should allow a classification of social simulations into categories of good, bad, or outright ugly.

As computational social scientists we need to better understand the microprocesses that compose the overall quality of social simulation:

- How is a problem chosen for investigation?
- How is the problem-space reduced by abstraction?
- How is the model designed?
- How well are the entities and relations understood?
- How is the simulation language chosen?
- How is the model implemented?
- How are verification and validation conducted?
- How are simulation runs actually conducted?
- How is the model being maintained?

Requiring additional quality criteria for social simulation models is not an argument against the unity of science. It is a plea for greater specificity and more rigor in the evaluation of quality in the field of social simulation.

From a methodological perspective, quality criteria could also help support the (virtual) experimental function of social simulations in terms providing ways for assessing the veracity of artificial worlds. Computational experiments could thus be framed within the context of a social simulation characterized by a set of quality features, taking all lifecycle stages into consideration, as in evaluating experimental results *conditional upon* the quality of the social simulation model. Such a function would enhance the value of social simulation as an experimental method and highlight its scientific usefulness.

Finally, the topic of quality in social simulations also motivates a broader discussion concerning similarities and differences between social simulation and other scientific approaches in science, such as statistical and mathematical models. While all scientific approaches share some of the same quality criteria, each has also unique quality criteria that are not applicable in other approaches. For social simulations there are aspects such as quality of code or visualization dashboards that are *sui generis* to the approach itself. The lifecycle approach to assessing quality in social simulations could also shed new light on parallel efforts in statistical and mathematical models of social systems, since there too we find a similar sequence of stages, from the formulation of research questions to analyzing and communicating model results, albeit with significant variations in technical details if not in the overall process.

4 Summary

Computational social science arises from a number of research traditions that have roots in The Enlightenment and even earlier origins in Aristotle's comparative analysis of social systems. Therefore, our existing standards of scientific quality and excellence have been inherited through the history and philosophy of science in terms of basic principles, such as formalization, testing, replication, and dissemination.

More specifically, the properties of Truth, Beauty, and Justice proposed for mathematical social science (Lave and March, 1993) in an earlier generation remain equally valid quality criteria for assessing social simulation models. But useful as such classic standards of quality may be, social computing adds new scientific features (e.g., emphasis on understanding complex adaptive systems, object-oriented ontologies, network structures that can evolve in time, nonlinear dynamics) that require development as new standards for quality evaluation. Social simulation models in particular (e.g., agent-based modeling) require further specific requirements for judging quality. This paper proposed a set criteria for discerning quality in social simulations, especially agent-based models, based on universal stages of simulation model development. These criteria are offered as an initial heuristic framework to consider and develop as a work-in-progress, not as a finalized set of fixed criteria.

Acknowledgements. Funding for this study was provided by the Center for Social Complexity of George Mason University and by ONR MURI grant no. N00014-08-1-0921. Thanks to members of the Mason-Yale Joint Project on Eastern Africa (MURI Team) for discussions and comments, and to Joseph Harrison for converting the earlier LaTeX article into LNCS style. The opinions, findings, and conclusions or recommendations expressed in this work are those of the author and do not necessarily reflect the views of the sponsors. This paper is dedicated to the memory of Auguste Comte (1798–1857), who lived at the Hotel de Saint-Germain-des-Pres, Paris, where this paper originated during the 2010 Quality Commons workshop.

References

1. Ambler, S.W.: The Elements of UML 2.0 Style. Cambridge University Press, Cambridge (2005)
2. Cioffi-Revilla, C.: Simplicity and Reality in Computational Modeling of Politics. Computational and Mathematical Organization Theory 15(1), 26–46 (2008)
3. Cioffi-Revilla, C.: On the Methodology of Complex Social Simulations. Journal of Artificial Societies and Social Simulations 13(1), 7 (2010)
4. Cioffi-Revilla, C.: Comparing Agent-Based Computational Simulation Models in Cross-Cultural Research. Cross-Cultural Research 45(2), 1–23 (2011)

5. Cioffi-Revilla, C., Gotts, N.M.: Comparative Analysis of Agent-Based Social Simulations: GeoSim and FEARLUS Models. Journal of Artificial Societies and Social Systems 6(4) (2003)
6. Cioffi-Revilla, C., Rouleau, M.: MASON RebeLand: An Agent-Based Model of Politics, Environment, and Insurgency. International Studies Review 12(1), 31–46 (2010)
7. Cleveland, W.S.: Visualizing Data. AT&T Bell Laboratories, Murray Hill (1993)
8. Cover, J.A., Curd, M. (eds.): Philosophy of Science: The Central Issues. W. W. Norton, New York (1998)
9. Doran, J.E.: Iruba: An Agent-Based Model of the Guerrilla War Process. In: Troitzsch, K.G. (ed.) Representing Social Reality: Pre-Proceedings of the Third Conference of the European Social Simulation Association (ESSA), pp. 198–205. Verlag Dietmar Foelbach, Koblenz (2005)
10. Epstein, J.M., Axtell, R.: Growing Artificial Societies: Social Science from the Bottom Up. The Brookings Institution, Washington, D.C. (1996)
11. Few, S.: Information Dashboard Design: The Effective Visual Communication of Data. O'Reilly, Sebastopol (2006)
12. Gilbert, N., Troitzsch, K.: Simulation for the Social Scientist, 2nd edn. Open University Press, Buckingham (2005)
13. Grimm, V., Revilla, E., Berger, U., Jeltsch, F., Mooij, W.M., Railsback, S.F., Thulke, H.-H., Weiner, J., Wiegand, T., DeAngelis, D.L.: Pattern-Oriented Modeling of Agent-Based Complex Systems: Lessons from Ecology. Science 310, 987–991 (2005)
14. Guttag, J.V.: Introduction to Computation and Programming Using Python. MIT Press, Cambridge (2013)
15. Hempel, C.G.: Aspects of Scientific Explanation. Free Press, New York (1965)
16. Hoffmann, M.J.: Entrepreneurs and the Emergence and Evolution of Social Norms. In: Urban, C. (ed.) Proceedings of Agent-Based Simulation 3 Conference, pp. 32–37. SCS-Europe, Ghent (2002)
17. Hoffmann, M.J.: Self-Organized Criticality and Norm Avalanches. In: Proceedings of the Symposium on Normative Multi-Agent Systems, pp. 117–125. AISB (2005)
18. Kaplan, A.: The Conduct of Inquiry. Chandler, San Francisco (1964)
19. Lave, C.A., March, J.G.: An Introduction to Models in the Social Sciences. University Press of America, Lanham (1993)
20. Luke, S.: Mersenne Twister (2011), `http://www.cs.gmu.edu/~sean/research/` (accessed May 30, 2011)
21. Meeker, B.F.: Some Philosophy of Science Issues in the Use of Complex Computer Simulation Theories. In: Szmatka, J., Lovaglia, M., Wysienska, K. (eds.) The Growth of Social Knowledge: Theory, Simulation, and Empirical Research in Group Processes, pp. 183–202. Praeger, Westport (2002)
22. Nikolai, C., Madey, G.: Tools of the Trade: A Survey of Various Agent Based Modeling Platforms. Journal of Artificial Societies and Social Simulations 12(2), 2 (2009)
23. Osgood, C.E., May, W.H., Miron, M.S.: Cross-Cultural Universals of Affective Meaning. University of Illinois Press, Urbana (1975)
24. Rouchier, J., Cioffi-Revilla, C., Gary Polhill, J., Takadama, K.: Progress in Model-to-Model Analysis. Journal of Artificial Societies and Social Simulation 11(2), 8 (2008)

25. Rosenberg, D., Grafton, A.: Cartographies of Time: A History of the Timeline. Princeton Architectural Press, New York (2010)
26. Sargent, R.G.: Verification and Validation of Simulation Models. In: Henderson, S.G., Biller, B., Hsieh, M.H., Shortle, J., Tew, J.D., Barton, R.R. (eds.) Proceedings of the 2007 Winter Simulation Conference. IEEE Press, Piscataway (2004)
27. Schelling, T.C.: Dynamic models of segregation. Journal of Mathematical Sociology 1(2), 143–186 (1971)
28. Simon, H.A.: The Sciences of the Artificial, 3rd edn. MIT Press, Cambridge (1996)
29. Taber, C.S., Timpone, R.J.: Computational Modeling. Sage Publications, Thousand Oaks (1996)
30. Thomas, J.J., Cook, K.A. (eds.): Illuminating the Path. IEEE Computer Society, Los Alamitos (2005)
31. Tufte, E.R.: Envisioning Information. Graphics Press, Cheshire (1990)
32. Wasserman, S., Faust, K.: Social Network Analysis. Cambridge University Press (1994)

A New Framework for ABMs Based on Argumentative Reasoning*

Simone Gabbriellini and Paolo Torroni

DISI, University of Bologna
Viale del Risorgimento, 2
40136, Bologna - Italy
{simone.gabbriellini,paolo.torroni}@unibo.it

Abstract. We present an argumentative approach to agent-based modeling where agents are socially embedded and exchange information by means of simulated dialogues. We argue that this approach can be beneficial in social simulations, allowing for a better representation of agent reasoning, that is also accessible to the non computer science savvy, thus filling a gap between scholars that use BDI frameworks and scholars who do not in social sciences.

Keywords: Agent based social simulation, behavioral models, abstract argumentation, social networks, opinion dynamics.

1 Introduction

ABMs within the social sciences can be classified into two streams of research: (*a*) a first stream that uses mathematical approaches; (*b*) a second stream that uses formal logics and BDI frameworks.

Analytical, generative and computational sociologists advocate ABMs to model social interactions with a finer-grained realism and to explore micro-macro links [20]. As a result, there are many proposals for ABMs of social phenomena, such as human hierarchies [26], trust evolution [19], cooperation [4], cultural differentiation [5] and collective behaviors [17]. All these models belong to the first stream and share at least two common features: (*a*) a network representation [27] to mimic social embeddedness; (*b*) a preference for mathematical, game theoretical or evolutionary computing techniques. In all these models, agents do, in fact, interact socially within a large population, but very little explicit reasoning is done. The second stream is focused on how social agents should reason, and it encompasses models of trust [11], cognitive representations[12] and norms evolution and evaluation [2]. These models usually rely on formal logics[1] and BDI frameworks [24] to represent agent opinions, tasks and decision-making capabilities. What emerges from this duality is that ABM in social sciences always

* This paper is an extended version of the short paper accepted by AAMAS 2013 [14].
[1] The relevance of logic in social simulations is an open issue, with both detractors and supporters [10].

B. Kamiński and G.Koloch (eds.), *Advances in Social Simulation*,
Advances in Intelligent Systems and Computing 229,
DOI: 10.1007/978-3-642-39829-2_3, © Springer-Verlag Berlin Heidelberg 2014

assume agent's reasoning capabilities (in the sense of information processing), but rarely this feature is explicitly modeled. Agents are pushed or pulled, with some degree of resistance, but such a representation of influence has already been challenged [22]. We hypothesize that BDI frameworks have not encountered a wide diffusion among social scientists because most BDI architectures are complex to use by non-computer-scientists. It is no coincidence that cognitive, AI and computer scientists use this approach instead of social scientists. On the other hand, cognitive and computer scientists do not implement agents that interact socially to any significant extent in simulations.

This paper aims at evaluating a new framework for agent-based modeling, which may be appealing for both streams of research in social simulation: it explicitly models agents reasoning capabilities *and* it can be applied to socially embedded and interacting agents. The approach we propose is built on well established theories from social, cognitive, and computer science: the "strength of weak ties" by Granovetter [18], the "argumentative nature of reasoning" by Mercier & Sperber [21] and "computational abstract argumentation" by Dung [9]. Computational abstract argumentation is a reasoning approach that formalizes arguments and their relations by means of networks, where arguments are nodes and attacks between arguments are directed links between such nodes. We believe that this formalization, while it offers a logical and computational machinery for agent reasoning, it is nevertheless friendly to social scientists, who are already familiar with network concepts. There is already a plea for the use of logic-related approaches in ABM [25], but we are not aware of any previous ABM that uses argumentation to investigate social phenomena. Our approach represents a framework in the sense that it leaves the modeler many degrees of freedom: different embedding structures can be accommodated, as well as different trust models and different ways of processing information. The only pivotal point is the representation of information and reasoning with abstract argumentation.

The paper proceeds as follows: in the next section, we discuss the concept of embeddedness [18] that will be used to connect our agents in a relational context; we then present a brief formalization of how agent reasoning and interacting capabilities unfold; we present an implementation of this idea by means of an ABM, along with its scheduling and discuss some experimental results; finally, we conclude and present some ideas for future work.

2 Weak Ties and Social Agents

In social simulations, embeddedness is almost always represented with (more or less explicit) network structures. Embeddedness could be something abstract, i.e. represented with relational networks, or spatial, i.e. represented with Von Neuman or Moore neighborhoods. In any case, these different kinds of embeddedness may all be explicitly represented by network topologies.

The basic idea of this social trait comes from Granovetter's hypothesis, which states that our acquaintances are less likely to be connected with each other than our close friends [18]. This tendency leads to social networks organized as densely knit clumps of small structures linked to other similar structures by bridges between them. Granovetter called this type of bridges "weak ties", and demonstrated their importance in permitting the flow of resources, particularly information, between otherwise unconnected clusters [17]. Embeddedness and bridges express a network topology which exhibits small-world features [28].

Building on Granovetter's "strength of weak ties" theory [18], sociological research on "small world" networks suggests that in a social network the presence of bridges promotes cultural diffusion, homogeneity and integration, but only under the assumption that relations hold a positive value [13]. This last concern is a trademark of the social simulation stream which uses a non-reasoning approach to agent modeling. We will show that our model does not need such a specification.

Following the experimental design by Flache & Macy [13], we use a "caveman graph" to represent a situation where clusters are maximally dense. We use this topology as a starting point to confront our results with a renown model in literature. We then allow for two kind of structural settings:

- a first one where each "cave", i.e. each cluster of the graph, is disconnected from the others, thus agents can interact within their own cluster only;
- a second one where a random number of bridges is added between caves, thus agents can interact occasionally with members of different caves. Even if our mechanism does not guarantee that all the caves become connected, on average the resulting networks exhibit small-world network characteristics.

Such a network structure is imposed exogenously to agents and kept static once generated. Random bridges play the role of weak links. By connecting previously unconnected densely knit caves, they play the role that acquaintances play in real life, and thus bridges are supposed to carry all the information beyond that available in a single cave. However, we do not impose a positive or negative value to links. Instead, links only represent the possibility of communication between any two pair of agents. The bit of information transmitted may have a positive or negative value, depending on the content exchanged: something that reinforces agent opinions or that radically changes them.

We call the stream of information exchanged between two agent a "simulated dialogue". The dialogue mechanism represents the micro-level assumption that governs our model and builds on Mercier & Sperber's work.

3 Agent Reasoning and Interaction

According to Mercier & Sperber's argumentative theory of reasoning [21], the function of human reasoning is argumentative and its emergence is best understood within the framework of the evolution of human communication. Reasoning developed as a "tool" to convince others by means of arguments exchanged in dialogues. We report a brief summary of a communication process according to Mercier & Sperber:

1. Every time an addressee receives a new bit of information, she checks if it fits what she already believes. If this is the case, nothing happens, otherwise, if the new information uncovers some incoherence, she has to react to avoid cognitive dissonance;
2. She faces two alternatives: (*a*) either to reject the new information because she does not trust the source enough, to start a revision of her own opinions. In that case, the addressee can reply with an argument that attacks the new information; (*b*) or to accept the new information because she trusts the source enough, to start a coherence checking and allow for a fine-grained process of opinion revision;
3. The source can react as well to the addressee's reaction: if the addressee decides to refuse the new information, the source can produce arguments to inject trust in the addressee, like exhibiting a social status which demonstrates competences on the subject matter. Otherwise, the source can produce arguments to persuade the addressee that the new information is logical and coherent, or to rebut the addressee's reply.
4. Both addressee and source may have to revise their own opinions while involved in such a turn-taking interaction, until: (*a*) addressee (or source) revises her own opinions; (*b*) they decide to stop arguing because they do not trust each other.

Such a turn-taking interaction between communicants is called a "dialogue". As said before, our agents argue through *simulated* dialogues. Before discussing how such a simulated role-taking process unfolds, we introduce how agents represent their knowledge by means of abstract argumentation. In computational abstract argumentation, as defined by Dung [9], an "Argumentation Framework" (AF) is defined as a pair $\langle \mathcal{A}, \mathcal{R} \rangle$, where $\mathcal{A}$ is a set of atomic arguments and $\mathcal{R}$ is a binary *attacks* relation over arguments, $\mathcal{R} \subseteq \mathcal{A} \times \mathcal{A}$, with $\alpha \rightarrow \beta \in \mathcal{R}$ interpreted as "argument α attacks argument β." In other words, an AF is a network of arguments, where links represent attack relations between arguments. Consider this simple exchange between two discussants, D_1 and D_2:

- D_1: My government cannot negotiate with your government because your government doesnt even recognize my government (a).
- D_2: Your government doesn't recognize my government either (b).
- D_1: But your government is a terrorist government (c).

Abstract argumentation formalizes these positions through a network representation, as shown in Figure 1. Once the network has been generated, abstract argumentation analyzes it by means of *semantics* [6], i.e. set of rules used to identify "coherent" subsets of arguments. Semantics may range from very credulous to very skeptical ones. Each coherent set of arguments, according to the correspondent semantics, is called an "extension" of $\mathcal{A}$. Some well-known semantics defined by Dung are the *admissible* and the *complete* semantics. To illustrate the rules imposed by these semantics, let us consider a set S of arguments, $S \subseteq \mathcal{A}$:

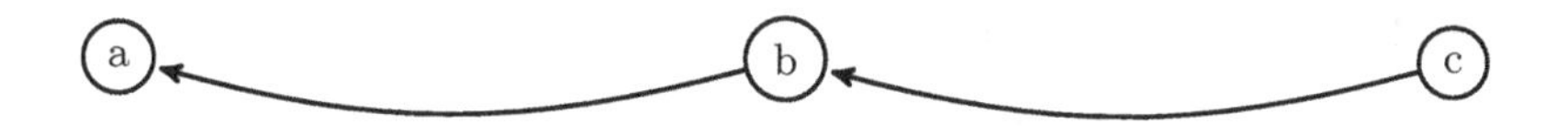

Fig. 1. Sample argumentation framework from Dung [9]

- S is *conflict-free* if $\forall \alpha, \beta \in S, \alpha \rightarrow \beta \notin \mathcal{R}$;
- an argument $\alpha \in S$ is *acceptable* w.r.t S if $\forall \beta \in \mathcal{A}$ s.t. $\beta \rightarrow \alpha \in \mathcal{R}$, $\exists \gamma \in S$ s.t. $\gamma \rightarrow \beta \in \mathcal{R}$;
- S is an *admissible* extension if S is conflict-free and all its arguments are acceptable w.r.t. S;
- S is a *complete* extension if S is admissible and $\forall \alpha \in \mathcal{A} \setminus S$, $\exists \beta \in S$ s.t. $\beta \rightarrow \alpha \in \mathcal{R}$.

In the words of Dung, abstract argumentation formalizes the idea that, in a debate, the one who has the last word laughs best. Consider again the very simple AF in Figure 1 and a credulous semantic like the complete semantic, which states that a valid extension is the one which includes all the arguments that it defends. It is easy to see that $\{a, c\}$ is a complete extension. a is attacked by b, but since c attacks b and does not receive any attack, c defends a, i.e. a is reinstated.

Our agents use a simulated dialogue process, introduced in [15], to exchange similar attacks between their AFs. A simulated dialogue $\mathcal{D}$ starts with an "invitation to discuss" from A (communicator) to B (addressee), by picking a random argument σ in her own extension. If B evaluates σ as coherent with her own AF, the dialog stops: A and B already "agree". On the contrary, if σ is not included in any of B's extensions, B faces an alternative: if she *trusts* A, she will revise her own opinions (i.e., by adding the new information to her AF and by updating her extensions); if instead B does not trust A, she will rebut α against σ and wait for a reaction from A. The exchange between A and B continues until one of the agents changes her mind (agreement is thus reached), or if both agents leave the dialogue because neither is persuaded.

For the sake of generality, we left several choice points open. Mainly, we do not commit to any specific argumentation semantics and we do not commit to any specific opinion revision mechanism. We also assume that agents rely on a trust model. Arguably, a realistic model of trust would to take into account the *authoritativeness*, *rank* and *social status* of the interlocutor [26]. To date, our dialogue model is orthogonal to trust, we define trust thresholds statically but different trust models can be accommodated in the future. Furthermore, in our model information is either accepted or rejected. Argumentative frameworks can handle situations in which human beings partially accept information by means of weighted argumentative frameworks [7] where a certain level of inconsistency between arguments is tolerated. Such argumentation semantics are called conflict-tolerant, whereby arguments in the same extension may attack each other [3]. Again, for the sake of simplicity, we use conflict-free semantics.

4 ABM and Arguments: NetArg

We used NetLogo [29] to develop NetArg[2] [14,16]: a model for simulating discussions between argumentative agents, along with a software module (a NetLogo add-on) that performes the computational argumentation analysis.

The model comprises a number of agents (100 in our experiments) distributed in 20 distinct caves. Every agent reasons from the same set of arguments, and she selects, with a fixed probability, one set of attack relations among the two ones available at the beginning, as shown the *AF*s in Figure 2, derived from a real debate in an online discussion forum about renewable energies[3]. We tested the model with different *AFs*, with random attacks and up to 10 arguments, as well as *AFs* taken from empirical contexts and we found results to be stable.

At each time step, each agent is asked to start a dialogue with one of her neighbors extracted at random (see Algorithm 1), who could be restricted to the same cave or not, depending on the presence of bridges. The random extraction assures that the probability to "argue" with members of the same cave is higher than with out-cave neighbors, according to the fact that bridges (weak ties) are less activated than strong ties.

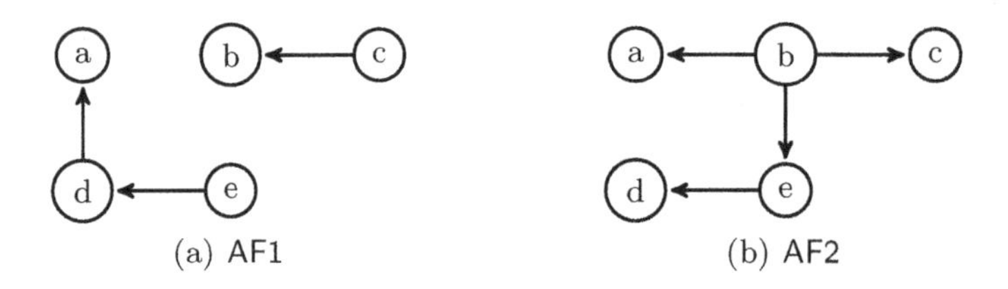

Fig. 2. The two argumentative frameworks distributed among the agents

The agent selected to start a dialogue picks one random argument in her extensions (i.e., an argument she believes in) and addresses the previously selected neighbor. The opponent replies, by following the dialogue procedure briefly sketched in the previous section.

In all our experiments, we opted for the complete semantics, which agents refer to when computing their extensions. At the beginning, each agent believes either $\{a, c, e\}$ or (b, d), the two possible extensions that a complete semantic returns from the *AFs* used. It is evident from the two plots in Figure 3 that, after some steps, agents adopt new opinions by means of dialogues: in (**a**) only two bars are present at time 0, each of which represents one of the two extensions available at the beginning (the distribution probability is set to 0.5 so they are equally distributed); in (**b**) more bars are present at time 50, i.e. more extensions are now available, because agents, by exchanging and accepting attacks, alter their own arguments network and thus new extensions are possible.

[2] The model can be downloaded from here:
`http://lia.deis.unibo.it/~pt/Software/NetARG-ESSA2013.zip`

[3] `http://www.energeticambiente.it`

Algorithm 1. Simulate an iteration of the model.

```
Require: N_I > 0 {N_I is the number of iterations}
Require: N_A > 0 {N_A is the number of agents}
  for I = 1 → N_I do
    for A = 1 → N_A do
      select a random agent B within A's neighbors
      initiate dialog with B
    end for
    record statistics
  end for
```

This opinion revision process gives raise also to a polarization effect at the population level. By polarization we mean that a population divides into a small number of factions with high internal consensus and strong disagreement between them. A perfectly polarized population contains two opposing factions whose members agree on everything with each other and fully disagree on everything with the out-group.

Using a modified version of the measure used by Flache & Macy [13], we measure the level of polarization P at time t as the variance of the distribution of the AF distances $d_{ij,t}$:

$$P_t = \frac{1}{N(N-1))} \sum_{i \neq j}^{i=N, j=N} (d_{ij,t} - \gamma_t)^2$$

where:

- N represents the number of agents in the population;
- $d_{ij,t}$ represents the AF distance between agents i and j at time t, i.e., the fact that agent i has an argument in her extension ($\bigcup_{\mathcal{E}}^{i}$) while the other does not, averaged across all available arguments ($|\mathcal{A}|$):

$$d_{ij} = \frac{|\bigcup_{\mathcal{E}}^{i} \setminus \bigcup_{\mathcal{E}}^{j} \cup \bigcup_{\mathcal{E}}^{j} \setminus \bigcup_{\mathcal{E}}^{i}|}{|\mathcal{A}|};$$

- γ_t represents the average distance value at time t.

In the next section, we present and discuss the results of three experiments made with NetArg.

5 Experimental Results

We present here the results of three different experiments that make use of the AFs in Figure 2. Each parameters combination has been ran 30 times and plots display averaged values for each combination. The first experiment that we discuss aims at testing if the model can reproduce Granovetter's theory about weak

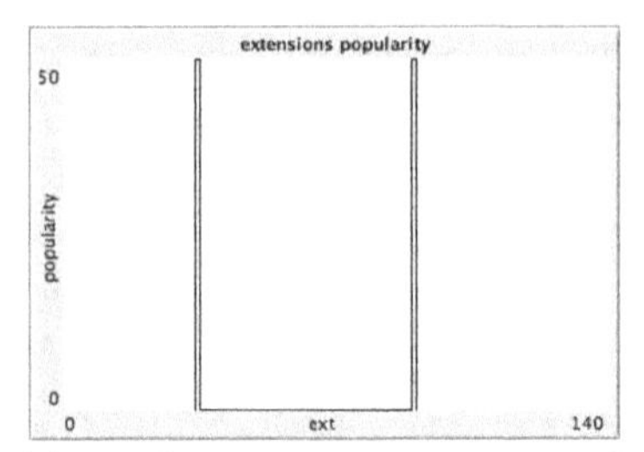

(a) NetArg model at time 0, with bridges allowed in the caveman graph.

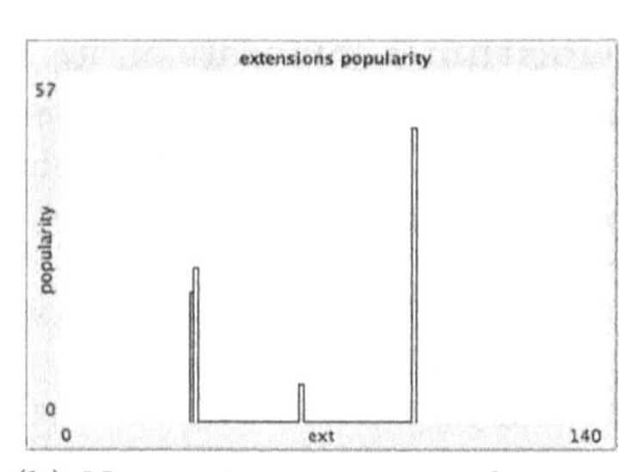

(b) New opinions emerge from interaction in dialogues at time 50.

Fig. 3. The distribution of extensions, at the population level, at time 0 and at time 50. In (b) it is evident the presence of newly formed extensions, not present at setup.

ties: does the presence of bridges lower polarization even with our argumentative agents? We set the AFs distribution fixed at 0.5 and allowed trust to take values: 0, 0.2, 0.5, 0.8 and 1. Results are shown in Figure 4. Dialogues enhance polarization because they give raise to new opinions sets, thus increasing opinion distance among the agents. With no bridges connecting caves (**a**), each cave quickly stabilizes at a local minimum. However, different caves will end up in different local minima, which results in a high polarization overall. Trust is able to lower the curve, but only until 0.8, because at 1 every agent changes her mind continuously so that polarization is even enhanced. In a sense, agents with total trust are "gullible" agents ready to believe anything. The instability arises if *all* agents are gullible, because there is no stable opinion. On the contrary, when bridges are present (**b**) polarization levels are lowered considerably. This time, caves can receive information from other caves, and this "small-world" topology lets the population exit from local minima. Increasing trust is more effective in this case, and values as low as 0.5 are able to lower polarization nearly to 0. We then control for different combinations of the two AF among the population, along with different level of trust. We can conclude that the model fits the predictions of Granovetter's theory: (1) the presence of bridges between caves fosters agreement and consensus, increasing the number of "like-minded" agents and (2) since only caves with bridges to other caves can receive new information, only connected caves learn new relations between arguments and change their minds.

In the second experiment we want to test if "majority wins", or if one AF is more "invasive" that the other, controlling for different distribution of the AFs among the agents. We distributed AF_1 among agents with different probabilities (0.2, 0.4, 0.6 and 0.8), controlling for different level of trust (0.2, 0.5 and 0.8). We replicated the experiment with and without bridges. Results are shown in Figure 5. In (*a*) no bridge is allowed, neither AF_1 (in white) nor AF_2 (in black) lose their positions (see the percentage for the initial distribution) to a significant extent. In (*b*) bridges are allowed, and the jump toward blacks is quite evident:

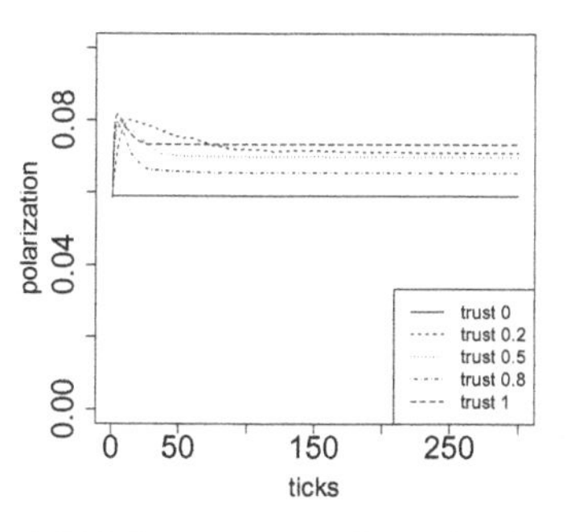

(a) Polarization without bridges

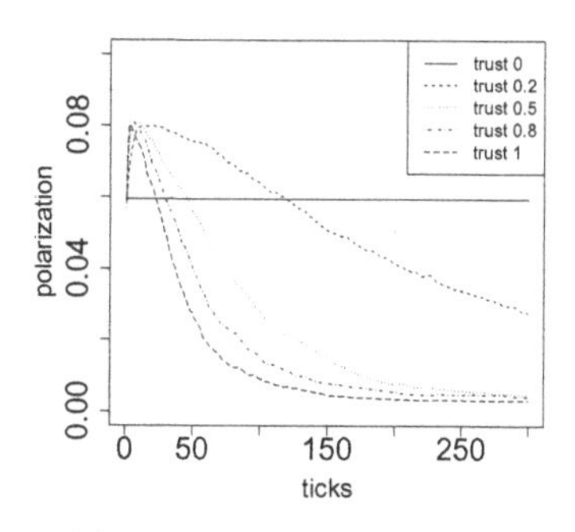

(b) Polarization with bridges

Fig. 4. Average polarization levels (over 100 runs) without bridges, and with bridges. AFs distribution is 0.5. Different levels of trust are shown.

no matter what the initial distribution is, even when AF_1 starts from 20% of the population, it still increase its audience if trust is high. AF_1 results much more aggressive toward AF_2 if bridges are present. A number of other new extensions arise, even if they are a strict minority. AF_2 contains more attacks, nevertheless is not able to win the population nor to defend itself from AF_1. More investigation toward AFs properties involved in ABM is needed in order to better understand this process.

In the third experiment, we check for AF resilience. Since AF_1 appears to be more aggressive, we label agents with AF_1 "innovators" and we explore if it is possible for a relatively small amount of innovators to convince the population to believe their extension, i.e. $\{a, c, e\}$. We can see in Figure 6 that AF_1 has a chance of winning over the whole population even if a low number of innovators

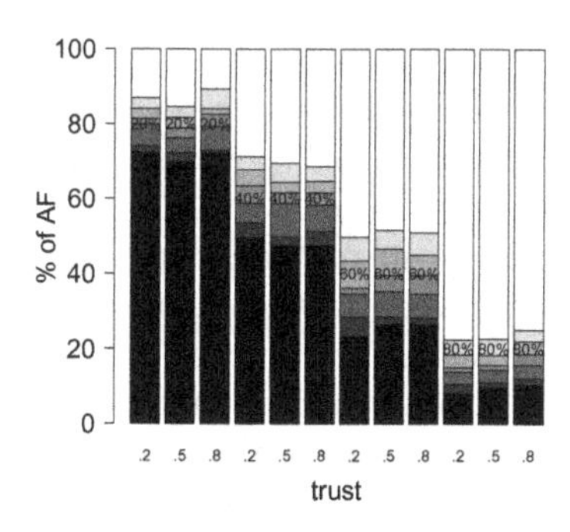

(a) Final diffusion of AFs without bridges.

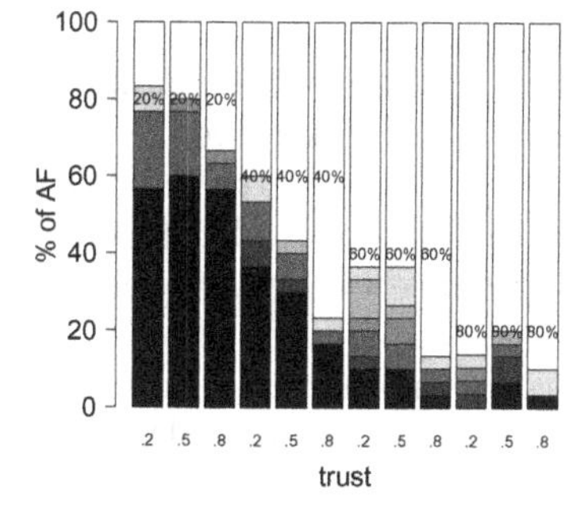

(b) Final diffusion of AFs with bridges.

Fig. 5. Diffusion of AFs. The percentages on the bars indicate the initial distribution when only two AFs (the black and the white) where present.

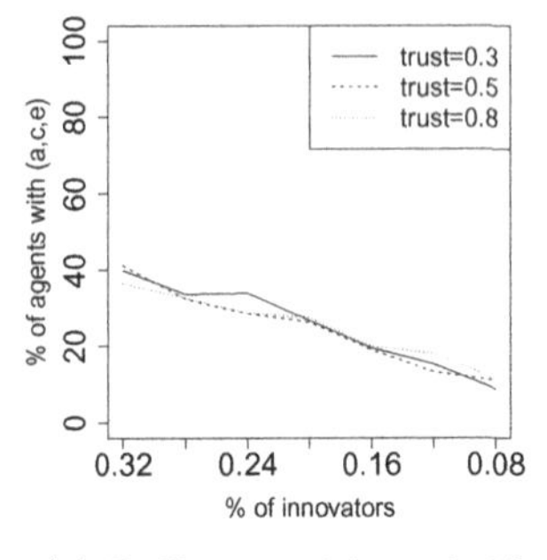

(a) Diffusion without bridges

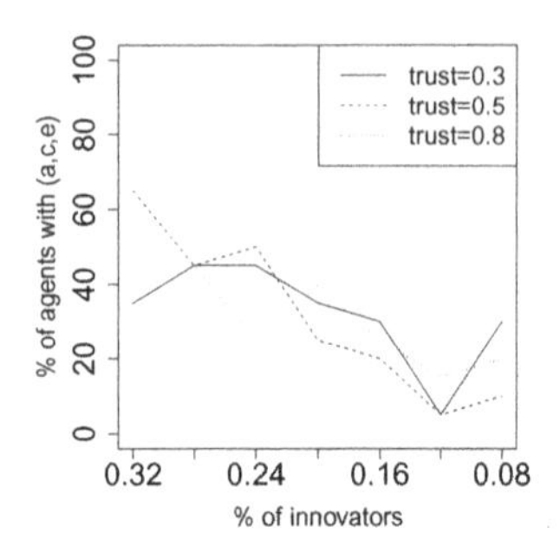

(b) Diffusion with bridges

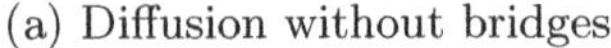

Fig. 6. The diffusion of a new argument among caves. On the x-axis, the initial percentage of innovators. On the y-axis, the percentage of agents that believes $\{a, c, e\}$ after the simulation.

is allowed (8% of total population). It is worthwhile noticing that, if bridges are not allowed, the proportion of agents who know $\{a, c, e\}$ at the end of the simulation is more or less equal to the beginning, and the level of trust does not influence agents much. When bridges are permitted, $\{a, c, e\}$ has a much higher probability to spread and, interestingly, results are much sharper: if $\{a, c, e\}$ reaches a tipping point, it wins the whole population, i.e., all agents change their minds and believe $\{a, c, e\}$, otherwise $\{a, c, e\}$ gets forgotten also by innovators. We conclude that bridges not only permit the diffusion of new ideas, but are the real key for innovations to happen, provided they succeed to overcome the threshold.

6 Conclusions

Using a network representation at different levels (social embedding and information), we have built a simple framework for social agents where reasoning is explicitly represented. We used abstract argumentation and argumentative theory of reasoning to build agents that exchange information through simulated dialogues. We demonstrated that our approach is, in principle, sufficient to reproduce two macro-behavior embedded in Granovetter's theory, i.e., the tendency to inclusion of weak ties and a competitive advantage for non-isolated caves. We showed that some argumentative frameworks are stronger than others and thus can, in principle, spread more efficiently when large audiences come into play. Finally, we also found that a small amount of "argumentative innovators" can successfully spread their opinions among a population, even at very low threshold.

As future work, we plan to further investigate patterns, strengths and weaknesses of AFs from a social science perspective (e.g., to understand which argumentation semantics better model human behavior, and if/why some opinions are stronger than others in a social debate). To accomplish this task, we will analyze real-world debates, like sustainable energy and political discussions within the e-Policy project.

There is a large literature on revising beliefs in artificial intelligence and knowledge representation. In particular, work by Alchourrón et al. [1] was influential in defining a number of basic postulates (known as *AGM postulates* in the literature) that a belief revision operator should respect, in order for that operator to be considered rational. Cayrol et al. [8] propose a framework for revising an abstract *AF* along these lines. However, considering our application, which is modeling possible outcomes of human debates, respecting the AGM postulates may not be a necessary requirement after all. We plan however to investigate the application of these and other methods, and evaluate which one performs best in simulating opinion diffusion in social networks.

To the best of our knowledge, our proposal is original in the social sciences, where argumentation has never been used for social simulation. It represents also a way for qualitative approaches to fit ABM formal requirements: for instance, discourse analysis results can be formalized as *AF* and fed into a simulator. Our approach envisages possible new grounds for cross-fertilization between the social and computer sciences, whereby surveys could be devised to retrieve arguments rather than numeric variables, possibly with the aid of argument extraction tools [23], and ABMs could be calibrated with empirically grounded *AF*s, to study the spreading of information, ideas and innovations with a finer-grained realism.

References

1. Alchourrón, C.E., Gärdenfors, P., Makinson, D.: On the logic of theory change: Partial meet functions for contraction and revision. J. Symb. Logic 50, 510–530 (1985)
2. Andrighetto, G., Villatoro, D., Conte, R.: Norm internalization in artificial societies. AI Communications 23, 325–333 (2010)
3. Arieli, O.: Conflict-tolerant semantics for argumentation frameworks. In: del Cerro, L.F., Herzig, A., Mengin, J. (eds.) JELIA 2012. LNCS, vol. 7519, pp. 28–40. Springer, Heidelberg (2012)
4. Axelrod, R.M.: The complexity of cooperation: agent-based models of competition and collaboration. Princeton University Press, Princeton (1997)
5. Axelrod, R.M.: The dissemination of culture: A model with local convergence and global polarization. J. Conflict Resolut. 41(2), 203–226 (1997)
6. Baroni, P., Giacomin, M.: Semantics of abstract argument systems. In: Argumentation in Artificial Intelligence. Springer (2009)
7. Bistarelli, S., Santini, F.: A common computational framework for semiring-based argumentation systems. In: ECAI 2010 - 19th European Conference on Artificial Intelligence, Lisbon, Portugal, pp. 131–136 (August 2010)
8. Cayrol, C., de Saint-Cyr, F.D., Lagasquie-Schiex, M.C.: Revision of an argumentation system. In: Brewka, G., Lang, J. (eds.) Principles of Knowledge Representation and Reasoning: Proc. 11th Int. Conf., KR 2008, September 16-19, pp. 124–134. AAAI Press, Sydney (2008)

9. Dung, P.M.: On the acceptability of arguments and its fundamental role in non-monotonic reasoning, logic programming and n-person games. Artif. Intell. 77(2), 321–357 (1995)
10. Edmonds, B.: How formal logic can fail to be useful for modelling or designing mas. Online resource, `http://cfpm.org/logic-in-abss/papers/Edmonds.html`
11. Falcone, R., Castelfranchi, C.: Trust and relational capital. Journal of Computational and Mathematical Organization 17(2), 179–195 (2011)
12. Falcone, R., Pezzulo, G., Castelfranchi, C., Calvi, G.: Contract nets for evaluating agent trustworthiness. In: Falcone, R., Barber, S.K., Sabater-Mir, J., Singh, M.P. (eds.) Trusting Agents. LNCS (LNAI), vol. 3577, pp. 43–58. Springer, Heidelberg (2005)
13. Flache, A., Macy, M.W.: Small worlds and cultural polarization. J. Math. Sociol. 35(1-3), 146–176 (2011)
14. Gabbriellini, S., Torroni, P.: Arguments in social networks. In: Proceedings of the Twelfth International Conference on Autonomous Agents and Multiagent Systems (AAMAS 2013). IFAAMAS (2013)
15. Gabbriellini, S., Torroni, P.: Ms dialogues: Persuading and getting persuaded. a model of social network debates that reconciles arguments and trust. In: Proc. 10th ArgMAS (2013)
16. Gabbriellini, S., Torroni, P.: NetArg: An agent-based social simulator with argumentative agents (demonstration). In: Proc. 12th AAMAS (2013)
17. Granovetter, M.: Threshold models of collective behavior. Am. J. Sociol. 83(6), 1420–1443 (1978)
18. Granovetter, M.: The strength of weak ties: a network theory revisited. Sociological Theory 1, 201–233 (1983)
19. Macy, M., Skvoretz, J.: The evolution of trust and cooperation between strangers: a computational model. Am. Sociol. Rev. 63, 638–660 (1998)
20. Manzo, G.: Variables, mechanisms and simulations: can the three methods be syntesized. Revue Française de Sociologic 48(5), 35–71 (2007)
21. Mercier, H., Sperber, D.: Why do humans reason? arguments for an argumentative theory. Behavioral and Brain Sciences 34(02), 57–74 (2011)
22. Palloni, A.: Theories and models of diffusion in sociology. Working Paper 11, Center for Demography and Ecology (1998)
23. Pallotta, V., Delmonte, R.: Automatic argumentative analysis for interaction mining. Argument & Computation 2(2-3), 77–106 (2011)
24. Rao, A., Georgeff, M.: Modeling rational agents within a BDI-architecture. In: Readings in Agents. Morgan Kaufmann (1998)
25. Reich, W.: Reasoning about other agents: a plea for logic-based methods. JASSS 7(4) (2004)
26. Skvoretz, J., Fararo, T.: Status and participation in task groups: A dynamic network model. Am. J. Sociol. 101(5), 1366–1414 (1996)
27. Wasserman, S., Faust, K.: Social network analysis: methods and applications. CUP, Cambridge (1999)
28. Watts, D.J., Strogatz, S.H.: Collective dynamics of small-world networks. Nature 393, 440–442 (1998)
29. Wilensky, U.: Netlogo. In: cCL and Computer-Based Modeling. Northwestern University, Evanston (1999)

Testing Model Robustness – Variation of Farmers' Decision-Making in an Agricultural Land-Use Model

Georg Holtz and Marvin Nebel

Institute of Environmental Systems Research, University of Osnabrück, Germany
{gholtz,mnebel}@uos.de

Abstract. Two empirically grounded agent-based models of agricultural land-use change are presented and compared with respect to factors driving groundwater use. The models are identical in most aspects, but one model uses a utility-approach to represent farmers' decision-making, while the other uses satisficing. The model-comparison exercise helps to distinguish substantial from implementation-dependent conclusions drawn from the model(s), and confirms the importance of model robustness tests.

Keywords: model comparison, robustness, land-use, utility, satisficing.

1 Introduction

The treatment of uncertainty is a major challenges for agent-based modelling ([1], [2]). An important procedure to enhance a model's credibility is testing the robustness of conclusions drawn from the model. Robustness means that there is no significant change in the results when supposedly insignificant changes are made in the model structure or specification of particular relationships in it. In many studies a model's sensitivity against random number sequences, (some) parameter values and sometimes also accessory (technical) assumptions ([3]) are tested. But often there is also a certain level of uncertainty related to a model's core assumptions, such as decision-making of agents. In case the agents represent humans, a wide range of theories from psychology, social psychology, economics, sociology and other (sub-) disciplines is available to describe their decision-making process and related behaviour ([4], [5]). A major difficulty and source of uncertainty hence relates to the "correct" choice of rules for agent behaviour ([6]). Often, this choice is made based on researchers' own background and available data, and involves some degree of arbitrariness.

More specifically, in the context of the case-study to be presented below, many approaches exist to represent farmers' decision making. Related to farming models [7] argue that one reason for the poor use of such models as decision support tools is the poor understanding by researchers of the actual process of decision-making by farmers. Economic approaches assuming the profit-maximising farmer for long have played an important role ([8], [9]). Since the 1990s, a considerable range of complementary factors has been identified in empirical studies, from socio-demographics and the

B. Kamiński and G.Koloch (eds.), *Advances in Social Simulation*,
Advances in Intelligent Systems and Computing 229,
DOI: 10.1007/978-3-642-39829-2_4, © Springer-Verlag Berlin Heidelberg 2014

psychological make-up of the farmer, over characteristics of the farm household and the structure of the farm business to the wider social milieu ([8]). However, little is known about the relative contributions of these various factors in varying contexts and a general model is lacking. In order to reduce the uncertainty of model outcomes with respect to the choice of (farmers') decision-making rules, a robustness analysis that compares model results under various assumptions would be ideal to substantiate the conclusions drawn from agricultural land-use models. In most cases such an analysis is not conducted due to time/budget constraints, although the studies cited above indicate that this would be important.

In this paper we present and compare two empirically grounded agent-based models of agricultural land-use change and associated groundwater over-use in the Upper Guadiana, Spain. The models have been implemented to reconstruct the history of agriculture in the Upper Guadiana Basin in order to identify characteristics of farmers which are important to understand land-use change in that region. The models use an identical general setting including options available to farmers (prices, costs related to crop area planted, labour intensity, yields, water needs, etc.) and regulations, and partly identical implementations of farmers with regard to the diffusion of innovations and path-dependency arising from historical developments of the farm. But they implement two different decision-making rules for farmers: one is based on utility ([10]) and one is based on satisficing ([11]). We present the different models and compare them with regard to similarities and differences of simulation results in order to draw lessons on substantial and implementation-dependent results.

The next section introduces the case study. Section 3 then outlines the similarities and differences between the two models and compares simulations to empirical data. Section 4 then compares the models with respect to the factors driving the dynamics. Section 5 draws the conclusions.

2 The Upper Guadiana Case Study

The Upper Guadiana Basin (UGB) is a rural area located in the Autonomous Region Castilla La Mancha in central Spain. The UGB is one of Spain's driest areas ([12]), the population density is very low and the UGB is classified as a rural area. Irrigation of farm land accounts for approximately 90% of total groundwater use while irrigation using surface water is hardly significant ([13]). During the last decades, the amount of irrigated farming has increased and farming practices have changed towards water-intensive crops, especially in the Mancha Occidental aquifer (MOA), the area's main aquifer. In the MOA around 17.000 farms exist, and it accounts for 90% of the UGB's groundwater extractions ([14]). Irrigated agriculture encompasses the possibility to plant water-intensive crops like high-yield cereals (maize, alfalfa) and melons, which do not grow in the region without irrigation. It further provides the possibility to achieve higher yields of traditional crops like winter wheat and barley as well as of vineyards and olives. All this led to a considerable expansion of irrigated agriculture in the UGB until the mid 1980s. The associated overexploitation of groundwater resources in the MOA endangered wetlands of high ecological value

([13], [15], [16]). To counteract this development regulations and subsidy schemes were introduced. Legal extraction of groundwater was considerably limited and subsidy schemes for water saving were temporarily introduced. However, many farmers did and still do disagree with the obligatory pumping restrictions introduced in the MOA and take more water than granted ([13], [17]). Aspirations to reduce groundwater extractions have also been hampered by the EU Common Agricultural Policy (CAP) which had considerable influence on the profitability of various kinds of crops, during some periods favouring water-intensive crops, especially maize ([18]). The overall developments have led to a non-sustainable situation (figure 1). Understanding the factors driving farmers' irrigation decisions is highly relevant in the context of future land-use in the UGB.

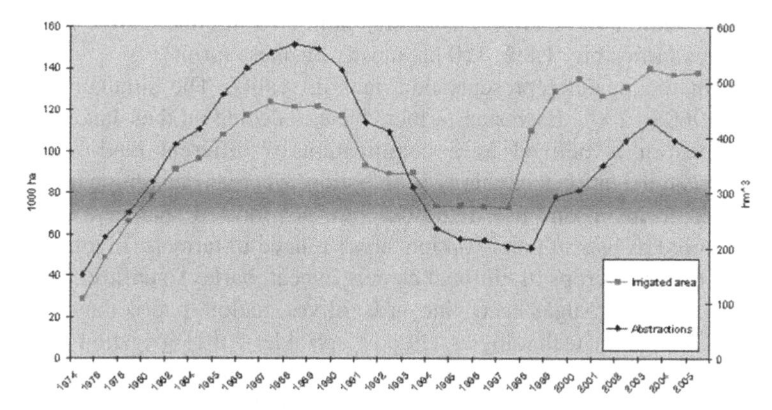

Fig. 1. Irrigated *area and water extractions in the UGB's main aquifer. The shaded area represents estimated renewable water resources. (Source:* [10]*)*

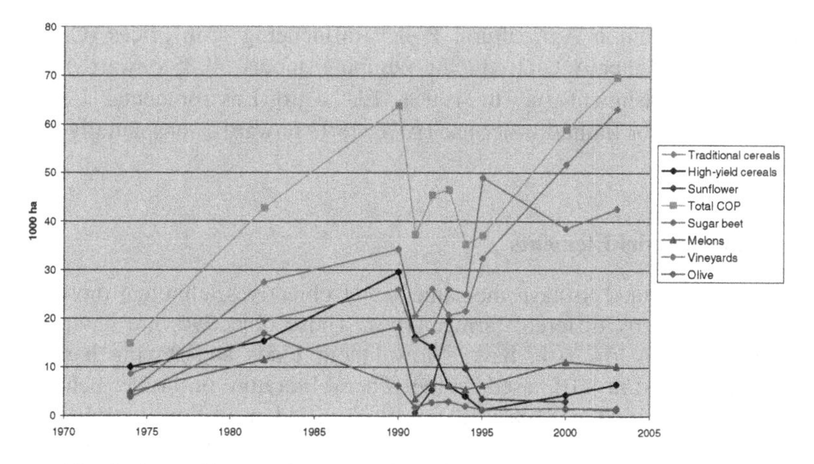

Fig. 2. Area *of irrigated crops in the UGB's main aquifer (Source:* [10]*)*

3 Model Description

This section first describes the general model setup and some aspects of farmers' decision-making which are identical for both models, which are presented afterwards. For a more detailed description see [10] or the model description provided at www.openabm.org.

3.1 General Setup

Each agent represents a farm. Agents are classified in three different types (part-time farms, family farms and business farms) and five different farm size classes (very small: 46,5% of all farms, 4 ha, mostly part-time farms; small: 35%, 8 ha, mostly family farms; medium1: 12%, 32 ha, mostly family farms; medium2: 4,7%, 70 ha, mostly business farms; big: 1,8%, 150 ha, mostly business farms).

One step in the model represents one year in reality. The simulation covers a period from 1960 to 2009. Each step farmers have to decide on their land-use pattern. A land-use pattern is defined as a combination of different land-uses. A crop combined with an irrigation technology is an option, whereas one option together with its related area forms a land-use. Land-uses are not located spatially, but land-use patterns are stored as lists of pairs {option, area} related to farmers. In total the model contains nine different crops: traditional cereals (wheat, barley), sunflower, high yield cereals (maize, alfalfa), sugar beet, vineyards, olives, melon, paprika and garlic, and four different irrigation technologies (flood-, sprinkler-, drip-irrigation or rainfed). Overall there is an amount of 23 options[1]. Prices, variable costs (rainfed/irrigated), yields (rainfed/irrigated), water needs for irrigation, risk and labour input needed (including scale effects) for crops, as well as efficiency of irrigation technologies are given exogenously and based on empirical data. Different regulations are implemented, which imply subsidies or penalties for specific crops and amounts of water use: EU- Common Agricultural Policy influencing crop prices (CAP, 1993 onwards), Spanish Water Act introducing pumping quotas (1985 onwards), Spanish Vine irrigation banishment (up to 1996), EU Agro Environmental Programme providing subsidies for limited water use (AEP, 1993 onwards, substantially modified in 2003).

3.2 Identical Model Elements

All farmers are assumed to have the same set of objectives, although those may be weighted differently by different farmer types. Those objectives are having a high gross margin, having low risk, having low labour loads and staying legal. These objectives are derived in [10] based on the general literature on farmer behavior and case-specific information and can be briefly justified as follows: profit and risk aversion are standard considerations when modeling western farmers. Further, in the Guadiana, many farmers take more groundwater than granted. Expecting that all

[1] Not all combinations of crop and irrigation technology are feasible.

farmers always accept formal rules is thus misleading. On the other hand, many farmers do follow formal rules, although it would be profitable not to do so and the chance of being controlled and fined is very low in the UGB. In order to incorporate this situation, the model considers a motivation to comply with formal rules, which is independent of the risk of getting a penalty for non-compliance. Another issue is that family farms and part-time farms which are mostly run with family labor have some natural limit on labor capacity. The labor force of big farms which belong to land-owners considering the land as capital is not restricted to family members. For those farmers, constraints on the applied labor force may however still arise from other sources like the availability of skilled workers or from the organizational structure of a farm. In the models the labor intensity of various land-use patterns can hence influence farmers' decision. Labor intensity also influences profits because family labor is considered available at no cost, but labor force beyond family labor causes costs. Several processes that are part of farmers' decision-making are identical in both models. In each time step:

1. For each farmer *f* a set of "considered options" is identified. The set includes those options that are combinations of crops and technologies both known to *f*. Further, unknown options are randomly considered, the probability increasing with usage of an option by other farmers, what induces a self-reinforcing diffusion process.
2. Based on *f*'s land-use-pattern in the previous step p_0, a set[2] of "considered patterns" p_i is developed. Each p_i is created through (randomly) iterating three basic operations on p_0 several times (the original p_0 is also considered further): a) add one considered option which is not yet part of this pattern and associate some area to it, forming a new land-use. The respective area is subtracted from a random other land-use. b) Remove one land-use (if not the last), the respective area is added to a random other land-use. c) Re-scale two land-uses: exchange some random area between two land-uses.
3. Land-use patterns p_i which are too different from p_0 are discarded. "Difference" is thereby evaluated based on two types of distances d between p_0 and p_i: a) a distance d_s related to skills representing uncertainty and learning efforts associated with a change in the area of land-uses (especially implementation of new crops and technologies) and b) distance d_c related to capital representing investments necessary and capital loss arising from the change in the area of land-uses. Both distances depend on the history of this farm because individual farms build up stocks of irrigation technologies and accumulate knowledge on crops based on their land-use decisions. For each land-use pattern p_i, f keeps p_i for further consideration randomly with a probability $p = (1-d)^{\alpha}$. This random filtering is done for both types of distances, α_s (skills) and α_c (capital) being respective parameters that can be adjusted to explore the influence on model behaviour. That is, those patterns that are "close" enough to *f*'s current land-use in both regards are likely kept for further consideration while more distant p_i are likely discarded.
4. Consequences arising from regulations (subsidies, penalties, legality) are evaluated for each of the remaining patterns.

[2] In the implementations used in this article for each farmer 1000 "considered patterns" are developed in each time-step.

The selection among the remaining land-use patterns differs between the models as described in the next sections.

3.3 The Utility Based Model (Holtz and Pahl-Wostl 2012)

This model version selects among different land-use patterns through the calculation of a utility for each land-use pattern and selection of the one that maximizes utility. Utility calculation is based on the four objectives identified above (profit, risk, labor, legality), i.e. it is not limited to monetary profit (see Table 1). The parameters of the utility function differ between farm types (cf. [10]).

Table 1. Calculation of utility $U(p_i)$ of a land-use pattern p_i (adapted from [10])

$U(pi) = G(g(p_i))*R(r(p_i))*W(w(p_i))*L(l(p_i))$ G: function of the influence of gross margin $g(p_i)$, R: function of the influence of risk $r(p_i)$ W: function of the influence of labour (work) load $w(p_i,)$, L: function of the influence of staying legal $l(p_i))$
$G(g) = g^\gamma$ $0<\gamma<=1$. γ is a parameter representing decreasing marginal utility of gross margin (for $\gamma<1$).
$W(w) = Max[0, Min(1, (1 - \frac{w - w_f}{\kappa - w_f})^\beta)]$ w is the amount of labor needed for a land-use pattern, κ sets an upper limit of the labor load that can be handled by a farmer f and w_f is the available family labor (it is $\kappa \geq w_f$). Note that if $w < w_f$ then $W(w)=1$ and if $w > \kappa$ then $W(w)=0$. $W(w)$ (and thus utility) decreases between w_f and κ. β determines the shape of this decrease.
$R(r) = (1 - r)^\rho$ The values of risk r associated with a land-use pattern as calculated in the model are associated with the standard deviation of gross margin due to short-term fluctuations of prices for products and inputs as well as variability of yields. r is in [0.0,1.0]. ρ determines the shape of utility regarding r.
$L(l) = Max(l, 0.5^\lambda)$ Being legal is binary: $l \in \{0 = illegal, 1 = legal\}$. If $l=1$ (legal behaviour) $L(l)=1$, thus utility is not reduced. If $l=0$ (illegal behaviour), λ varies the impact of illegal behaviour on utility.
In total the utility function is hence: $U(p_i) = g^\gamma \cdot (1 - r)^\rho \cdot Max(0, Min(1, 1 - \frac{w - w_f}{\kappa - w_f})^\beta)) \cdot Max(l, 0.5^\lambda)$

Figure 3 shows the simulation results that best fit[3] the empirical data presented in figures 1 and 2. In the simulation results the COP[4] area rises until 1985 to a similar level as observed empirically (cf. figure 2). The subsequent drop in irrigated COP area after groundwater extractions have been legally limited is reproduced by the model, but underestimated. The model does not capture the empirically observable increase of irrigated COP area after 1995; instead a higher level of horticultural crops

[3] The model was fitted manually. For a sensitivity analysis see [10].

[4] COP is an abbreviation for cereals, oilseeds and proteins. In this model COP cover traditional cereals, high-yield cereals and sunflowers.

is maintained. Horticultural crops also initially rise to a similar level in the model and empirically. However, empirically only melons play a role while in the model the main share is garlic. The observed difference could be explained through small errors in data which artificially favor garlic over melons, as discussed below. A more significant shortcoming of the model is its inability to reproduce the empirical drop in the area of horticulture after 1990. Irrigated vineyards rise to an approximately similar level in the model and in empirical data but an intermediate drop (in official data) of irrigated vineyards around 1990 is not captured by the model. The total irrigated area and the water extractions resemble the development in areas of irrigated crops discussed above regarding COP and horticultural crops and also the shortcomings of the model to capture certain aspects of empirical data. The model captures the developments of irrigated area and water extractions before 1985 but does not reproduce a drop around 1985-1990 and a subsequent re-rise but simulation results instead show a slight decline. The area subscribed to the voluntary AEP may be an explanation for some of the observed differences (cf. [10] for discussion of this issue).

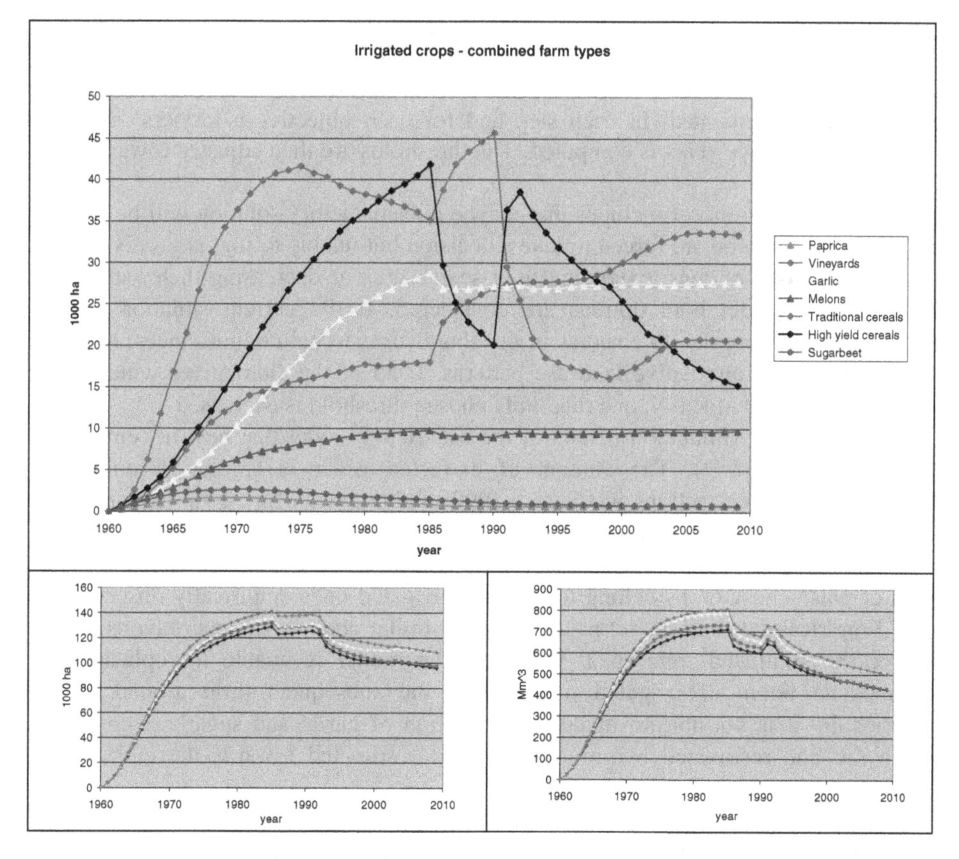

Fig. 3. The above figure shows irrigated crops. The small figures show the total irrigated area (left) and water extractions (right), including some parameter variations. (Source: adapted from [10])

3.4 The Satisficing Model (Nebel 2011)

Satisficing is a classic strategy in the decision-making literature ([19]) that forms an alternative to optimizing strategies and goes back to Herbert Simon ([20], [21]). In this concept alternatives are evaluated sequentially, in the order in which they appear in the choice set. "Instead of hopelessly searching for the best" ([22], p. 17) a decision maker continues seeking a solution only until he finds one that is "good enough" ([23]). The satisficing model version uses a multidimensional aspiration level as a basis for evaluating different land-use patterns. Four thresholds (one for each of the four objectives profit, risk, labor and legality) are defined. A solution is considered satisfactory if and only if all thresholds are reached.

Given that this model simulates a span of 50 years, fixed underlying aspiration levels cannot be assumed. This model hence uses flexible thresholds that are influenced by comparison with others and former experiences of a farmer. First, a yearly increase of the aspiration level for economic returns is implemented. Between 1960 and 2009 the gross domestic product of Spain increased almost every year ([24]) and it is assumed that farmers in the UGB want to be part of this economic growth. Additionally Simon ([21], p. 883) states that "people tend to aspire for a future that is little better than the present". Based on these considerations in each step each farmer's profit threshold is increased. Second, an adjustment of thresholds to past outcomes is implemented. In each step and for each objective a farmers' average value in the past five years is computed. The thresholds are then adjusted towards this average.

The satisficing concept includes that maybe no satisfactory solution will be found. Decision makers, who are forced to take a decision but unable to find any satisfactory solution have two options: extending their search area or decreasing their aspiration level. In this model both options are considered. If the current situation is not satisfactory anymore due to changes in thresholds or external circumstances, farmers start to search for alternative land-use patterns. If all of the alternatives which they consider are not satisfactory, one randomly chosen threshold is decreased.

Figure 4 shows simulation results of the satisficing model that best fit[5] empirical data (figure 2 and 3). The amount of extracted water is similar to empirical observations with extractions just before 1990 of about 600 Mm^3; a subsequent drop to 200 Mm^3 followed by an increase of extractions up to 400 Mm^3. Simulation results show mainly three irrigated types of crops: vineyards, traditional cereals and high yield cereals. Those crops belong to the most irrigated ones empirically observed as well. Empirical data and model results show a similar area of irrigated melons at the end of the simulated time span. However, the model is unable to replicate the intermediate boom of irrigated melons and the subsequent drop around 1990. Additionally it is not able to model the irrigation of garlic and sugarbeet correctly which are underestimated before 1990 and overestimated towards the end of the simulation time.

In the satisficing model the effect of laws and policies is immense. The water act (1985) and the CAP (1993 onwards) reduce irrigation and water extraction to a large

[5] The model was fitted manually. For a sensitivity analysis see [11].

extend. 1997 the irrigation of vineyards becomes legal, which has a great influence on model results. Within 12 years the area of irrigated vineyards increases to 50.000 ha, while irrigation of most other crops stays comparatively low. However, this finding is detached from reality. Though empirical data show an increase of irrigated vineyards in the mid-1990's, this is observable already before its irrigation becomes legal. After 1995 the area of irrigated vineyards remains relatively constant between 40.000 and 50.000 ha.

Another feature plays a prominent role when looking at empirical data: around 2003 irrigation of traditional cereals increased massively up to 60.000 ha. This model is not able to capture this feature. All simulations with any parameter combination show only moderate irrigation of traditional cereals after 1995.

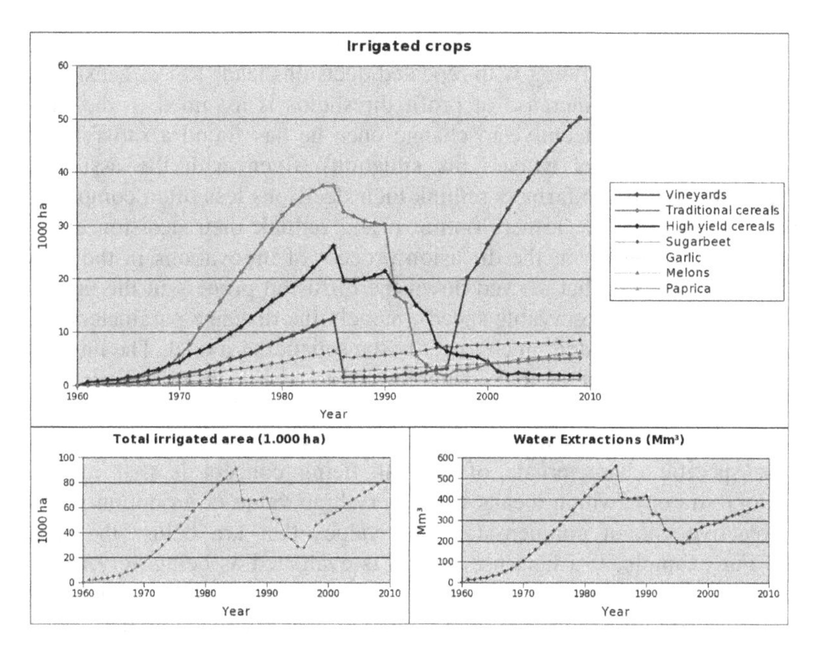

Fig. 4. "Best fit" results of the satisficing model. The above figure shows irrigated crops. The small figures show the total irrigated area (left) and water extractions (right). (Source: adapted from [11])

4 Model Comparison

The satisficing model performs significantly better than the utility-based model: it manages to cover the drop and re-rise in irrigated area and water extracted; and the areas of irrigated crops are approximated more closely, especially in the phase of fast changing regulations after 1985. Both models face similar limitations regarding the empirically observed increase of irrigated traditional cereals after the mid 1990's, which is not reproducible by any of the models. Furthermore, results of the utility-based model are more sensitive to data variations. This becomes apparent when comparing results for melons and garlic. Both of those crops are horticultural crops with similar characteristics: returning high profits, but requiring high labor loads and having high water needs (resulting in illegal behavior). The data used for this model slightly favor irrigated garlic especially for small farmers due to little higher profits and less risks. This advantage by a narrow margin has strong effects in the utility-based model: at the end of the simulation the area of irrigated garlic is almost twice as large as that of irrigated melons, whereas the amount of irrigated garlic and irrigated melons is almost equal throughout the simulation in the satisficing version. If farmers don't optimize they are almost indifferent between options that contain almost equal characteristics (such as melons and garlic). The cautious conclusion from the comparison of model performance is that the most important factors influencing farmers' decision-making in the satisficing model may be closer to reality than those which determine farmer behavior in the utility-based model. The following analyses the differences.

When comparing the models a striking difference can be observed in the model dynamics. The use of satisficing with repeated decisions leads to less sensitive model dynamics. If no constant increase of profit thresholds is assumed in the satisficing model, a farmer has no incentive to change once he has found a satisfying pattern (unless external influences worsen his situation). Even with the assumption of increasing profit thresholds farmers rethink their decisions less often compared to the utility-based model, where farmers optimize and rethink their decision every year. This leads to a slow-down of the diffusion process of innovations in the satisficing model. Thus, parameters that slowed down the diffusion process in the utility-based model to an empirically observable speed[6], namely the distance parameters α_s and α_c (cf. section 3.2), are estimated much lower in the satisficing model. The interpretation in the utility-based model is that switching costs and learning efforts delay the diffusion process, while the satisficing model suggests that farmers not actively looking for improvements are the main reason for slow diffusion.

Another specific characteristic of the satisficing concept is that it is a non-compensatory strategy, which means that a very high value of an option with regard to one objective cannot compensate other values that are below the respective threshold. For example, if a land-use pattern is evaluated as being very risky, other attributes such as very high profit cannot compensate this, the option will not be chosen. Thus satisficing favors "all-round options", which contain no negative outlier. For example, the area of irrigated garlic is significantly higher in the utility-based model since it returns higher profits and therefore compensates for high labor loads and for an illegal amount of necessary groundwater extractions. The satisficing model shows less irrigated garlic due to the fact that irrigated garlic includes high labor loads and illegal behavior. As a consequence of the compensatory nature of the utility-based model the maximum work-load κ (cf. section 3.3) played a significant role to fit the model to comparably small areas of horticultural crops. These provide a very high profit which overcompensates for illegal behavior, high risk and high labor loads in the utility-based model. Setting κ to a value that does not allow for larger areas of horticultural crops due to non-availability of labor was virtually the only possibility to roughly approximate empirical data. The interpretation then is that the available labor on (family and part-time) farms sets a limit to the area of irrigated horticultural crops.

[6] Cf. the sensitivity analysis in [10].

In contrast, the satisficing model evaluates all scales equally: a chosen land-use pattern must exceed all thresholds. As a consequence κ does not play a similar significant role for limiting the area of horticultural crops. Instead being legal or at least "being not too illegal" plays a much more important role in the satisficing model[7].

5 Conclusions

Two similar agent-based models of agricultural land-use were implemented to investigate the factors that drive the groundwater (over-)use in the Upper Guadiana. The models differed (only) with respect to the decision-making algorithm of farmers among sets of possible future land-uses. One followed a utility-based approach while the other implemented satisficing. The different decision-making algorithms lead to differing conclusions on the reasons for the observed empirical historical development, and would thus lead to different policy recommendations to influence future development in the Guadiana. The utility-based model stresses switching costs, learning efforts and the available labour force as main factors that limit a stronger shift to horticultural crops which would provide more cash per drop and thus would allow reducing groundwater extractions without strong economic losses. In contrast, the satisficing model highlights the routine (non-optimizing) behaviour of farmers as the main reason. Although none of the models reproduces the historical development very closely the satisficing model captures the overall behaviour better and outperforms the utility-based model, what indicates that the factors highlighted by this model might be closer to the drivers in the real system.

The model comparison demonstrates the importance of model robustness analyses which go beyond parameter variations, but include variation of uncertain assumptions, such as actor rationality.

References

1. Windrum, P., Fagiolo, G., Moneta, A.: Empirical Validation of Agent-Based Models: Alternatives and Prospects. Journal of Artificial Societies and Social Simulation 10(2) (2007)
2. Heckbert, S., Baynes, T., Reeson, A.: Agent-based modeling in ecological economics. Annals of the New York Academy of Sciences 1185, 39–53 (2010)
3. Galán, J.M., Izquierdo, L.R., Izquierdo, S.S., Santos, J.I., del Olmo, R., López-Paredes, A., Edmonds, B.: Errors and Artefacts in Agent-Based Modelling. JASSS 12(1) (2009)
4. Jackson, T.: Motivating Sustainable Consumption - a review of evidence on consumer behaviour and behavioural change. A report to the Sustainable Development Research Network (2005)
5. An, L.: Modeling human decisions in coupled human and natural systems: Review of agent-based models. Ecological Modelling 229, 25–36 (2012)

[7] For a more elaborate explanation of the treatment of the binary scale "legality" in the satisficing model see [11].

6. Hare, M., Pahl-Wostl, C.: Model uncertainty derived from choice of agent rationality - a lesson for policy assessment modeling. In: Giambiasi, N., Frydman, C. (eds.) Simulation in Industry: 13th European Simulation Symposium, pp. 854–859. SCS Europe Bvba, Ghent (2001)
7. Matthews, R., Gilbert, N., Roach, A., Polhill, G., Gotts, N.: Agent-based land-use models: a review of applications. Landscape Ecology 22, 1447–1459 (2007)
8. Edwards-Jones, G.: Modelling farmer decision-making: concepts, progress and challenges. Animal Sciences 82, 783–790 (2006)
9. Janssen, S., van Ittersum, M.K.: Assessing farm innovationand responses to policies: A review of bio-economic farm models. Agricultural Systems 94, 622–636 (2007)
10. Holtz, G., Pahl-Wostl, C.: An agent-based model of groundwater over-exploitation in the Upper Guadiana, Spain. Regional Environmental Change 12(1), 121 (2012)
11. Nebel, M.: Implementation and analysis of 'satisficing' as a model for farmers' decision-making in an agent-based model of groundwater over-exploitation, University of Osnabrück (2011)
12. Lopez-Gunn, E.: Policy change and learning in groundwater policy: a comparative analysis of collective action in la Mancha, Spain. King's College, London (2003)
13. Llamas, R., Martínez-Santos, P.: NeWater Report: Baseline Condition Report Upper Guadiana Basin (2005)
14. Acreman, M.: GRAPES project: Technical Report. Groundwater and River Resources Programme on a European Scale (GRAPES). Report to the European Union ENV4-CT95-0186. Institute of Hydrology, Wallingford, UK (2000)
15. Martínez-Santos, P., Llamas, R., Martínez-Alfaro, P.E.: Vulnerability assessment of groundwater resources: A modelling-based approach to the Mancha Occidental aquifer, Spain. Environmental Modelling & Software 23(9), 1145–1162 (2008)
16. Martínez-Santos, P., de Stefano, L., Llamas, R., Martínez-Alfaro, P.: Wetland Restoration in the Mancha Occidental Aquifer, Spain: A Critical Perspective on Water, Agricultural and Environmental Policies. Restoration Ecology 16(3), 511–521 (2008)
17. Ross, A., Martinez-Santos, P.: The challenge of groundwater governance: case studies from Spain and Australia. Regional Environmental Change 10(4), 299–310 (2010)
18. Varela-Ortega, C.: Policy-driven Determinants of Irrigation Development and Environmental Sustainability: A Case Study in Spain. In: Molle, F., Berkoff, J. (eds.) Irrigation Water Pricing: The Gap Between Theory and Practice. CAB International (2007)
19. Payne, J.W., Bettman, J.R.: Preferential Choice and Adaptive Strategy Use. In: Gigerenzer, G., Selten, R. (eds.) Bounded Rationality - The Adaptive Toolbox. MIT Press (2001)
20. Simon, H.A.: A behavioural model of rational choice. The Quarterly Journal of Economics 59, 59–118 (1955)
21. Simon, H.A.: Satisficing. In: Greenwald, D. (ed.) The McGraw-Hill Encyclopedia of Economics, pp. 881–886. McGraw-Hill (1993)
22. Simon, H.A.: Invariants of human behaviour. Annu. Rev. Psychol. 41, 1–19 (1990)
23. Gotts, N.M., Polhill, J.G., Law, A.N.R.: Aspiration Levels in a Land Use Simulation. Cybernetics and Systems 34(8), 663–683 (2003)
24. Maddison, A.: Angus Maddison homepage (2009), http://www.ggdc.net/maddison/oriindex.htm

Agent-Based Dynamic Network Models: Validation on Empirical Data

Richard Oliver Legendi[1,2] and László Gulyás[1,2]

[1] AITIA International, Inc.,
Czetz János u. 48-50, H-1039 Budapest, Hungary
[2] Eötvös Loránd University, Regional Knowledge Centre
Iranyi Daniel u. 4., H-8000 Szekesfehervar, Hungary
{rlegendi,lgulyas}@aitia.ai
http://mass.aitia.ai

Abstract. Inspecting the dynamics of networks opens a new dimension in understanding the mechanisms behind complex social phenomena. In our previous work, we defined a set of elementary dynamic models based on classic random and preferential networks. Focusing on edge dynamics, we defined several processes for changing networks with a fixed set of vertices. We applied simple rules, including the combination of random, preferential and assortative mixing of existing edges. Starting from an empty initial network, we examined network properties (like density, clustering, average path length, number of components and degree distribution) of both snapshot and cumulative networks for various lengths of aggregation time windows. In this paper, we extend our analysis with a comparison to results obtained from empirical data from two selected data sets. The knowledge of the baseline behavior of the abstract elementary dynamic network models helps us to identify the important, idiosyncratic properties of the empirical networks.

Keywords: dynamic networks, social network analysis, empirical data.

1 Introduction

Methodological analysis of social networks is an important field of research in modern sociology. However, there is a significant interest towards social network analysis in several additional scientific disciplines, and applications emerged in economics [7], [8], biology [9], epidemiology [10] and counter-terrorism [11] for example. Since real-world systems happen in time, the dynamism of the network indicates an interesting research direction for which we may find several examples in the literature, for example, by epidemologists [12], social scientists [13], [14] and computer scientists [15].

Working with dynamic networks (e.g., collecting longitudinal samples of networks or trying to model the evolution of networks), one realises that sampling always involves the act of aggregation. This problem is unavoidable, but poorly studied and usually ignored (see Figure 1). Therefore, we pay special attention

B. Kamiński and G.Koloch (eds.), *Advances in Social Simulation*,
Advances in Intelligent Systems and Computing 229,
DOI: 10.1007/978-3-642-39829-2_5,

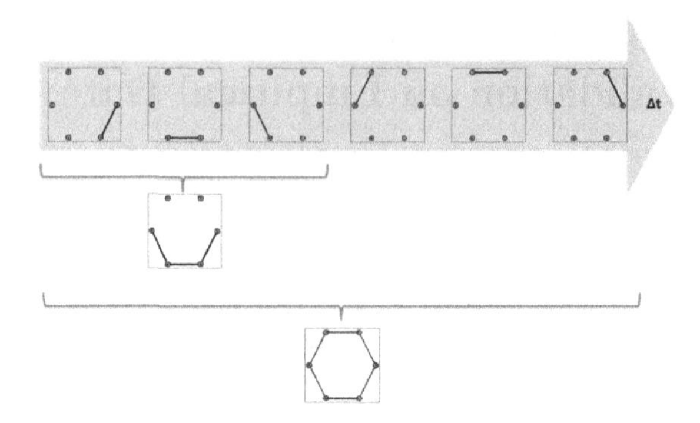

Fig. 1. The importance of the sampling window length

to the length of the cumulation time window and its effect to the aggregated network's properties.

In the following, the term *snapshot network* will refer to a network at time t : $G_t(V_t, E_t)$. The aggregated network from t_0 to t, $\bar{G}_t = (\cup_{i=t_0}^t V_i, \cup_{i=t_0}^t E_i)$ will be called a *cumulative network*. Clearly, snapshot networks and their aggregations with various lengths could be very different. See Figure 1 as an illustration.

In our previous works (see [1], [2], [3], [5] and [4] as an extensive summary) we evaluated the properties of various elementary dynamic network models. In this paper, we extend these studies by comparing the results of artifical simulations to a selected set of empirical data available in the social science literature. Our goal is to validate the previous theoretical results against data obtained from a real-world system.

The paper is structured as follows. The next section briefly discusses our previous results and conclusions regarding the elementary dynamic network models. Section 3 describes the chosen empirical datasets and reports our results, while the last section outlines the directions for future work and concludes the paper.

2 Dynamic Network Models

Constructing a static network for further analysis in general requires either a snapshot of a network in a specific point of time, or some kind of aggregation (e.g. collecting longitudinal samples of networks). However, both approaches may miss important facts and tendencies of network dynamics: the ongoing change of a network may be among the most interesting properties to observe. For instance, dissemination of information, social norms, innovation, or even diseases may behave in an inherently different way if we use different time windows for our investigation (e.g. the dynamic interpretation lets us observe individuals switching affiliations periodically in community identification [16]).

In our previous studies [4] we defined elementary models of dynamic networks based on the classic network models for static networks. We also analyzed models [3] where edges appear periodically in each $k * t$ time step: random links are created with a random appearance and lifetime, where edges are present in the snapshot network only in timestep t (cf. Lee et al. [19]).

The classic models we had before our eyes were the Erdős-Rényi (ER) [18] model of random networks and the preferential attachment model of Albert and Barabási [17], yielding scale-free networks (PA). For the dynamic ER model, we have created two variations: one where edges appear with the same probability and one with relinking. For the dynamic PA model, we introduced three different versions depending on how the two ends of new edges are selected (randomly, assortatively or preferentially), and each of them has two separate submodels defining if the used weights are specified by the snapshot (SPA) or by the cumulative network (CPA). The following subsection describes these models briefly.

2.1 Elementary Models of Dynamic Networks

ER1 We start from an initial graph G_0 created by the static ER model with density p_0. In each time step, we add all non-existing edges with probability p_A and delete existing ones with probability p_D (with double buffering).

ER2 Starting from an initial G_0 graph created by the static ER model with density p_0, in each time step we add k_A randomly selected non-existing edges, and delete k_D random existing ones (if possible).

CPA/SPA We start from an initial graph G_0 created by the static ER model with density p_0. In each time step we remove k_D random edges and add k_A edges using preferential attachment based on the cumulative or snapshot network. In detail, we select a starting node for the edge randomly (nodes are selected uniformly). The target node is then selected with probability linearly proportional to the number of edges the node currently has in the cumulative or snapshot network. In case the edge thus created already exists, we repeat the process.

AssortativeCPA/SPA We start from an initial graph G_0 created by the static ER model with density p_0. In each time step, we remove k_D random edges and add k_A edges using assortative mixing based on the cumulative or snapshot network. In detail, when creating a new link, we select the starting node u randomly, and the end node v based on the following w weights[1]:

$$w(u,v) = \begin{cases} 1/|deg(u) - deg(v)| & \text{if } deg(u) \neq deg(v) \\ 1 & \text{otherwise} \end{cases}$$

DoubleCPA/SPA We start from an initial graph G_0 created by the static ER model with density p_0. In each time step, we remove k_D random edges and add k_A edges by selecting both the starting and end node based on the preferential weights of the cumulative or snapshot network.

2.2 Evolution of Structural Properties

Figure 2 shows a selection of our results as an illustration for all the models over the entire time period (i.e., until they reach a stable point, the complete network).

[1] Note that isolated vertices and nodes with only one link difference have the same weights.

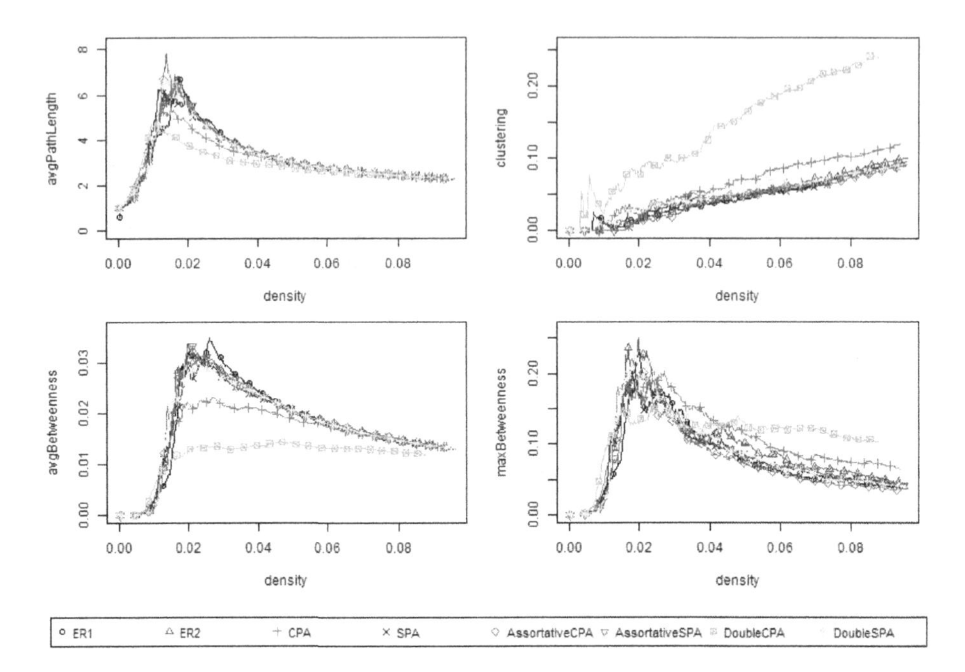

Fig. 2. Evolution of structural properties of the cumulative networks. The charts contain data obtained from 500 time step simulations where $N = 100, G_0 = 0$. Other parameters configured to add and remove one edge in each time step ($pA = 0.2e^{-3}, pD = 1, kA = 1, kD = 1$). Each plot represents the average of three simulation runs for each of three random seeds defined for all the models (27 runs in total).

It demonstrates the evolution of structural properties of the cumulative networks for all the different models over time as a function of density. Three universality classes may be identified: most of the models show similar behaviour, except the ones with preferential attachment (namely the CPA and DoubleCPA models) which dampen the density dependence of the measured values.

2.3 Evolution of Degree Distributions

The previously mentioned difference between the found universality classes is also reflected in the evolution of the degree distributions of the models. It is sensitive to the length of the aggregation window as well and inherently different for the snapshot and cumulative networks. Considering that the snapshot networks are very sparse[2], we omit their degree distributions because of space limitation. However, the degree distribution of the cumulative networks are interesting from several aspects. The same dynamic network may produce a normal, lognormal or even power law distribution for different aggregation lenghts. We also found

[2] Which may be similar to the snapshots of real-world networks in a given instant of time, based on the used time frame.

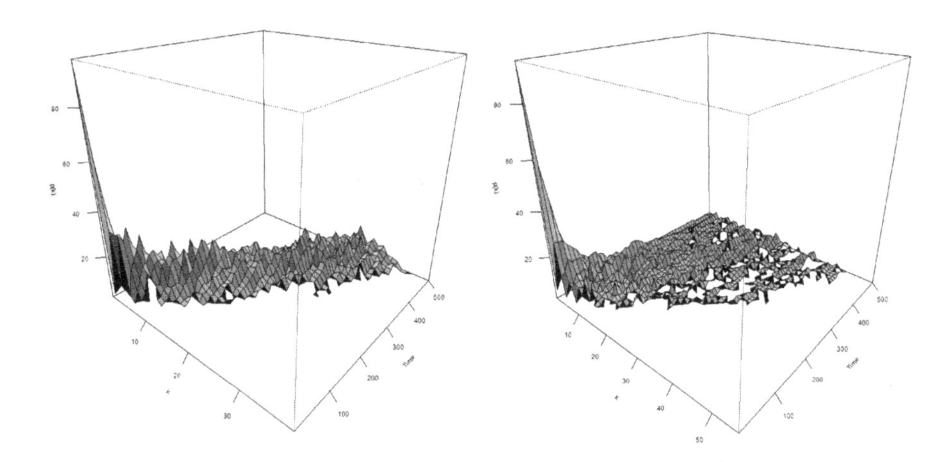

Fig. 3. Typical degree distribution results for the elementary dynamic network models. Frequency is shown on the z axis, while how the distribution changes over time can be read from the left to the right on each chart. The picture on the left hand side obtained by an ER1 model shows the case we found for most of the models. The picture on the right hand side obtained by a DoubleCPA model shows a more spread out distribution which is the characteristic of the DoubleCPA and CPA models.

that it can be stabilized to an extent by using bidirectional preferential attachment. Figure 3. demonstrates a typical result for most the models and for the preferential ones.

3 Comparison against Empirical Data

After we got an overall insight on the evolution of properties of the introduced models, the question is how far are these theoretical results from the properties of real-world systems? In order to answer that, we turned our attention to the evaluation of publicly available and well-known empirical temporal network datasets. In the following sections we briefly describe these datasets and the results we obtained through the approach of dynamic analysis.

3.1 The Gulf Dataset

The Gulf dataset *"covers the states of the Gulf region and the Arabian peninsula for the period 15 April 1979 to 31 March 1999. The Kansas Event Data System used automated coding of English-language news reports to generate political event data focusing on the Middle East, Balkans, and West Africa. These data are used in statistical early warning models to predict political change. The ten-year project is based in the Department of Political Science at the University of Kansas; it has been funded primarily by the U.S. National Science Foundation.*

There are two versions of the data: a set coded from the lead sentences only (57,000 events), and a set coded from full stories (304,000 events)" [20].

The dataset covers the connection of states within the Gulf region and the Arabian peninsula over almost 20 years. It is annotated with timestamps thus making it perfect for temporal network analysis. For the current comparison, we used the dataset coded from the lead sentences, for which the networks are also available preprocessed both in monthly and daily granularity [21].

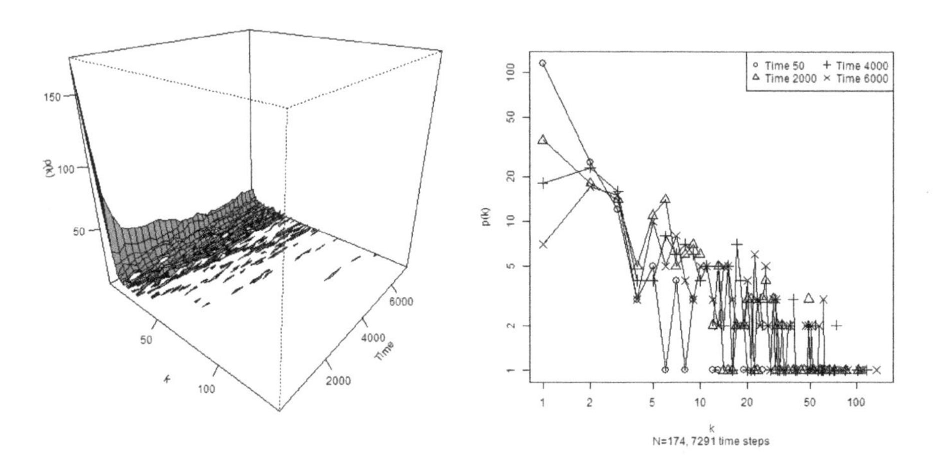

Fig. 4. Degree distribution evolution for the cumulative network constructed for the Gulf dataset. The degree distribution P(k) is shown on the z axis, while how the distribution changes over time can be read from left to right on the picture on the left hand side. The picture on the right hand side shows slices of the 3D chart in four selected points on a log-log scale.

Results and Discussion. The network created from the dataset contains 174 nodes and 57,131 edges. We made simulations with identically parameterized versions of the introduced dynamic network models for comparison. The appendix contains the concrete results for different structural properties for the dataset both in daily and monthly granuality as a reference. The properties of the cumulative network does not extend a significant difference between the daily and the monthly versions, hence in the following we consider the monthly version for the discussion.

The first important thing to note is that the structure of the degree distribution is similar to the preferential elementary dynamic networks. They show the same spread out structure that characterizes the DoubleCPA and CPA models (illustrated on Figure 4).

Comparing the results obtained from the theoretical models we found that some of the statistics are identical, however, there are notable differences in a few cases illustrated in Figure 5. The exceptionally high clustering values, the very low average but extremely high maximal betweenness suggests the existence of a highly connected core in the network. Other properties like average degree is linearly increasing in an identical way for all the models and for the dataset.

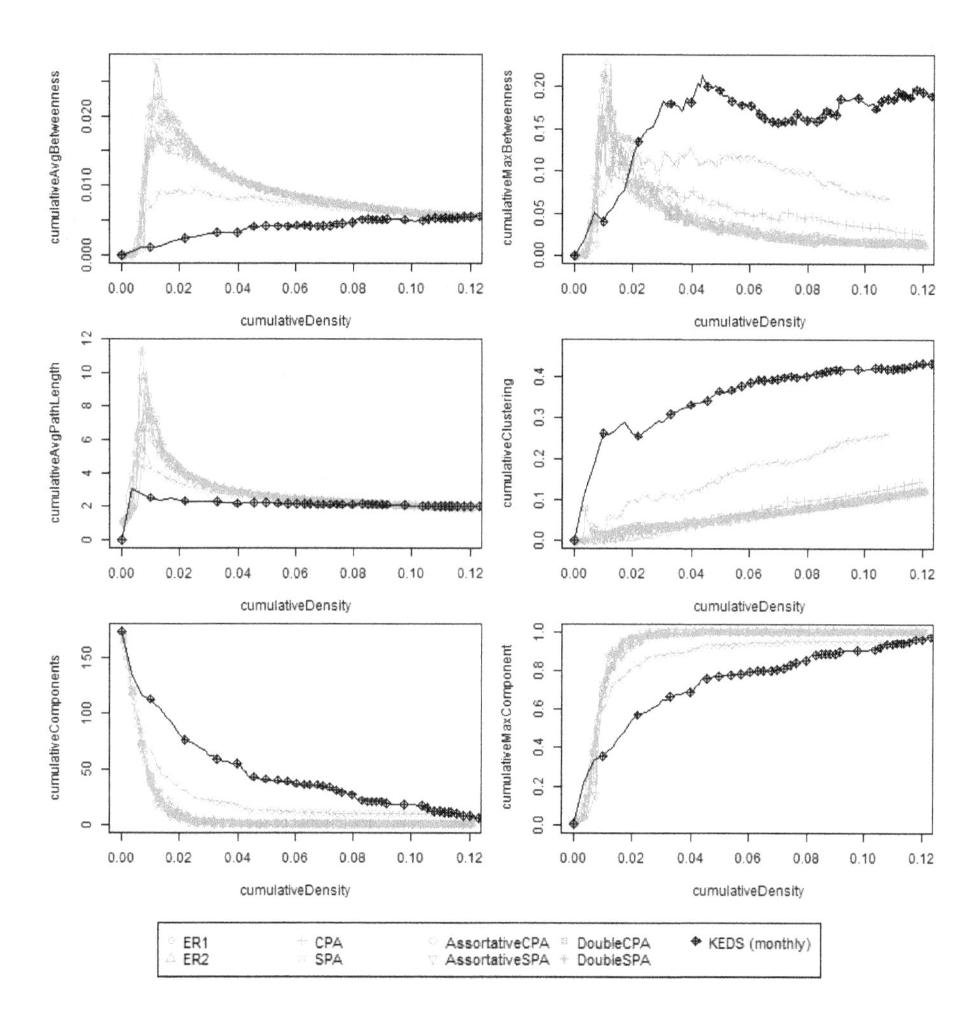

Fig. 5. Comparison of simulation results and the empirical data from the monthly Gulf dataset. The charts show the evolution of different statistics of the cumulative network (average and maximum betweenness, average path length, clustering, number of components and proportional size of the largest component) as a function of density. The empirical dataset is highlighted, the specific elementary dynamic network values can be read from Figure 2.

3.2 Sexual Network of Internet-Mediated Prostitution

Another dataset we investigated is Rocha's, Liljeros' and Holme's temporal, bipartite sexual network dataset [22] constructed to analyze Internet-mediated prostitution. The dataset describes a large, but very sparse network, containing sexual contacts between 6,624 anonymous escorts and 10,106 sex buyers over eight years. *"The community studied is a Brazilian, public online forum with free registration that is financed by advertisements. In this community, male members grade and categorize their sexual encounters with female escorts, both using anonymous nicknames. The forum is oriented to heterosexual males."*

Results and Discussion. The density change of the cumulative network shows an almost linear increasment to about ~ 0.003. This finding is important because (i) density has a direct influence on other statistics, thus this linear increasment may also be observed on several depending values (like in the total number of edges, proportional size of the maximal component, average betweenness, average degree and the number of components also deacreasing in a linear tendency), and (ii) this increase also indicates that the cumulative network is in the "initial" state where the statistics have not yet reached their peek. The clustering of the network is constant zero due to the bipartite nature of the dataset.

Other statistics shown on Figure 6 however show identical evolution over time to the results we got for the different dynamic network models. Average path length increases quickly and starts decreasing slowly after a high peek. The maximal betweenness also shows the same initial increasing tendency we described before.

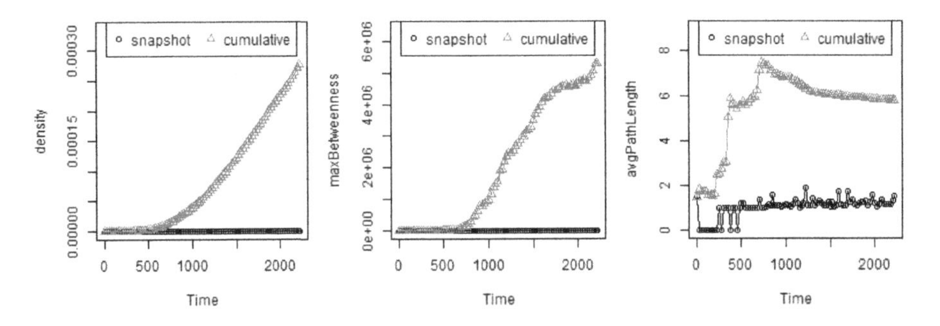

Fig. 6. Simulation results by using the dataset of Rocha et al. with N=16,730 and 2,235 simulation time steps. Charts contain every 25th measure points of the evolution of density, maximum betweenness and average path length over time.

The degree distribution (depicted on Figure 7) also shows a familiar structure we obtained for larger dynamic network simulations: the number of isolated vertices is decreasing as a "spread out" pattern is emerging in the distribution. An interesting aspect of the dataset is that it keeps the scale-free property for

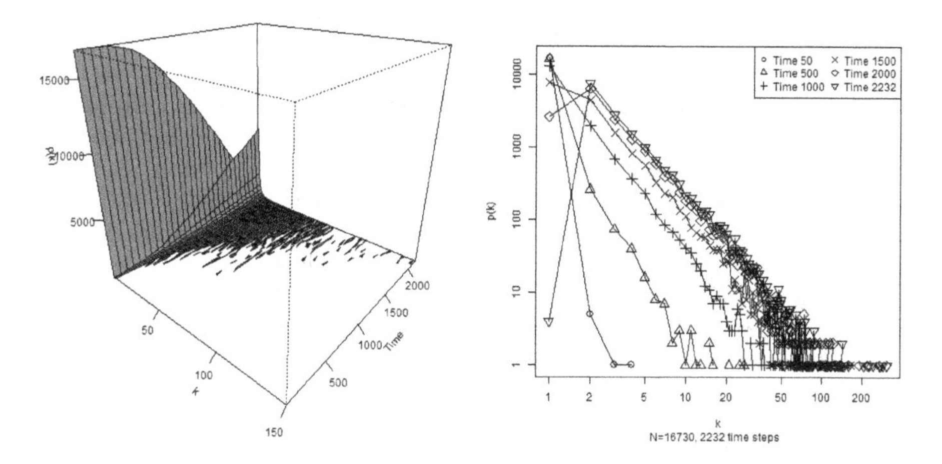

Fig. 7. Degree distribution evolution for the cumulative network constructed for the dataset of Rocha et al.

the first 2,000 time steps, and the log-log degree distribution starts transforming drastically in the last 200 time steps. This indicates that additional accumulated data might alter the powerlaw nature of the degree distribution of the cumulative network.

4 Conclusion and Future Works

In this paper we have studied how elementary dynamic network models behave compared to various empirical networks. Starting from an empty initial network, we examined network properties (like density, clustering, average path length, number of components and different views about the degree distribution) of both snapshot and cumulative networks for various lengths of aggregation time windows.

The results reported in this paper summarize the first phase of a longer research plan. We found that the baseline properties of the elementary dynamic network models show identical trends, in the analysed statistics, to the ones exposed by the empirical data sets. The knowledge of the baseline behavior of the abstract elementary dynamic network models helps us to identify the important, idiosynchratic properties of the empirical networks. However, we found notable differences.

In the first case, the Gulf dataset show exceptionally high clustering values, very low average but extremely high maximal betweenness. The reason for that is the existence of a highly connected core in the network. We would like to extend our investigation with further datasets and – as the Gulf dataset emphasized – perform experiments with a stable core that is highly connected and/or unchanged during the simulation. We also found that time granuality (i.e., monthly or daily aggregation) does not yield significant difference in the results, and it can be described with the size difference in the snapshot network.

Our findings emphasize the importance of preferential attachment in real-world systems, since the elementary dynamic network models incorporating the mechanisms of preferential attachments are the closest in behaviour to the trends found in the empirical data.

In the second case, we found that, based on the change in its density (and other related statistics, like the average centrailty values), the sexual network dataset is still in the beginning of the evolution of its cumulative network. Another interesting finding is that the degree distribution of the network is almost stable for during the first 2,000 time steps it is exposing a power law pattern, but in the next 200 a drastical change takes place in it deforming the distribution.

Acknowledgments. This work was partially supported by the European Union and the European Social Fund through project FuturICT.hu (grant no.: TÁMOP-4.2.2.C-11/1/KONV-2012-0013).

References

1. Legendi, R.O., Gulyas, L., Kampis, G.: Properties of Elementary Random and Preferential Dynamic Networks. Poster Presented at European Conference on Complex Systems 2011, Vienna, Austria (2011)
2. Legendi, R.O., Gulyas, L.: Cumulative Properties of Dynamic Social Networks. In: Proceedings of the 7th European Social Simulation Association Conference, Montpellier, France (2011)
3. Legendi, R.O., Gulyas, L.: Effects of Time-Dependent Edge Dynamics on Properties of Cumulative Networks. In: Proceedings of the Satellite Meeting EPNACS 2011 within ECCS 2011, Vienna, Austria (2011)
4. Gulyas, L., Legendi, R., Kampis, G.: Elementary Models of Dynamic Networks. Advances in Dynamic Networks. Special Issue of European Physical Journal (forthcoming, 2013)
5. Legendi, R.O., Gulyas, L., Kampis, G.: Effects of Preferential Attachment on Properties of Elementary Dynamic Networks. Poster Presented at ELTE Innovation Day, Budapest (2012)
6. Gulyas, L., Horváth, G., Cséri, T., Szakolczy, Z., Kampis, G.: Betweenness Centrality Dynamics in Networks of Changing Density. Presented at the 19th International Symposium on Mathematical Theory of Networks and Systems (MTNS 2010) (2010)
7. Tesfatsion, L., Judd, K.L.: Handbook of Computational Economics. Agent-Based Computational Economics: A Constructive Approach to Economic, ch. 16, vol. 2, pp. 831–880. Elsevier (2006) ISBN 1574-0021
8. Goyal, S.: Introduction to Connections: An Introduction to the Economics of Networks. Princeton University Press (2007)
9. Keeling, M.J., Eames, K.T.D.: Networks and epidemic models. Journal of the Royal Society, Interface / the Royal Society 2(4), 295–307 (2005) ISSN 1742-5689, doi:10.1098/rsif.2005.0051

10. Isella, L., Stehle, J., Barrat, A., Cattuto, C., Pinton, J.-F., Van den Broeck, W.: What's in a crowd? Analysis of face-to-face behavioral networks. J. Theor. Biol. 271, 166–180 (2010), doi:10.1016/j.jtbi.2010.11.033
11. Carley, K.: Destabilizing Terrorist Networks. In: Proceedings of the 8th International Command and Control Research and Technology Symposium. Conference held at the National Defense War College, Washington DC. Evidence Based Research, Track 3. Electronic Publication (2003)
12. Read, J.M., Eames, K.T.D., Edmunds, J.W.: Dynamic social networks and the implications for the spread of infectious disease. Journal of the Royal Society Interface the Royal Society 5(26), 1001–1007 (2008)
13. Skyrms, B., Pemantle, R.: A Dynamic Model of Social Network Formation. In: Proceedings of the National Academy of Sciences (2004)
14. Kossinets, G., Watts, D.J.: Empirical Analysis of an Evolving Social Network. Science 311(5757), 88–90 (2006), doi:10.1126/science.1116869
15. Tizghadam, A.: On Congestion Control in Mission Critical Networks. In: Proceedings of the Second IEEE International Workshop on MissionCritical Networking (2008)
16. Tantipathananandh, C., Berger-Wolf, T.Y., Kempe, D.: Knowledge Discovery and Data Mining. In: A Framework for Community Identification in Dynamic Social Networks, pp. 717–726 (2007), doi:10.1145/1281192.1281269
17. Barabási, A.-L., Albert, R.: Emergence of Scaling in Random Networks. Science 286(5439), 509–512 (1999), doi:10.1126/science.286.5439.509
18. Erdős, P., Rényi, A.: On random graphs, I. Publicationes Mathematicae (Debrecen) 6, 290–297 (1959)
19. Lee, S., Rocha, L.E.C., Liljeros, F., Holme, P.: Exploiting temporal network structures of human interaction to effectively immunize populations. Quantitative Biology - Populations and Evolution (November 2010)
20. The Kansas Event Data System: Gulf data set, `http://web.ku.edu/~keds/data.dir/gulf.html` (accesssed February 25, 2013)
21. Pajek datasets: KEDS - The Kansas Event Data System, `http://vlado.fmf.uni-lj.si/pub/networks/data/KEDS/keds.htm` (accesssed February 25, 2013)
22. Rocha, L.E.C., Liljeros, F., Holme, P.: Information dynamics shape the sexual networks of internet-mediated prostitution. Proc. Natl. Acad. Sci. 107, 5706–5711 (2010)

Appendix: Daily/Monthly Gulf Dataset Comparison

The following charts demonstrate the evolution of the properties for both the dynamic cumulative and snapshot networks of the Gulf dataset. The left column contains the results obtained from the daily granuality, while the right column contains the same property for the monthly version. The properties of the cumulative network does not show notable differences. On the other hand, the snapshot network properties do show considerable differences. Since the snapshot network of the monthly dataset is about $\sim$ 30 times larger, this is not surprising. This difference results in a higher overall rating for the properties, and larger connected components.

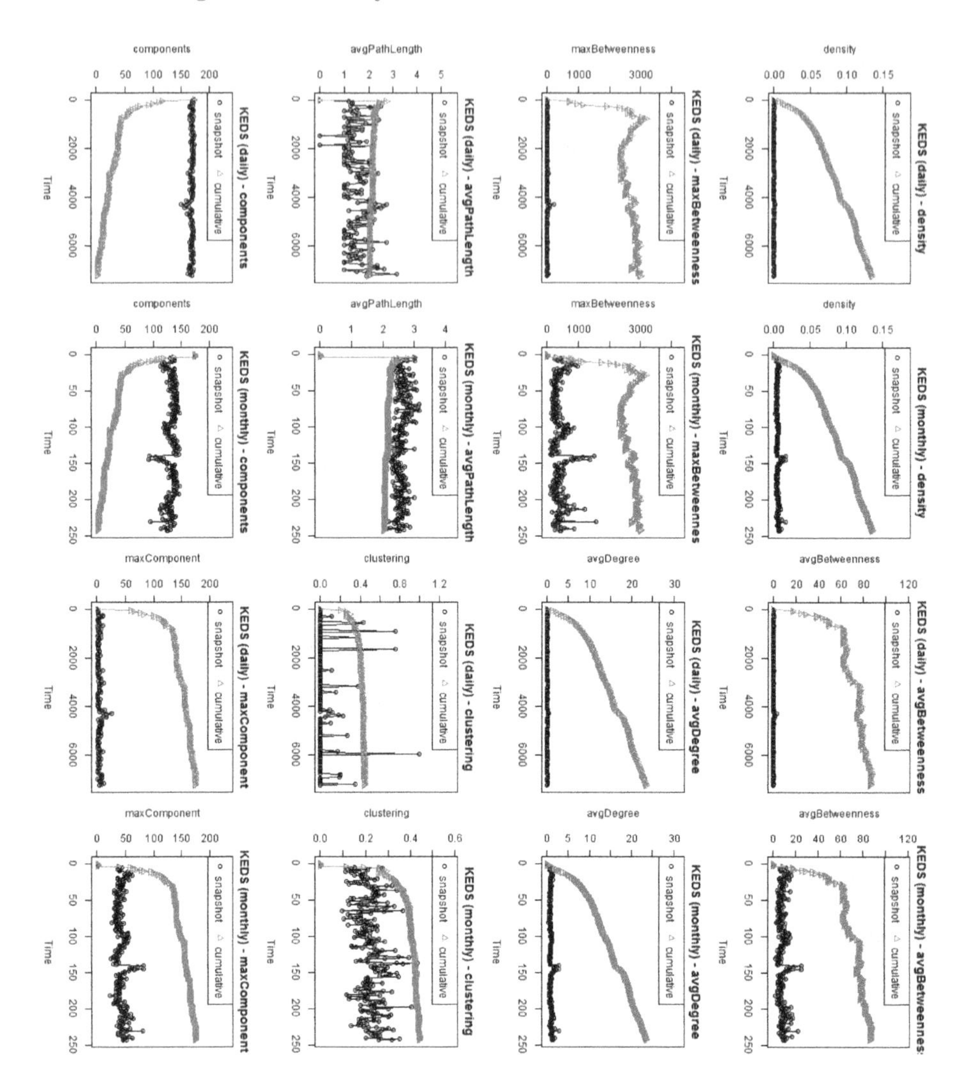
KEDS (daily) - density
KEDS (daily) - maxBetweenness
KEDS (daily) - avgPathLength
KEDS (daily) - components
KEDS (monthly) - density
KEDS (monthly) - maxBetweennes
KEDS (monthly) - avgPathLength
KEDS (monthly) - components
KEDS (daily) - avgBetweenness
KEDS (daily) - avgDegree
KEDS (daily) - clustering
KEDS (daily) - maxComponent
KEDS (monthly) - avgBetweennes
KEDS (monthly) - avgDegree
KEDS (monthly) - clustering
KEDS (monthly) - maxComponent
snapshot
cumulative
Time

Analyzing the Problem of the Modeling of Periodic Normalized Behaviors in Multiagent-Based Simulation of Social Systems: The Case of the San Jerónimo Vegetable Garden of Seville, Spain

Flávia Cardoso Pereira dos Santos[1], Thiago Fredes Rodrigues[1], Henrique Donancio[1], Glenda Dimuro[2], Diana F. Adamatti[1], Graçaliz Pereira Dimuro[1], and Esteban De Manuel Jerez[2]

[1] Universidade Federal do Rio Grande (FURG) - Brasil
{faflasan,trodrigues02,henriquedonancio, dianaada,gracaliz}@gmail.com
[2] Depto de Expresión Gráfica y Arquitectónica, Universidad de Sevilla, Sevilla, Espanha
glenda.dimuro@gmail.com

Abstract. This paper presents some results obtained through the modeling of a multiagent system for the simulation of production and social management processes of an urban ecosystem: the San Jerónimo Vegetable Garden (HSJ) of Seville, Spain. The social organization of HSJ is based on the performance of periodic routines by the organizational roles, and also on periodic norms that regulates their behaviors. For the modeling this kind of periodicity, that are commonly observed in social system, we used a combination of tools to offer a suitable solution, as the MOISE+ (part of JaCaMo platform) and the MSPP (Modeling and Simulation of Public Policies) *framework*. Although those tools separately present limitations for the modeling of periodic actions and norms associated to the performance of those actions, they can be used in a combined way, where the norms are specified in the MSPP *framework*, which support periodicity, and the normalized routines in the MOISE+ model.

1 Introduction

This work addresses, in an interdisciplinary approach, some preliminary results obtained in the modeling of a multiagent system for the simulation of the social production and management of a urban ecosystem – a joint effort for interrelating knowledge, seeking collective interpretations, adopting as case study the current tendency of (re)approaching the countryside to the city through urban vegetable gardens. The chosen organization is the project of social vegetable gardens conducted at the San Jerónimo Park (Seville/Spain), driven by the NGO "Ecologistas en Acción".

The general objective is to develop a MAS-based simulation tool for the analysis of the current reality of the project, allowing discussions on the adopted social management processes, and also for investigating how possible changes in actions, behaviors, and roles assumed by the agents in the organization, especially from the point of view of their participation in the decision making processes, may transform this reality, from

B. Kamiński and G.Koloch (eds.), *Advances in Social Simulation*,
Advances in Intelligent Systems and Computing 229,
DOI: 10.1007/978-3-642-39829-2_6,

the social, environmental and economical point of view, then contributing for the sustainability of the project.

In previous works [1,2,3], we presented the first phase of the MAS organization modeling, developed using the organizational model MOISE+ [4,5], identifying the organizational roles of the San Jerónimo vegetable garden (in Spanish: Horta San Jerónimo - HSJ) and their routines, the social interactions, the regulative and constructive norms.

The social organization of HSJ is based on the performance of periodic routines by the organizational roles, and also on periodic norms that regulates their behaviors.

Due to the difficulty found in the modeling of the periodic routines that characterize the behaviors of the different organizational roles identified in the HSJ social organization, as well as their normative aspects, we decided to explore more the JaCaMo platform (http://jacamo.sourceforge.net/), which is a framework for Multiagent Programming that combines three separate technologies, namely Jason (for programming autonomous agents, Cartago (for programming environment artifacts) and MOISE (for programming multiagent organizations).

However, we observed that the MOISE+ model lacks facilities for specifying routines. So, we decided to analyze another related tool, namely the MSPP (Modeling and Simulation of Public Policies) framework[6,7], which allowed us to specify the periodicity in role behaviors in the HSJ social organization.

Then, for the modeling of such kind of periodicity commonly observed in social system, we found that a combination of tools can offer a suitable solution, namely, the MOISE+ of JaCaMo platform and the MSPP framework. Although those tools separately present limitations for the modeling of periodic actions and norms associated to the performance of those actions, they can be used in a combined way, such that the norms are specified in the MSPP framework, which support periodicity, and the normalized routines in the MOISE+ model.

The aim of this work is to present an analysis, through the modeling of the HSJ social organization, of the results obtained with the application of the MOISE+ organizational model [2] and the MSPP framework, pointing to the limitations found with respect to the modeling of periodic normalized routines in a social system. Then, the proposed combined solution is discussed.

The paper is organized as follows. Section 2 explain the social project of HSJ, presenting some examples of our preliminary modeling of the HSJ social organization. Section 3 presents the Jacamo Plataform and the MSPP framework, with a discussion on their use for the modeling of periodic routines. In Section 4, we present the restrictions of Moise+ and how to solve them with the MSPP framework. Sections 5 presents the conclusion and further works.

2 About the San Jerónimo Urban Vegetable Garden Social Organization

The San Jerónimo Urban Vegetable Garden (in Spanish: Horta San Jerónimo - HSJ) is an initiative of the NGO Ecologists in Acción to promote social participation in organic farming practices through the use of urban vegetable gardens to recreation, and conducting activities related to environmental education [8,9,10].

Currently, it covers an area of about 1.5 hectares of the San Jerónimo Park (public space of municipal property), around 42 units (with sizes ranging from 75 to 150m^2), that are divided into individual cultivable plots (primarily cared for by retiree gardeners, but not exclusively), school gardens (devoted to elementary students from neighborhood schools) and assigned plots to other associations for scientific experiments (as the Andalusian Seed Network and Andalusia Platform Free of Transgenics).

The gardens are bound to gardeners and the right to use (and not property) of the plot occurs for a period of two years (although extendable), in which the gardener agrees to follow the guidelines and rules set out in regulation set by the NGO "Ecologistas en Acción" [11]. Also, they participate in promoted courses and workshops on environmental education.

All plots are considered homelike, but have a "holder" or guardian gardener. However, there is no impediment to appoint a helper or volunteer (other than family) to help in cultivating the garden. Its main features are the fact of being a nonprofit social garden, that is, the production is dedicated to the consumption of those who cultivate (to sale is illegal) and is supported economically by municipal funding and collaboration of the participants.

The HSJ is governed by rules and is the ONG "Ecologistas en Acción" who verifies compliance with these standards. When a rule is disobeyed, punishment / penalty should be applied, according to the rules of the garden. The gardener responsible may be expelled from the parcel that is under his responsibility.

2.1 Initial Modeling of the HSJ Social Organization

In order to organize and establish the behaviors of the different roles and roles' routines of HSJ organization, as well as the frequency of these routines, we used the so-called ellipses, a kind of Venn diagram of set theory. The use of ellipses helps us to analyze the periodicity of the roles' routines, helping the understanding of the agents' behavior, as well as the identification of interactions between them and the environment. As an example, Figure 1 shows the ellipses of the routines of the NGO's Secretariat, described as:

- **Daily routines:** to receive candidates' documentations desiring to participate in the project, called the Aspiring Vegetable Gardener, registering them in the waiting list; to receive transfer request to plot possession.
- **Monthly routines:** to receive monthly fees paid by the Vegetable Gardener to cover costs with water (drip), pest control material, use of common tools, etc.; to inform the meetings; to register Auxiliar Vegetable Gardener (informed by Vegetable Gardener).
- **Biennial routine:** to receive a request from a gardener to continue in the project.
- **Seasonal routine:** to sent the received documentation for the NGO Administration.

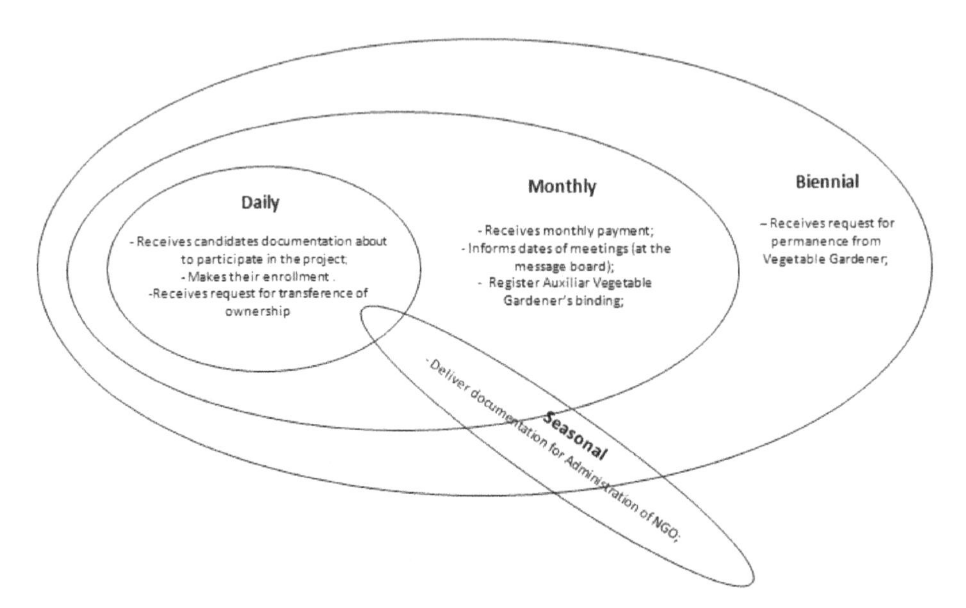

Fig. 1. Ellipses of the routines of the NGO Secretariat

After that, we use the MOISE$^+$[4,5] organizational model to model the organization of HSJ. The modeling consists of the specification of three dimensions: the structural, where roles and inheritance links and groups are defined; the functional, which establishes a set of comprehensive plans and missions to achieve; and the deontic, which is the dimension responsible for defining what role or permission is required to accomplish each mission.

The representation of the organization of HSJ in the MOISE+ organizational model is shown in Figure 2. Observing the relationship between the roles, an example of inter-group communication is the one observed between the gardener (included in the group "plot") and the technician (included in the group "NGO"). More information about the MOISE+ modeling of HSJ can be seen in [2].

In the sequence, to allow a clear visualization of the interactions between the class instances, we use UML activity diagrams. An Activity Diagram is a diagram defined by the Unified Modeling Language (UML), representing the flows driven by processes. It is essentially a flow chart that shows the flow of control from one activity to another. Usually this involves the modeling of sequential steps in a computational process.

In our work, we use these diagrams to visualize the interactions between roles of the HSJ organization.

Figure 3 is an activity diagram showing interactions between the the roles of Auxiliar Vegetable Gardener, Vegetable Gardener and Technician. In the following we explain it briefly.

Initially, the agent that assumes the Auxiliar Vegetable Gardener role arrives at the NGO's building, and it listens to a lecture. This lecture is given by another agent,

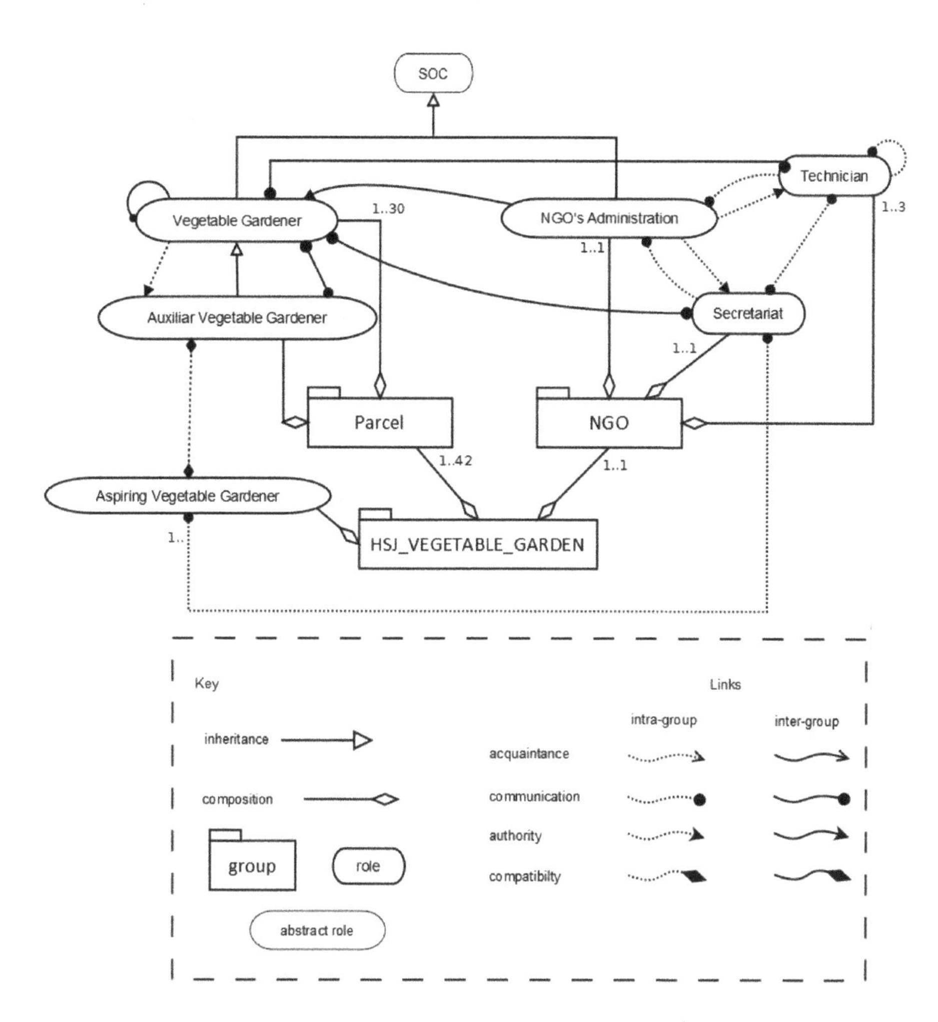

Fig. 2. Structural Specification of HSJ with MOISE$^+$ model

playing the Technician role, and teaches some rules on how to harvest adequately. This activity is executed as soon as the Technician agent perceives that everyone has arrived at the hall. The interaction between each role is accomplished through oral communication, in which the latter agent talks to everyone.

Following this interaction, another one is performed between the Auxiliar Vegetable Gardener and the Vegetable Gardener agents. The former one requests authorization to use a cabinet, using oral communication. Finally, the latter answers the request giving the other agent permission to use the requested cabinet, giving it the key to access that object.

Many other interactions follow these, as Figure 3 presents.

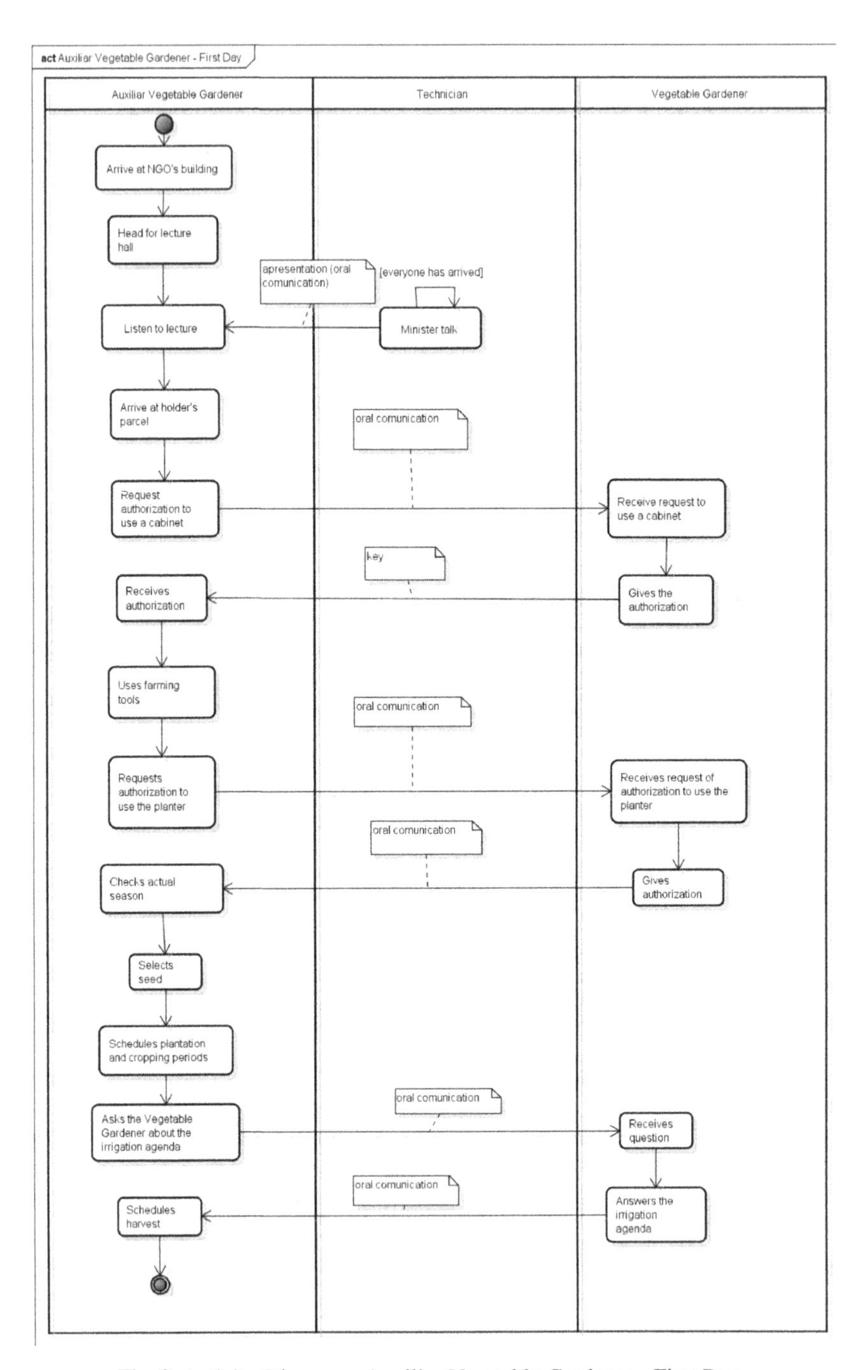

Fig. 3. Activity Diagram - Auxiliar Vegetable Gardener - First Day

3 The JaCaMo Plataform and the MSPP Framework for the Modeling of Role Rotines in HSJ

For some roles identified when studying the San Jerónimo vegetable garden, we observed that their routines are repeated periodically. In the JaCaMo plarform, the modeling of such role routines cannot be easily done, since there are no native tools in the platform that allow this specification.

Currently, the allowed processes in the MAS organization, in terms of the goals that must be achieved, can be described through the MOISE+ model. This tool presents a good abstraction level to specify these objectives, as well as the definition of an hierarchy between them. However, a periodic routine involves the achievement of periodic goals (e.g., in periods of one month, one week, one day), and MOISE+ model does not have structures to do represent such periodicity.

Moreover, in social systems, there are situations in which norms must be applied, imposing sanctions over not allowed actions when they are performed by the agents. In HSJ, many actions related to imposed norms were identified, as "sell garden's harvest", "irrigate with hoses" and "use of chemicals in the garden".

As there is no way in MOISE+ for defining the periodicity of actions (in the achievement of goals), also there is no direct mean to define norms, their basic attributes (name, periodicity, applying role) and the sanctions.

Aiming to offer a modular way of describing periodic norms, simplifying our modeling of the social system comprised by the San Jerónimo garden, we use the MSPP (Modeling and Simulation of Public Policies) framework [6,7].

This tool complements the MOISE+ model, offering another abstraction layer. In MSPP model, routines can be modeled, having the norms defined in the framework, while, in the the MOISE+ model, the normalized actions are specified, constituting the routines.

In the following, we will briefly explain the JaCaMo platform and the MSPP framework, highlighting the components of both platforms. JaCaMo [12] is a multiagent systems development framework that combines three technologies: Jason, CArtAgO e MOISE+.

3.1 Jason

Jason is an *AgentSpeak-L* interpreter, and provides a platform to develop multiagent systems, supporting agent communication based in the speech act theory. There are many BDI systems *ad hoc* implementations, however an important characteristic of the *AgentSpeak-L* language is its theoretic base. Jason is implemented in Java (multiplatform) and is available as *Open Source* under the GNU LGPL license [13].

3.2 CartAgO

CartAgO (Common ARTifact infrastructure for AGents Open environments) is a multiagent systems virtual environment development and simulation framework. Through this tool is possible to implement virtual environments as a computational layer encapsulating the facilities and non-autonomous services exploited by agents during runtime.

CartAgO is based in the Agents & Artefacts (A & A) meta-model [14] to model multiagent systems. This model introduces a high-level metaphor, based in the idea that human workers work in a cooperative way with its environment: agents are computational entities that do some type of goal-oriented task (analogous to human workers), and artifacts are the resources and tools dynamically created, handled and shared by agents to support its activities, both individual and collective (as in the human context).

Therefore, it is possible to develop artifacts that are instantiated in the environment and can provide services to agents, also able to do communication with external services (e.g., web-services). CartAgO is a Java-based, Open Source technology and is available in [14].

3.3 MOISE+

The MOISE+ [4] organizational model was created to model the MAS's organization and is the specification of three dimensions: the structural (where roles, inheritance links and groups are defined), the functional (where a set of global plans are defined, with missions to achieve those goals) and the deontic (where specifies which role has to commit to which mission).

Therefore, MOISE+ is a multiagent organizational model, based in notions as roles, groups and missions, which allows the system to have an explicit organization and that a platform can be used to make agents achieve its tasks defined in the organization.

In the following, we present a brief introduction to the model's functional dimension. It is composed by a set of Social Schemes, that is, a set of goals, structured by means of plans. In the formal specification of MOISE+, the set of all social schemes is denoted by *SCH* and a scheme *sch* is represented by the tuple

$$sch = (\mathcal{G}, \mathcal{P}, \mathcal{M}, mo, nm) \tag{1}$$

where $\mathcal{G}$ is the set of goals in the scheme, $\mathcal{P}$ the set of plans that builds the goals decomposition tree, $\mathcal{M}$is the set of missions, that is, a set of global goals that can be bound to a role, mo: $\mathcal{M} \mapsto \mathbb{P}\ (g)$ is a function that determines the set of goals in each mission, nm: $\mathcal{M} \mapsto \mathbb{N} \times \mathbb{N}$ determines the maximum and minimum number of agents that must commit to each mission.

A *scheme* is a global goal decomposition tree, whose root is the goal of the entire *scheme*. The decomposition is made by plans (denoted by the '=' operator), that point a way of achieving a role. For instance, on plan "$g0 = g1; g2; g3$" the role $g0$ is decomposed in three plans, indicating that it will be achieved only if plans $g1, g2$ and $g3$ are also achieved.

3.4 The Modeling and Simulation of Public Policies Framework

The framework for insertion of public policies becomes concrete in the form of artifacts in the model CArtAgO. This framework includes two types of normative artifacts, namely: NormObrig and NormPrb, modeling rules of obligation and prohibition, respectively.

There are also special agents (BDI agents for the Jason platform) to perform / verify those norms. They are: the government agent, which is responsible for emit norms; the

Table 1. Comparing the analyzed results

NORM	TYPE	PERIODICITY	MOISE	MSPP FRAMEWORK
Monthly payment	Obligation	Monthly	Not Support	Support
Acquire cupboard	Right	First Day	Support	Support
Cultivate garden	Obligation	Monthly	Not Support	Support
Irrigate	Obligation	Daily (reactive)	Not Support	Support
Request technical support	Permission	Daily	Not Support	Support
Evict violator gardener	Obligation	Monthly	Not Support	Support

social agents, which are subject to the norms and seek to reach proper objects; and the government officials detectors and / or effectors, which are responsible for detecting the compliance of a policy issued by the government agent as well as regulatory environment resources, and apply possible sanctions to actions that characterize the breaches of the rules.

The framework presupposes the adoption of these four types of agents interacting to promote the policy cycle.

The norms are structured in the following way [6,7]:

- **Id:** the identifier of the norm;
- **Recipient:** specifies the role to which the norm applies;
- **Action:** specify an action to be held by the agent who takes the role to which the norm was addressed;
- **Condition:** specifies contextual a condition necessary for the application of the norm;
- **Periodicity:** Specifies the event that should occur (month, week, or a specific action) to verify the condition;
- **Exception:** specifies a condition in which the norm does not apply;
- **Sanction:** specifies the sanction to be imposed in case of violation of the norms.

The social agents, effectors and detectors agents are constantly observing the norms. They acquire knowledge of any changes in the system. Their knowledge of the norms is added through beliefs that define any action that is prohibited, obligatory.

Once committed a breach of these norms, it is up to the detector agent and / or effector agent to seek in the artifact the appropriate sanction and eventually to apply it. Public resources available in the environment and up to the same shall also be willing as a CArtAgO artifact to establish interaction in the system.

4 Comparing MOISE+ and the MSPP Framework

In this section, we present a comparision between the MOISE+ modelling and MSPP Framework, in the context of HSJ social organization. The first column of Table 1 shows a sample of norms observed in the HSJ regulation. In the second column the types of the norms are shown, namely, *obligation* (i.e., a norm that must be accomplished), *right* (i.e., a norm that represents a right of an agent playing a certain role) and *permission* (i.e., a norm that indicates an allowed action). In the third column the periodicity of the normative actions is presented. In the last two columns we indicate whether or not the MOISE+ model and MSPP Framework support such normative action.

For example, observe in Table 1 that only the norm "Acquire cupboard" is supported by the MOISE+ organizational model, since there is no periodicity in this action, i.e., it occurs just once. The gardener, when acquiring its plot in HSJ, has the right to have a cupboard to store utensils, tools or materials for general use. In MOISE+, this task's execution occurs through the realization of this mission on the first day and ends by indicating its satisfiability level, which indicates that the goal has already been achieved, in this case, "satisfied".

Also in Table 1, the "Monthly Payment" norm states that a gardener should pay a monthly tax if he/she has a cultivation parcel. This norm is an obligation one, with a monthly periodicity. This type of norm can not be modeled by MOISE+, since it does not have structures to natively allow the description of norms and routines, as described in Section 3. For that, we propose the use of the MSPP framework, since it better suits this task, having structures (the Periodicity parameter) that allow such normalization of periodic actions, as discussing in section 3.4.

Thus, the analysis of routines in the San Jerónimo Vegetable Garden allows us to observe that due to the periodicity of the the routines and norms, as shown in Table 1, using only the MOISE+ model it is not possible to complete the modeling of the organization. However, the use the MSPP framework proved to be adequate, meeting the need that the system requires. Then, we propose to use those tools in a combined way, such that the norms are specified in the MSPP framework, which support periodicity, and the normalized routines in the MOISE+ model.

5 Conclusion and Further Works

The JaCaMo platform has resources that make it a fully-fledged platform for modeling multiagent systems, including social systems. However, for the case study of this paper, some limitations could be observed related to MOISE+ model, with respect to the modeling of routines, namely, periodic routines and norms. In the case study of HSJ, several actions are periodic, which is very common in social systems and even in other types of systems.

The specification of routines could not be done in MOISE+, making necessary the use of another tool. In this paper, we propose to use the MSPP framework. The use of this framework enabled the specification of normative actions of the organizational roles, offering a more modular specification like other aspects pertaining to a MAS (organization, population and environment, already treated in JaCaMo). Besides that, it allows its modification according to conditions evaluated by agents, in their routines. For example, irrigation is not necessary if it has rained.

We emphasize that the features of the framework complements those provided by the MOISE+ model, where goals and actions can be described hierarchically. The description of the periodicity is made in the framework and the actions are specified in MOISE+.

In Figure 4, a proposal for future work is presented. We intend to integrate different types of artifacts (Organizational, Regulatory, Communication and Physical), for the domain of HSJ.

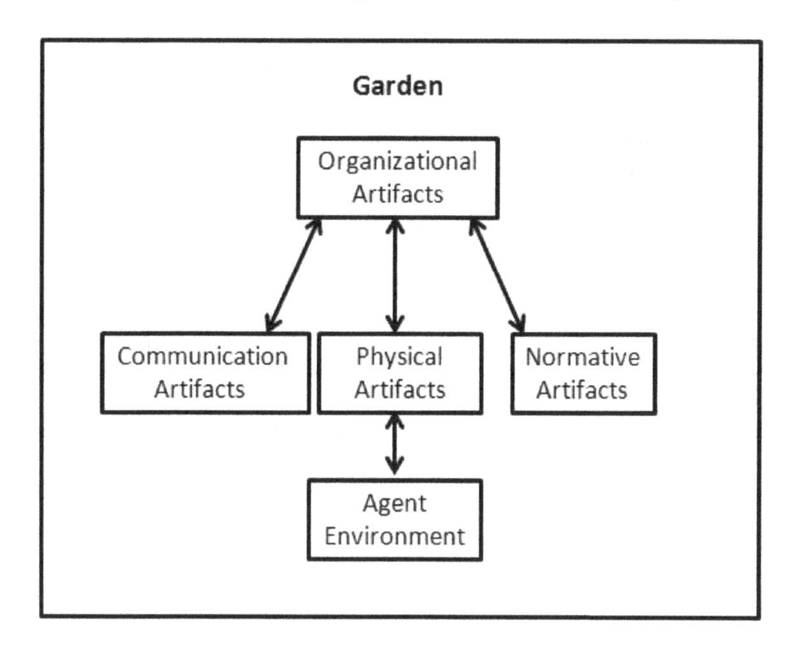

Fig. 4. Proposed Architecture: artifacts

The Organizational artifacts make the communication between the roles in the organization, defined in the MOISE+ model, and among the population of agents. The Communications Artifacts mediate communication, forwarding messages to their recipients, overseeing its sending order. The Normative Artifacts demonstrate how the relationships and interactions occur between roles with relation to the norms of the HSJ. The Physical Artifacts are abstractions over the environment, symbolizing physical objects and services that agents use to accomplish their tasks.

Therefore, it is intended to apply an approach in the scenario of the HSJ, where the artifacts are connected to the environment and each role use them to find what is needed for the simulation of the activities in the the garden.

References

1. Santos, I., Rodrigues, T.F., Dimuro, G.P., Costa, A.C.R., Dimuro, G., De M. Jerez, E.: Towards the modeling of the social organization of an experiment of social management of urban vegetable gardens. In: Lugo, G., Hübner, J. (eds.) Proceedings of the 2011 Workshop and School of Agent Systems, their Environment and Applications (WESAAC), pp. 98–101. IEEE, Los Alamitos (2012)
2. Santos, F.C.P., Dimuro, G., Rodrigues, T.F., Adamatti, D.F., Dimuro, G.P., Costa, A.C.R.: Modelando a organização social de um SMA para simulação dos processos de produção e gestão social de um ecossistema urbano: o caso da Horta San Jerónimo da cidade de Sevilla, Espanha. In: Hübner, J.F., Brandão, A.A.F., Silveira, R., Marchi, J. (eds.) Anais do WESAAC 2012 - VI Workshop-Escola de Sistemas de Agentes, seus Ambientes e Aplicações, Florianópolis, pp. 93–104, UFSC (2012)

3. Santos, F.C.P., Rodrigues, T.F., Dimuro, G., Adamatti, D.F., Dimuro, G.P., Costa, A.C.R., De Manuel Jerez, E.: Modeling role interactions in a social organization for the simulation of the social production and management of urban ecosystems: the case of San Jerónimo vegetable garden of Seville, Spain. In: Dimuro, G.P., Adamatti, D.F., Coelho, H., Sichmam, J.S., Balsa, J., Tedesco, P., Costa, A.C.R. (eds.) 2012 Third Brazilian Workshop on Social Simulation (BWSS), pp. 136–139. IEEE, Los Alamitos (2012)
4. Hübner, J.F.: Um Modelo de Reorganização de Sistemas Multiagentes. PhD thesis, Universidade de São Paulo, São Paulo (2003)
5. Hübner, J.F., Sichman, J.S., Boissier, O.: A model for the structural, functional, and deontic specification of organizations in multiagent systems. In: Bittencourt, G., Ramalho, G.L. (eds.) SBIA 2002. LNCS (LNAI), vol. 2507, pp. 118–128. Springer, Heidelberg (2002)
6. Santos, I., Rocha, A.C.R.: Toward a framework for simulating agent-based models of public policy processes on the jason-cartago platform. In: Proceedings of the Second International Workshop on Agent-based Modeling for Policy Engineering in 20th European Conference on Artificial Intelligence (ECAI)- AMPLE 2012, pp. 45–59. Springer, Berlin (2012)
7. Santos, I., Rocha, A.C.R., Dimuro, G.P., Mota, F.P.: A framework for simulating agent-based models of public policy cycles on the Jason-CArtAgO platform (to appear, 2013)
8. Dimuro, G.: La produccion social del habitat en ecosistemas urbanos (PhD Proposal, advisor: Esteban de Manuel Jerez) (2009)
9. Dimuro, G., Jerez, E.M.: La comunidad como escala de trabajo en los ecosistemas urbanos. Revista Ciencia y Tecnologia 10, 101–116 (2011)
10. Dimuro, G., Jerez, E.M.: Comunidades en transicion: Hacia otras practicas sostenibles en los ecosistemas urbanos. Cidades Comunidades e Territorios 20-21, 87–95 (2010)
11. Dimuro, G., Dimuro, G., Costa, A.C.R., Pinheiro, T.V.T., Grol, C.V., Rodrigues, T., Santos, F.C.P.: Modelagem do sistema multiagente para simulação de processos de gestão social em ecossistemas urbanos, estudo de caso: Horta san jeránimo. Technical Report 1, FURG/Universidad de Sevilla (2011)
12. Bordini, R.H., Hübner, J.F.: JaCaMo project, `http://jacamo.sourceforge.net/` (accessed in September 2012)
13. Bordini, R.H., Hubner, J.F., Wooldridge, M.: Programming Multi-Agent Systems in AgentSpeak using Jason. Wiley, New Jersey (2007)
14. Ricci, A., Santi, A., Piunti, M.: CArtAgO (common atifact infrastructure for agents open environments) (2013)

Saddle Points in Innovation Diffusion Curves: An Explanation from Bounded Rationality

Lorena Cadavid and Carlos Jaime Franco Cardona

Universidad Nacional de Colombia – sede Medellín, Medellín, Colombia
{dlcadavi,cjfranco}@unal.edu.co

Abstract. Empirical evidence shows that mostly complete and successful processes of innovation diffusion are S-shaped. However, some diffusion processes exhibit a non-perfect S-curve, but show a saddle point, which is displayed as a double-S. The reasons behind this phenomenon have been little studied in the literature. This paper addresses the emergence of the double-S phenomenon in the innovation diffusion process and provides an explanation for it. In order to do that, the authors develop an agent-based simulation model to representing the diffusion of two innovations in a competitive market considering elements of bounded rationality. The results show saddle points appear as a result of three characteristics: (1) the heterogeneity in the population, (2) the presence of asymmetric information and (3) the satisfaction criterion for selection.

Keywords: Innovation Diffusion, Agent-based Modeling, Bounded Rationality.

1 Introduction

The spread of an innovation over markets is known as innovation diffusion, a process by which an innovation is communicated through certain channels over time among members of a social system [1]. Empirical studies show that successful and complete processes of innovation diffusion take S shape [2], as many natural phenomena; hence, theoretical studies attempt to find the rate and amount of adopters in a specific population during a period of time [3].

Nevertheless, some diffusion processes exhibit a non-perfect S-curve, but show a saddle point, which is displayed as a double-S. Figure 1 shows the diffusion curves for telephone, automobile, air conditioning and clothes washer in U.S.A, as some of the examples of this phenomenon. Evidence form Europe is collected by Goldenberg, Libai, and Muller [4], as well as recent S-shape diffusion data are presented by Tellis and Chandrasekaran [5].

B. Kamiński and G.Koloch (eds.), *Advances in Social Simulation*,
Advances in Intelligent Systems and Computing 229,
DOI: 10.1007/978-3-642-39829-2_7, © Springer-Verlag Berlin Heidelberg 2014

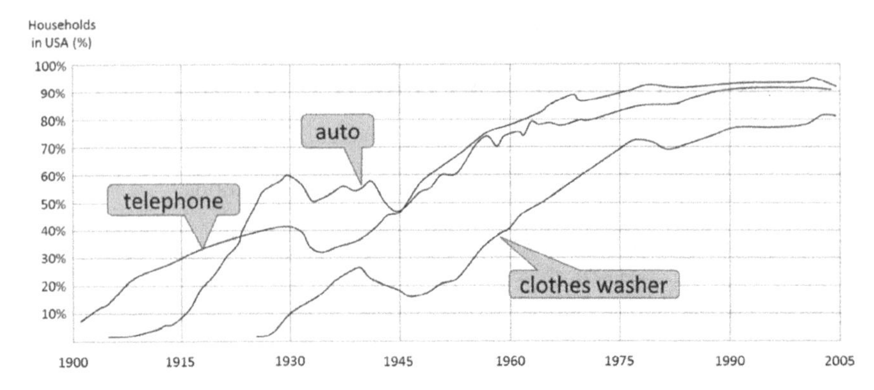

Fig. 1. Double-S phenomenon – the USA case (Source: the authors, based on Walton and Rockoff [2])

It is possible to see, a few periods after diffusion takeoff point, how diffusion slows down; even, it is evident a decrease in amount of adopters during that deceleration time ("dis-adoption" decisions). This phenomenon does not make sense in light of classic innovation diffusion theory [6]; according to it, the more individuals become adopters, the stronger word-of-mouth force and, therefore, the more adopters market will have.

Several questions arise from this issue. Why individuals or households stop adopting the innovation after diffusion takes off? What other "force", besides the word-of-mouth, affecting the adoption decision? And what happens during that non-diffusion period, which makes diffusion takes off again after some time?

While numerous studies try to reproduce the classic S-shaped diffusion phenomena, little attention has been paid in the double-S diffusion phenomena, and the reasons behind this emergent behavior remain still confusing for researchers and practitioners.

Therefore, this paper addresses the emergence of the double-S phenomenon in the innovation diffusion process and provides an explanation for it based on bounded rationality guideless. The authors use agent based modeling as a tool for this purpose, and find the emergent phenomenon after the interaction among individuals, who must choose between two identical innovations and have imperfect information and limited calculation capacity.

This research has an explanatory scope. It is not aimed to policy evaluation, since the double-S phenomenon cannot be considered good or bad by itself; however, uncover reasons which underlie the emergence of the phenomenon can be useful for several managerial purposes.

2 Diffusion of Innovations

An important part of the innovation diffusion literature can be classified in two dominant research families: (1) those that characterize the mechanisms and patterns of diffusion (innovation diffusion), and (2) those that seek to understand and characterize the structure of decision-making and individual processes of adoption of innovation (innovation adoption) [3]. These families are presented below.

2.1 Innovation Diffusion Modeling

Mathematical or computational representation of the process of diffusion of an innovation is known as innovation diffusion modeling, and has been a topic of academic and practical relevance since the 60s, when the first models emerged [7]. It is possible distinguish two trends in innovation diffusion modeling: (1) modeling through aggregation of collective behavior and (2) modeling through individual behavior representation [8, 9].

Diffusion models at the aggregate level have produced a vast literature in the last 50 years in the field of innovation diffusion [7, 10, 11]. These models provide an empirical generalization based on the mathematical formulation of differential equations.

However, literature has identified several limitations for these models. Perhaps the most notable of them is that they do not explicitly consider the individuals heterogeneity and the complex dynamics of social processes that shape the phenomenon of diffusion, so they explain only a limited set of theoretical issues [12, 13]. Hence, recently research has been focused to an individual modeling, which considers both, the individual heterogeneity as the complexity of interaction networks between them.

Diffusion models at the individual-level are composed of two main elements: (1) modeling of adoption decision-making of analysis units (individuals, households and organizations, etc.), and (2) modeling of social interaction between those units. Interaction or social influence is associated with the way in which individuals are interrelated, while the modeling of decision-making allows adoption decisions of individuals are explicitly incorporated into diffusion models [14].

2.2 Innovation Adoption Modeling

Adoption decisions at individual-level modeling began in the 70s, and has been a major field of research over the past 20 years [10]. Approaches have been mainly based on bounded rationality [15], because the asymmetric information and the partial understanding of market dynamics by actors in the diffusion process [9, 16].

Kiesling et al. [11] find that adoption models used in individual diffusion models are a kind of heuristics, that is to say, mechanisms that enable individuals to go directly to the conclusions of your thoughts without following a detailed process of reasoning [17, 18]. These models can be grouped into five categories; the simplest of these corresponds to the threshold heuristics, while the most complex refers to psychosociological approaches.

Implementation of heuristics implies the concept of heterogeneity in individuals, which is particularly important when this decision rule is considered. Overall, the population heterogeneity refers to variation in the degree of consumer innovation, patience, price sensitivity and needs; among those, heterogeneity in the propensity to adopt is a common approach to incorporate the heterogeneity of individuals [11]. It is supposed these differences allow the product takes place over a market, and their distribution in a specific social system determines the shape of the pattern of diffusion [10, 14].

3 The Double-S Phenomenon

Commonly, diffusion models predict a monotonic increase in sales after the launching of product and up to the peak of growth. Despite that, evidence shows that in some markets, a sudden decrease in sales may follow an initial rise. This special point in diffusion curves is referred as a slowdown phenomenon [19] or "saddle" point (concept proposed by Goldenberg, Libai and Muller [4].

After a literature review, Tellis and Chandrasekaran [5] point out three possible explanations for this phenomenon, two of which can be considered as external explanations and one of them as an internal one: economic contraction and important technological advance (external) and discontinuity in the transition between the early and late markets (internal). The authors find, after using a discrete-time split population survival model for several products in several countries, that all the reasons influence the S-shape phenomenon in some extent.

In a previous work, Golder and Tellis [20] argue that this phenomenon is due to informational cascades, which happen when people collectively adopt a behavior with increasing momentum, and declining individual evaluation of the merits of such behavior after, because of their tendency to infer information from the behavior of prior adopters.

Nonetheless, although their argument lies in individual behavior, they propose a hazard model to determine the impact of explanatory variables such as price declines, income declines, and market penetration on the time to saddle point; whereby their hypothesis is not completely tested at the individual level (as in the Tellis and Chandrasekaran [5] study).

At individual level, Goldenberg, Libai, and Muller [4] use cellular automata to describe the process by which internal communication breaks down between the early adopters and early majority, which could lead to a saddle point in the diffusion curve. Nevertheless, cellular automata models do not allow modeling socioeconomic characteristics of population, so conclusions from them are very limited.

Peres et al. [10] point out that one of the further research line on innovation diffusion is still to explain saddle formation and incorporate the findings into the diffusion framework.

4 Methodology

In order to analyze the emergence of the double-S phenomenon in the innovation diffusion process, it was developed an agent based simulation model, which represents the diffusion of two identical innovations in a market. The model was developed in NetLogo language and environment [21]. The agents are individuals that are interconnected according to a specific network structure.

Considered innovations correspond to theoretical exercises, which were simulated for a population of 100 individuals (which compose one only agent breed, besides observer) in a horizon of 20 periods. In each simulation step, only those individuals who have not chosen one of the innovations use the heuristic to make a decision. Also, it is assumed that the decision-making process is not reversible, i.e. once one of the individual adopted one of the innovations, he/she cannot change his/her mind about it (this is equivalent to a model of first purchases, as suggested by Bass [6] in their original research).

The diffusion processes described above is based on bounded rationality guidelines, and consists of a heuristic at individual level, which is composed by three modules according to Todd and Gigerenzer [22]: (1) information search, (2) alternatives evaluation and (3) selection criterion. The individual's awareness about innovations is determined by the individual exposure to each one of the innovations, which depends on advertising for innovation, and increases as more individuals in the network adopt one or another innovation. Figure 2 shows the decision rule schema for potential adopters.

The assessment made for every individual about each innovation is a weighted average between the influence of word-of-mouth (which depends on the number of previous adopters directly connected to the individual) and the advertising of the particular innovation. Thus, in every period of simulation, each individual monitors the information provided by the environment, and can be found in one of four possible scenarios: (1) individual is not informed of any innovation, (2) individual knows only innovation number 1, (3) individual knows only innovation number 2 or (4) individual knows both innovations.

Each one of these situations involves a different procedure in the heuristic regarding the selection criterion, namely:

- If the individual does not know any innovation, there is no decision to consider.
- If the individual knows only one innovation, it is verified that the value of this innovation meets or exceeds the satisfying level of individual (threshold of satisfaction - TS); in this case, the innovation is adopted. If this value is below its threshold of satisfaction, the individual does not adopt the innovation and adjusts its threshold of satisfaction value downward.
- If the individual knows both innovations, it can happen:
    - None of the innovations meets or exceeds the threshold of satisfaction of the individual; in this case, the individual does not adopt any innovation and decreases its threshold of satisfaction for the next step.
    - Only one of the innovations exceeds the threshold of satisfaction of the individual; in this case, the individual adopts this alternative.
    - Both innovations exceed the threshold of satisfaction of individual; in this case, the individual does not adopt any innovation and increases its threshold of satisfaction for the next step.

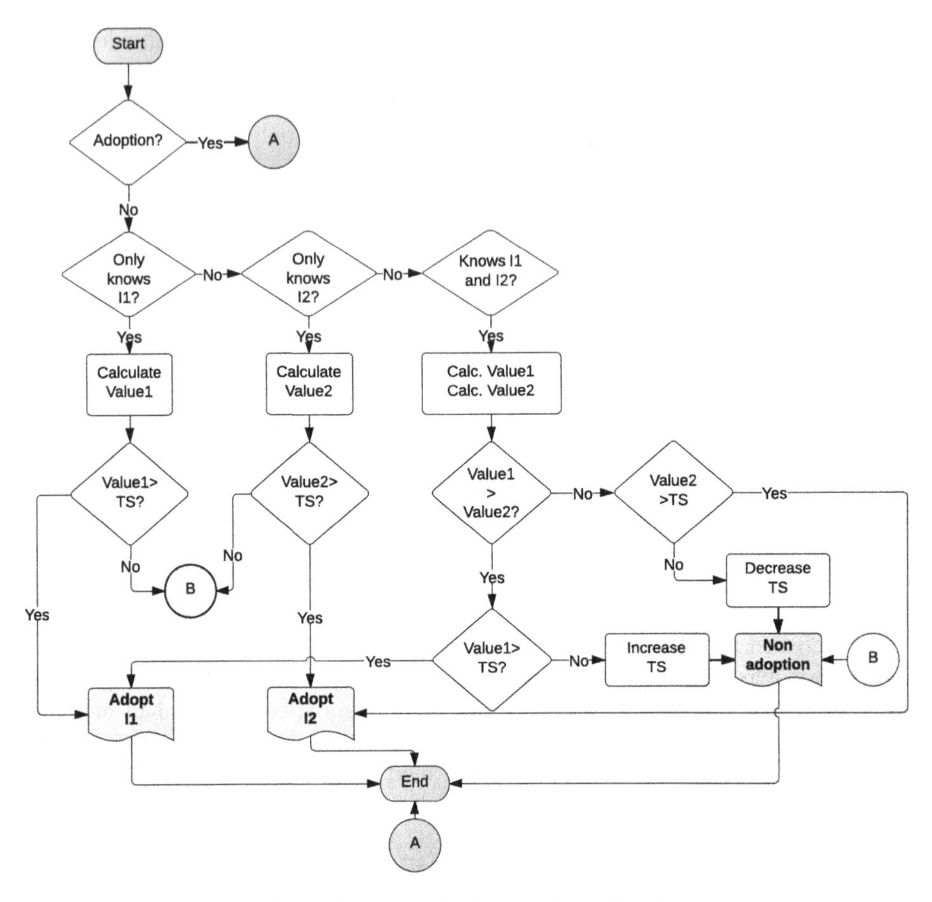

Fig. 2. Decision rule (Source: the authors)

5 Experimentation and Results

It was built a baseline scenario (BS) for evaluating the impact of the social network and population heterogeneity in the diffusion curves. For experimentation, heterogeneity was varied through the standard deviation of the normal distribution for modeling the threshold of satisfaction of individuals; also, the network structure was varied from a fully connected to a small-world network, which was implemented through generative algorithm developed by Watts and Strogatz [23]. As a result, Figure 3 shows the analyzed scenarios.

Every scenario was simulated 1000 times, in order to catch the variation of the stochastic variables; for that reason, the results present the average diffusion curves after the runs. Figure 4 shows the innovation diffusion curves for the considered scenarios. Note that, because it was considered two identical innovations, diffusion curves for both of them are almost equals between them.

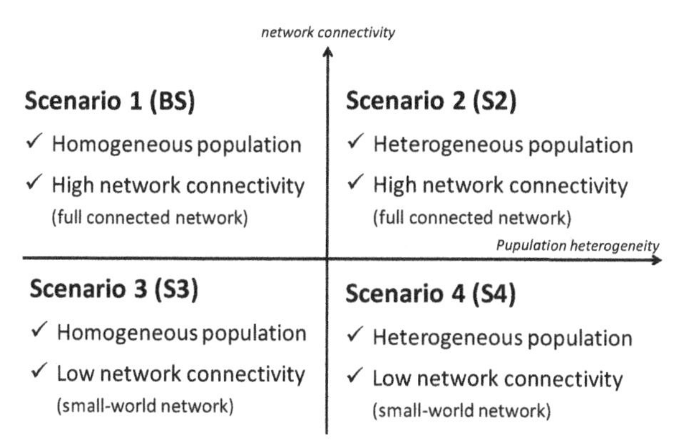

Fig. 3. Scenarios (Source: the authors)

Shape of the curves is similar between BS and S3, and between S2 and S4, and variation into each one of those groups is related with the speed of diffusion. In this sense, it can be argued that the network topology has implications for the speed at which diffusion of two innovations occur: the small-world structure network causes the diffusion occurs at a lower speed, taking longer to saturate the market.

This is because in a fully connected network, every adoption decision increases the value of the innovation for every one of the individuals in the network, making easier its own adoption decision; while, in a small-world network, adoptions increase the innovation value only for those individuals connected with the adopter, so that this information does not spread efficiently over the network.

Unlike the network topology, population heterogeneity change, not the diffusion rate, but the shape of diffusion curves. Thus, greater heterogeneity causes an earlier takeoff point in diffusion curves, so the earlier periods present adopters at higher rates than those observed for homogeneous population in the same periods.

Further, it is possible to see a double-S shape curve for both, S2 and S4. According the logic considered in the model, the saddle points appear as a result of three characteristics: (1) population heterogeneity, (2) asymmetric information and (3) satisfaction criterion for selection.

Unlike in homogeneous population, the presence of extreme points in the thresholds of satisfaction of individuals in heterogeneous population ensures the existence of individuals whose thresholds are below the valuations of innovations they know.

Given that in the first periods there are many individuals who know only one of the innovations, the threshold adjustment process is not necessary and, therefore, adoptions make an early appearance. These adoption decisions increase the value of innovation for other individuals, so the diffusion process is energized.

However, when the number of individuals who know only one of the innovations decreases, the diffusion makes slower, because of individuals must now wait until their thresholds of satisfaction reach suitable levels to make an adoption decision. This deceleration takes the shape of saddle point, which is observed in the graphs. Once satisfaction levels reach values around the valuations of the innovations, diffusion is accelerated again until market gets saturated, so the double-S shape in the curve is done.

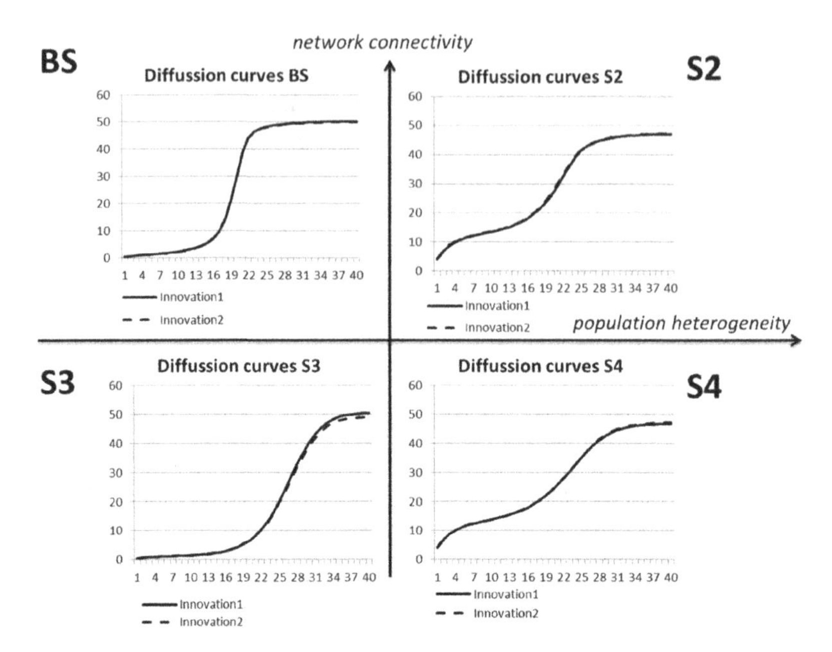

Fig. 4. Diffusion curves of scenarios (Source: the authors)

6 Conclusions and Recommendations

This paper proposed an alternative explanation for the double-S shape observed in several complete and successful innovation diffusion processes, which is based on the rationality of individuals instead of technological change, macroeconomic events or interactions among people.

For the analyzed phenomenon, the heterogeneity in population has a high impact on the diffusion curves, changing its shape pattern. In this sense, competing diffusion that happens in heterogeneous populations with asymmetric information tends to have saddle points that give a double S-shaped; while homogeneous populations allow the formation of simple S-shaped curve typically analyzed in simple diffusion models.

As well, it was shown that the network structure affects the speed of the diffusion, but not the shape of it; i.e, the network structure has not any incidence on the S-shape phenomenon.

This analysis differ from previous analysis for the same phenomenon, because it models explicitly the individual rationality. It was possible because the use of an agent based model, and perhaps because of the recent appearance of this technique in the innovation diffusion field, this analysis could not have been explored by previous research.

At the methodological level, this research shows the potential of using agent based models to address more questions about diffusion of innovations, especially those related with the individual rationality; thus, the individual level modeling can allow the integration between adoption and diffusion models through the explicit modeling of making adoption decision process.

References

1. Rogers, E.M.: Diffusion of Innovations. Collier Macmillan Publishers, London (1983)
2. Walton, G.M., Rockoff, H.: History of the American economy. South-Western Pub., Canadá (2010)
3. Montalvo, C., Kemp, R.: Cleaner technology diffusion: case studies, modeling and policy. Journal of Cleaner Production 6, S1–S6 (2008)
4. Goldenberg, J., Libai, B., Muller, E.: Riding the Saddle: How Cross-Market Communications Can Create a Major Slump in Sales. Journal of Marketing 66, 1–16 (2002)
5. Tellis, G.J., Chandrasekaran, D.: Getting a Grip on the Saddle: Chasms or Cycles. Working paper series 52, 463–479 (2012)
6. Bass, A.: new product gowth for model consumer durables. Management Science 15, 215–227 (1969)
7. Mahajan, V., Muller, E., Bass, F.: New Product Diffusion Models in Marketing: A Review and Directions for Research. The Journal of Marketing 54, 1–26 (1990)
8. Goldenberg, J., Efroni, S.: Using cellular automata modeling of the emergence of innovations. Technological Forecasting and Social Change 68, 293–308 (2001)
9. Georgescu, S., Okuda, H.: A Distributed Multi-Agent Framework for Simulating the Diffusion of Innovations. Journal of Power and Energy Systems 2, 1320–1332 (2008)
10. Peres, R., Muller, E., Mahajan, V.: Innovation diffusion and new product growth models: A critical review and research directions. International Journal of Research in Marketing (2010)
11. Kiesling, E., Günther, M., Stummer, C., Wakolbinger, L.M.: Agent-based simulation of innovation diffusion: a review. Cent. Eur. J. Oper. Res. (2011)
12. Berger, T.: Agent-based spatial models applied to agriculture: a simulation tool for technology diffusion, resource use changes and policy analysis. Agricultural Economics 25, 245–260 (2001)
13. Rahmandad, H., Sterman, J.: Heterogeneity and network structure in the dynamics of diffusion: Comparing agent-based and differential equation models. Management Science 54, 998–1014 (2008)
14. Kemp, R., Volpi, M.: The diffusion of clean technologies: a review with suggestions for future diffusion analysis. Journal of Cleaner Production 21, S14–S21 (2008)
15. Simon, H.A.: A behavioral model of rational choice. The Quarterly Journal of Economics 69, 99–118 (1955)
16. Collantes, G.O.: Incorporating stakeholders' perspectives into models of new technology diffusion: The case of fuel-cell vehicles. Technological Forecasting and Social Change 74, 267–280 (2007)
17. Manson, S.M.: Bounded rationality in agent-based models: experiments with evolutionary programs. International Journal of Geographical Information Science 20, 991–1012 (2006)
18. Secchi, D.: Extendable Rationality. Estados Unidos (2011)
19. Chandrasekaran, D., Tellis, G.J.: A critical review of marketing research on diffusion of new products. Review of Marketing Research, 39–80 (2007)

20. Golder, P., Tellis, G.J.: Growing, Growing, Gone: Cascades, Diffusion, and Turning Points in the Product Life Cycle. Marketing Science 23, 207–218 (2004)
21. Willensky, U.: NetLogo. The Center for Connected Learning (CCL) and Computer-Based Modeling. Northwestern University (2012)
22. Todd, P.M., Gigerenzer, G.: Bounding rationality to the world. Journal of Economic Psychology 24, 143–165 (2003)
23. Watts, D.J., Strogatz, S.H.: Models of the Small World. Nature 393, 440–442 (1998)

Estimation of Production Rate Limits Using Agent-Based Simulation for Oil and Gas Plants Safety

Yukihisa Fujita[1], Kim Nee Goh[2], Yoke Yie Chen[2], and Ken Naono[3]

[1] Central Research Laboratory, Hitachi, Ltd., Yokohama-shi, Kanagawa, Japan
yukihisa.fujita.hg@hitachi.com
[2] Computer and Information Sciences Department,
Universiti Teknologi PETRONAS, 31750 Tronoh, Perak, Malaysia
{gohkimnee,chenyokeyie}@petronas.com.my
[3] Hitachi Asia Malaysia Sdn. Bhd., Kuala-Lumpur, Malaysia
knaono@has.hitachi.com.my

Abstract. Safe production is one of the most important issues in oil and gas plants. Oil and gas companies lose enormous amounts of money and trust once they experience accidents such as explosions and oil spills. They have to make clear production rate limits to mitigate the human related risks that cause accidents. In this study, we model plant workers using an agent-based simulation to estimate the limits. The proposed model represents that plant workers solve problems with risks, and that the probability of the occurrence of a problem is proportional to three main factors, the production rate, the plant size, and the plant equipment degradation rate. The experimental results show that the proposed model can estimate the limits for different sizes of the three factors. The results imply that the limit estimation made by the proposed model is very crucial for plant operations to mitigate human-related accidents.

Keywords: agent-based simulation, oil and gas plants, production rate estimation, plant worker, risk mitigation.

1 Introduction

Petroleum is one the most important and useful energy resources in the world. According to the International Energy Agency, the worldwide energy demand has continuously increased [1,2]. Oil and gas companies have to meet the demands. However, resource production has risks that cause accidents such as explosions and oil spills. Once they experience such accidents, they lose enormous amounts of money and trust [3,4]. Therefore, safe production of oil and gas is one of the critical issues for these companies.

There are three production process stages, upstream, midstream, and downstream for produce oil and gas from the available resources. The upstream process contains searching for the resources, drilling exploratory wells, and

B. Kamiński and G.Koloch (eds.), *Advances in Social Simulation*,
Advances in Intelligent Systems and Computing 229,
DOI: 10.1007/978-3-642-39829-2_8, © Springer-Verlag Berlin Heidelberg 2014

operating the wells. The midstream process contains the gathering, refining, processing, storing, and transporting of oil and gas. The downstream process contains retailing and distributing it. Some of the important operations in these processes, such as the exploration and refinery, are executed at plants. The important factors to mitigate the risks in these plants are manual operations done by plant workers. Although operations in these plants are highly automated, there are still many manual operations. The plant workers manipulate plant equipment in order to perform these operations.

The plant workers play a critical role in safe production. According to Bayerl and Lauche [5], which investigated and observed the activities of oil production, the task performed by plant workers is "to keep the plant running by exporting gas at a maximum rate." However, according to investigations conducted by the U.S. Department of the Interior [6], human error caused 41% of all the accidents at offshore plants near the U.S. coast in 2000. From these data, we can say that the manual operations done by the plant workers have some human related risks that may cause accidents while the plant workers are trying to safely run the plants. In general, it is unrealistic to eradicate the human related risks as long as there are manual operations. One of the ways to mitigate these risks is to raise the level of the safety awareness of the plant workers. Sneddon et al. [7] suggested that oil and gas companies should provide training programs that focus on the cognitive skill of each plant worker in order to improve their awareness. O'Dea and Flin [8] clarified that the human factors, such as communication and job experience, are the critical requirements of leadership for safe production. These studies focus on the activities of the plant workers to mitigate the human related risks. However, while it is possible to reduce the risks, it is not possible to reduce them to zero. Therefore, we need to consider the human related risk tolerance from the viewpoint of the human activities to ensure safer production.

We propose an estimation method for the production rate limit for the human related risk tolerance in this study. The production rate limit is the maximum production rate of plants with no accidents. We assume that the amount of problems with risks is proportional to three factors, the production rate, the plant size, and the plant equipment degradation rate. Using this assumption, we constructed an Agent-based Simulation (ABS) model to estimate the limits. The simulation represents that the plant workers solve the problems generated with a probability decided by focusing on the three factors. We also show that the problems produce accidents over time. Using the simulation, we calculated the maximum production rate of the plants with no accidents.

The rest of the paper is organized as follows. First, Section 2 lists the related work in the area of production estimation and ABS. Next, Section 3 explains our proposed model. Then, Section 4 shows our experimental results. Finally, the meanings of the current results and the limitations of the model are discussed in Section 5, and this paper is concluded in Section 6.

2 Related Work

One of the ways to estimate and optimize production is to create a mathematical formulation of the plant equipment efficiency. Pongsakdi et al. [9] developed a production planning model by taking into consideration the uncertainty of product demand and price, and clarified that their model can sugget solutions to reduce financial risks and improve profits. Göthe-Lundgren et al. [10] proposed an optimization model in which the storage capacity minimizes the production cost, and clarified that an increase in the storage capacity reduces the total production cost. These models provide useful decision making information for efficiently operating plants. Additionally, there are methods that focus on controlling the equipment in the plants [11]. The control methods, which are called Model Predictive Controls (MPCs), calculate the efficiency of the plant operations based on the input and output of each equipment, and provide feedback to future control strategies. For example, Meum et al. [12] applied a nonlinear MPC to reservoir production. The simulation results showed that the method using a nonlinear MPC can increase the reservoir recovery performance. The studies we explained above provide the methods for estimating the production rate, amount, and efficiency from the viewpoints of equipment specification and equipment control. However, as mentioned in the previous section, the human factor has to be considered in order to ensure safer production.

Agent-based simulation is one of the solutions to consider such human factors. It is used to represent the emergent phenomena based on the individuals and for analyzing complex systems. One of the complex systems is the human behavior within organizations such as companies. Typical ABS usage is to measure the organizational performance. For example, OrgAhead [13] focuses on the relationship between the organizational structure and task processing performance, and finds the optimal organizational structure to achieve a maximum performance using machine learning techniques. The task represents product developments, military operations, and so on. SimVision[1] [14] focuses on the relationship between the information flow among employees and the task structure represented using the Program Evaluation and Review Technique [15], and estimates the organizational performance. These simulations can estimate the organizational performance for the given tasks. On the other hand, they are inadequate at representing the human related risks causing accidents. The plant workers have to tackle the problems in the plants, not the given tasks. Moreover, while the workers represented in the above models do their work to achieve their goals as fast as possible, the plant workers solve problems caused by daily production to keep the plant running and safe. In other words, while the workers represented in the above models are actively performing their duties, the plant workers passively perform theirs. Therefore, for our purpose, we have to construct a new model that has the following features, the human factors deciding the daily production rate, and the plant workers passively solving their problems.

[1] SimVision is a registered trademark of ePM, LLC in the U.S.A. and other countries.

3 Plant Operation Model

3.1 Overview

Figure 1 shows our proposed model, the "Plant Operation Model." This model consists of three components, the plant environment, the event, and the worker agent. The plant environment represents the work place of the worker agents and outputs the production rate as the results from the daily production. The event represents an incident that can grow from a risk to a fatal accident. The worker agent represents one plant worker who solves the problems in the plants. In this model, the plant environment generates the events in proportion to the current production rate, and the worker agent tackles and processes the generated events. The remaining events, which are not processed by the worker agents, reduce the production rate and grow worse over time. In a worst case scenario, the plant production rate becomes zero, which means the plant stops production due to a fatal accident. We can represent two plant features using this model, the human factors deciding the daily production rate and the plant workers passively solving their problems. We describe the implementation to estimate the production rate limits for the human related risk tolerance in the sections that follow.

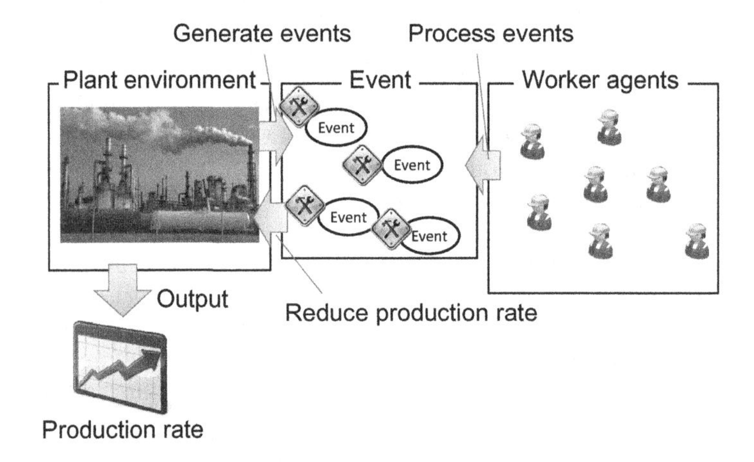

Fig. 1. Overview of plant operation model

3.2 Plant Environment

In this implementation, the plant environment has the following parameters:

- production rate r,
- maximum event generation quantity M, and
- maximum event generation probability P.

Production rate r is a real number with a $[0, 1]$ interval, and is calculated at every time step. It represents how efficiently the plant runs. Maximum event generation quantity M is a fixed positive integer, and maximum event generation probability P is a fixed real number with a $[0, 1]$ interval. These two parameters represent how often problems occur in the plant. They are respectively decided based on the plant size and the plant equipment degradation rate. In this model, the plant environment generates an event with a probability $r \times P$. This event generation process is iterated M times every time step. In other words, the higher the production rate, plant size and plant equipment degradation rate, the more problems that occur. This represents our assumption that the amount of problems with risks is proportional to three factors, the production rate, the plant size, and the plant equipment degradation rate. The generated event is assigned to one randomly selected agent. This represents when the plant worker finds equipment degradation that should be repaired.

3.3 Event

An event represents an incident that can grow from a risk to a fatal accident. One event e has the following parameters:

- time t_e,
- reduction rate R_e,
- required skill S_e, and
- required workload w_e.

Time t_e is the number of steps passed after generation of this event without processing by the worker agents. Reduction rate R_e is a fixed real number with a $[0, 1]$ interval. Suppose that E is a set of events that are being processed or not processed yet, the production rate r is calculated by using the following equation:

$$r = \prod_{e \in E} (1 - R_e). \qquad (1)$$

This means, the more problems that remain, the less the production rate. Required skill S_e is L-dimensional vector $S_e = \{s_i | 1 \leq i \leq L, s_i \in \{0, 1\}\}$. The number of dimensions L represents the number of skill types required in the plants. Each element of the vector represents the skill necessary to handle the equipment and solve problems. Required workload w_e, which is a positive integer, is reduced by 1 when the worker agent processes event e. When w_e becomes 0, event e is removed from the simulations. This means the plant worker solved the problem.

3.4 Worker Agent

The worker agents represent the plant workers. They process the events using their skills. The skills of a worker agent is represented by L-dimensional vector $S_w = \{s_j | 1 \leq j \leq L, s_j \in \{0, 1\}\}$, which is the same as required skill S_e.

In this implementation, the worker agents form a hierarchical structure. They can be classified as a "manager" or "operator." The managers have relationships with their subordinates and superior managers. The operators have no subordinates. The operators tackle and process assigned events. When the operators cannot process the events due to a lack of skill, they report the events to their superiors. The managers assign the events to their subordinates when they can process the events, otherwise the managers report the events to their superiors. This mechanism represents the typical cooperation within organizations. For this mechanism, the skills of the worker agents classified as managers equals the sum of their subordinates' skills to know whether the subordinates can process the events or not. For example, the skill of a manager is $\{1, 1, 1, 0\}$ when the manager has three subordinates who have $\{0, 0, 1, 0\}$, $\{0, 1, 1, 0\}$, and $\{1, 0, 0, 0\}$ skills, respectively.

In each time step, the worker agents select one event that has the highest t_e from the assigned events. Then, the worker agents select the appropriate action for this step from the following:

(a) processing,
(b) reporting, or
(c) assigning.

When there are no events assigned to the worker agents, the worker agents do not select any actions. Each action is explained in the following paragraphs using Fig. 2.

(a) Processing. This action can be selected by only the worker agents classified as operators. The worker agent selecting this action processes the selected event when its skill S_w satisfies the required skill S_e for the event, this means all the elements of vectors S_w, S_e satisfy the following condition; the ith element of S_w equals 1 when the ith element of S_e equals 1. The required workload of the selected event is reduced by 1. In the example described in Fig. 2(a), the worker agent whose skill set satisfies the required skill processes the event. As a result, the required workload of the event is reduced from 3 to 2.

(b) Reporting. This action can be selected by only the worker agents who can select neither processing nor assigning actions. The worker agent selecting this action assigns the selected event to its superior. When there are two or more superiors, one of them is randomly selected. In the example described in Fig. 2(b), worker agent B reports the event to superior A because the skill set of worker agent B does not satisfy the required skill for the event. The reported event is assigned to worker agent A.

(c) Assigning. This action can be selected by only the worker agents classified as managers. The worker agent selecting this action assigns the selected event to its subordinates. The worker agent assigns the events to one of its subordinates when the subordinate can process the event solely. The worker agent divides the event into sub-events and assigns sub-events to its subordinates when there are no subordinates who can process the event solely. The event is divided into sub-events with the minimal number of worker agents who can process them.

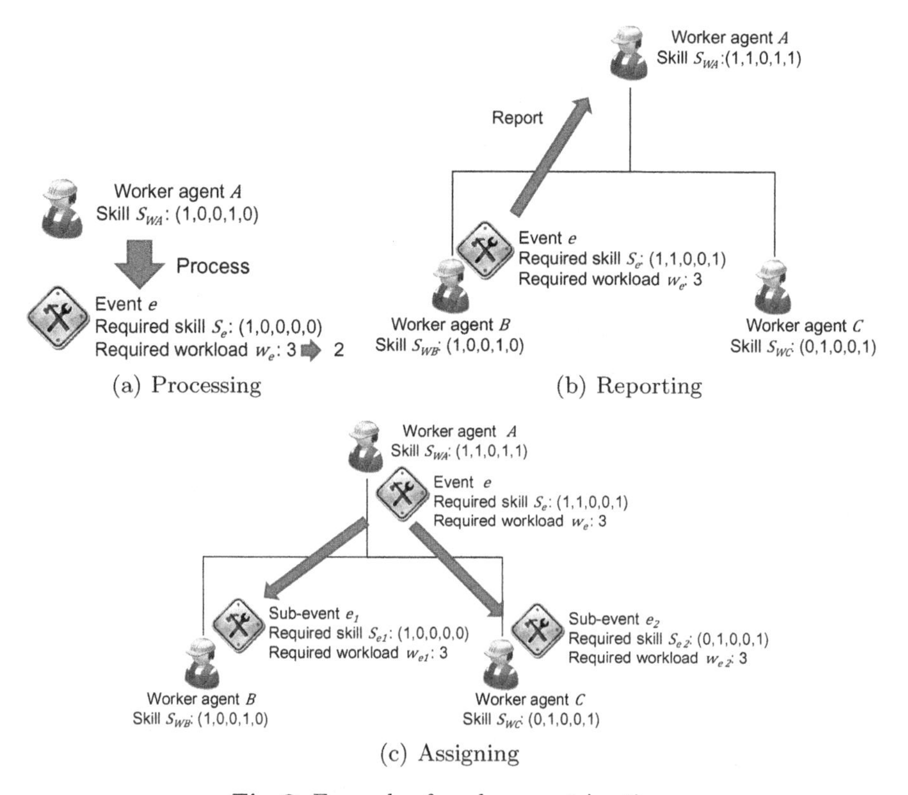

Fig. 2. Example of worker agents' actions

In the example described in Fig. 2(c), worker agent A divides event e into two sub-events e_1 and e_2, and assigns them to subordinates B and C, respectively.

3.5 Time Step

In each time step, this simulation processes as follows:

(1) the plant environment generates events and assigns them to worker agents,
(2) where each worker agent selects one action based on the assigned events, and
(3) each worker agent acts as selected.
(4) The events are removed when their required workload $w_e = 0$,
(5) and time t_e of the events, which are not processed by the worker agents, are increased, and then,
(6) the production rate r is calculated.

Since production rate r is recalculated every time step, the probability of event generation is different for each time step. The number of events removed in one time step does not exceed the number of worker agents. Additionally, the remaining events worsen over time. We represent the tradeoff between the production rate and problem occurrence, and calculate the production rate limit for safe production by simulation using this model.

4 Experiment

4.1 Purpose and Conditions

In this experiment, we evaluated the production rate for various parameter settings, and found the production rate limit for the given settings.

Table 1. Parameter settings

Parameter	Value
No. of agents	33
No. of skill types L	50
Maximum event generation quantity M	5, 10
Maximum event generation probability P	0.10, 0.15, 0.20
Required workload w_e	$\lceil \exp(0.05t_e) \rceil$
Required skill S_e	refer to **Table 2**
Reduction rate R_e	refer to **Table 2**
No. of steps	6000
No. of iterations	50

Table 2. Event settings

Status	Conditions	Parameter	Value
Risk	$0 \leq w_e < 5$	No. of 1s in S_e	5
		Reduction rate R_e	0.05
Minor accident	$5 \leq w_e < 10$	No. of 1s in S_e	7
		Reduction rate R_e	0.15
Major accident	$10 \leq w_e < 15$	No. of 1s in S_e	22
		Reduction rate R_e	0.45
Fatal accident	$15 \leq w_e$	No. of 1s in S_e	40
		Reduction rate R_e	1.0

Table 1 lists the parameter settings for this experiment. We defined required workload w_e as a monotonically increased function that takes on t_e. We classified the events into four types according to the current required workload w_e as listed in Table 2. These settings represent that the problems worsen over time. The number of skill types, which the worker agents have, was decided by using a uniform distribution of between 7 to 15, which means the worker agents have 15 to 30 % of the necessary skills. The worker agents formed a tree structure, as shown in Fig. 3. We evaluated the results by using the production rates in percentages. All the results were obtained by using an averaging of 50 runs.

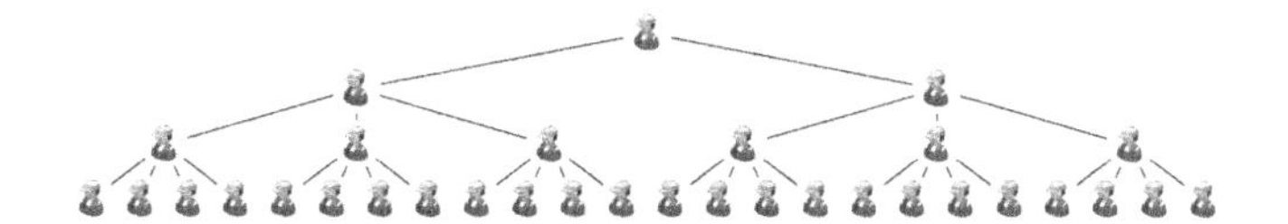

Fig. 3. Tree structure used in experiment

4.2 Results

First, Table 3 lists the final states for each maximum event generation quantity M and maximum event generation probability P. The processed events represents the number of events processed through 6000 steps. The risk, minor accidents, major accidents, and fatal accidents are the numbers of each remaining event after 6000 steps. As presented in Table 3, the greater M and P become, the lower the observed production rate. On the other hand, the number of processed events is increased due to an increase in both parameters. This means that the worker agents have the performance capacity to process more events. However, minor and major accidents occurred when $P = 0.20$ and $M = 10$. This suggests that the number of generated events exceeded the performance capacity of the worker agents.

Table 3. Final states of simulation runs for each maximum event generation quantity M and maximum event generation probability P. The production rate was calculated as a moving average of 100 steps. The processed events means the number of events processed through 6000 steps, and the risk, minor accidents, major accidents, and fatal accidents means the numbers of each remaining event at 6000 steps.

P	M	Production rate [%]	Processed events	Risk	Minor accidents	Major accidents	Fatal accidents
0.1	5	88.3	2651.7	2.3	0.0	0.0	0.0
0.15	5	82.9	3729.7	3.6	0.0	0.0	0.0
0.2	5	78.4	4683.8	5.0	0.0	0.0	0.0
0.1	10	78.1	4674.4	5.2	0.0	0.0	0.0
0.15	10	67.2	6152.1	7.6	0.0	0.0	0.0
0.2	10	56.0	6821.6	8.9	0.6	0.1	0.0

Next, Fig. 4 shows the time-series of the moving averages for 100 steps of the production rate with $M = 10$. As shown in Fig. 4, there are no drastic changes in the production rate. However, with the greater P, a wider range of changes appeared. In other words, an increase in P disrupts the production safety.

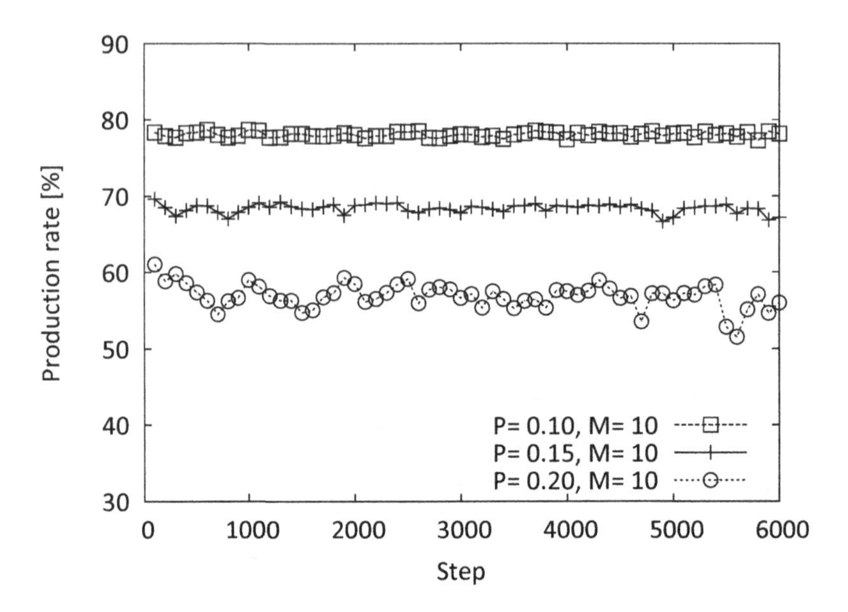

Fig. 4. Time-series of moving averages in 100 steps of production rate with $M = 10$. The data is plotted from 100 steps.

5 Discussion

We obtained the parameter settings that lead to accidents from the results of our experiments. In other words, the performance capacity of plant workers is exceeded in specific plant size and plant equipment degradation rate. While accidents occur when $P = 0.20$ and $M = 10$, they do not occur when $P = 0.15$ and $M = 10$. In this model, problems are generated when the probability is $r \times P$. Therefore, we can estimate the production rate limits from the parameters and the results of the production rate. The worker agents can safely process events when $P = 0.15$, $M = 10$ and the production rate is around 70%, that is, the worker agents can keep the plants safe when the probability is less than roughly 0.105 (0.15×0.7) when $M = 10$. We can calculate the production rate limit using the probability. In this case, the production rate limits are 0.7 when $P = 0.15$ and 0.525 when $P = 0.20$. We can mitigate the accidents if we reduce the production rate to make the probability less than the calculated value before the accidents occur. The current model has no method for controlling the production rate for such intention. We need to implement such a control method to use this model for real world applications.

In this experiment, we varied only two parameters, and the other parameters were fixed. Since different plants have different production rate limits, we have to set appropriate values to model the parameters when we apply the model to real world plants. Although real world plants have many parameters, they are represented by a few important parameters such as M and P. Therefore, we need to determine the values of such parameters. On the other hand, such parameter values are changed over time in real world plants. Moreover, we do not consider some of the more important factors such as the energy demands and financial status. This means that the proposed model can be applied to estimate the production rate limits only in the short term. To estimate the limit long term, we need to consider such factors.

The proposed model has limitations because it has both explicit and implicit assumptions like other simulation models. One of those limitations is that the model cannot represent the human cognitive factors. In the model, the worker agents always report the events if they have an insufficient skill set to process them. In other words, one assumption of the model is that plant workers solve problems only if they have sufficient skill. However, the plant workers cannot get root causes of problems from the phenomena that appear. As a result, they try to solve the problems even though they have an insufficient skill set. Processing with an insufficient skill set sometimes results in human-error causing accidents. Therefore, we need to consider such cognitive factors and introduce them into the model.

Another limitation is that the model cannot represent the communication difficulties among the plant workers. In real world plants, there are communication difficulties since the plants are located in highly dangerous and specific places like offshore exploration plants. In such dangerous places, using electrical equipment cannot be approved due to the risks of explosion. Therefore, they have to use traditional radio and wired phones with explosion-proof cases to communicate with each other. Moreover, the plant workers on offshore platforms have to collaborate with their colleagues at onshore plants because there are a limited number of plant workers on offshore rigs due to the difficult working environments. There is the potential for communication failures such as the misunderstanding of requests. These failures also present accident risks. Therefore, we also need to consider such communication problems.

6 Concluding Remarks

We proposed our plant operation model in this paper as the estimation method for the production rate limit for safe production. We modeled plant workers using agent-based simulation techniques. The proposed model represents that plant workers solve problems with risks, and that the probability of the occurrence of a problem is proportional to three main factors, the production rate, the plant size, and the plant equipment degradation rate. Then, we presented the experimental results. They showed that the proposed model can estimate the production rate limits for safe production. It is important to measure them for safe production. Finally, we discussed the meanings of the results and the limitations of the model.

Our future work is as follows: introduce other important factors into the model, and propose a new management method or technologies to improve the plant efficiency. As mentioned above, our model has representation limitations for the human cognitive factors and communication difficulties. These limitations are also critical for safe production. Therefore, they are the most important factors to focus on in our future work. Proposing new management methods or technologies is another important future work. To improve the plants efficiency with safe production, we need to create better management methods or new technologies.

References

1. International Energy Agency: Key World Energy Statistics 2012 (2012)
2. International Energy Agency: World Energy Outlook 2012 (2012)
3. Paté-Cornell, M.E.: Learning from the piper alpha accident: A postmortem analysis of technical and organizational factors. Risk Anal. 13(2), 215–232 (1993)
4. Gill, D.A., Picou, J.S., Ritchie, L.A.: The Exxon Valdez and BP oil spills: A comparison of initial social and psychological impacts*. Am. Behav. Sci. 56(1), 3–23 (2012)
5. Bayerl, P.S., Lauche, K.: Technology effects in distributed team coordination high-interdependency tasks in offshore oil production. Computer Supported Cooperative Work 19(2), 139–173 (2010)
6. U.S. Department of the Interior, Minerals Management Service, Engineering and Operations Division: Incidents Associated with Oil and Gas Operations, Outer Continental Shelf 2000. OCS Report, MMS 2002-016 (2002)
7. Sneddon, A., Mearns, K., Flin, R.: Situation awareness and safety in offshore drill crews. Cognition, Technology & Work 8(4), 255–267 (2006)
8. O'Dea, A., Flin, R.: Site managers and safety leadership in the offshore oil and gas industry. Safety Sci. 37(1), 39–57 (2001)
9. Pongsakdi, A., Rangsunvigit, P., Siemanond, K., Bagajewicz, M.J.: Financial risk management in the planning of refinery operations. Int. J. of Prod. Econ. 103(1), 64–86 (2006)
10. Göthe-Lundgren, M., Lundgren, J.T., Persson, J.A.: An optimization model for refinery production scheduling. Int. J. Prod. Econ. 78(3), 255–270 (2002)
11. Qin, S.J., Badgwell, T.A.: A survey of industrial model predictive control technology. Control Eng. Pract. 11(7), 733–764 (2003)
12. Meum, P., Tøndel, P., Godhavn, J.M., Aamo, O.: Optimization of smart well production through nonlinear model predictive control. SPE Intelligent Energy Conference and Exhibition, SPE 112100, 1–11 (2008)
13. Carley, K.M., Svoboda, D.M.: Modeling organizational adaptation as a simulated annealing process. Sociol. Methods & Res. 25(1), 138–168 (1996)
14. Jin, Y., Levitt, R.E.: The virtual design team: A computational model of project organizations. Computational & Mathematical Organization Theory 2(3), 171–195 (1996)
15. Malcolm, D.G., Roseboom, J.H., Clark, C.E., Fazar, W.: Application of a technique for research and development program evaluation. Oper. Res. 7(5), 646–669 (1959)

Moral Guilt: An Agent-Based Model Analysis

Benoit Gaudou[1,2], Emiliano Lorini[1], and Eunate Mayor[1]

[1] UMR 5505 CNRS IRIT, Toulouse, France
[2] University of Toulouse, Toulouse, France
benoit.gaudou@ut-capitole.fr,
{emiliano.lorini,eunate.mayor}@irit.fr

Abstract. In this article we analyze the influence of a concrete moral emotion (i.e. moral guilt) on strategic decision making. We present a normal form Prisoner's Dilemma with a moral component. We assume that agents evaluate the game's outcomes with respect to their ideality degree (*i.e.* how much a given outcome conforms to the player's moral values), based on two proposed notions on ethical preferences: Harsanyi's and Rawls'. Based on such game, we construct and agent-based model of moral guilt, where the intensity of an agent's guilt feeling plays a determinant role in her course of action. Results for both constructions of ideality are analyzed.

1 Introduction

Few aspects of human evolution have been more controversial that the explanation of human ethics and morality. Especially when natural selection theory reached its peak, a question started to be posed more and more often: is cooperation compatible with this phenomena? Or is it so that only selfish behavior can survive under such circumstances? And what about cooperation and other-regarding behavior?

According to Dawkins, all factors that lead to the evolving of instincts that favor other-regarding behavior can be summarized into four main types: "We now have four good Darwinian reasons for individuals to be altruistic, generous or 'moral' towards each other. First, there is the special case of genetic kinship. Second, there is reciprocation: the repayment of favors given, and the giving of favors in 'anticipation' of payback. [...] [T]hird, the Darwinian benefit of acquiring a reputation for generosity and kindness. And fourth, [...] there is the particular additional benefit of conspicuous generosity as a way of buying unfakeably authentic advertising."[11]. For Dawkins, through most of our prehistory, humans lived under conditions that would have strongly favored the evolution of other-regarding tendencies. The social side of our species, motivates that, whether kin or not, individuals would tend to meet again and again throughout their lives, favoring other-regarding behaviors.

Economic theories based in the self-regarding assumption have stated that, except for sacrifice on behalf of others (what we call 'altruism'), the rest is just long-run material self-interest, such theories abstract from reciprocity and other non-self-regarding motives which can guide individuals' behavior. Thus, although cooperation among purely self-regarding agents in indefinitely repeated games with sufficiently low discount rate is a widely accepted theoretical result, this narrow interpretation challenges observations of our everyday life. Indeed, there is compelling evidence that individuals adhere

B. Kamiński and G.Koloch (eds.), *Advances in Social Simulation*,
Advances in Intelligent Systems and Computing 229,
DOI: 10.1007/978-3-642-39829-2_9,

to norms of fairness in both experimental and real world situations that are non-repeated or infrequently repeated [2]. People do repay gifts and take revenge in interactions with complete strangers, even in those cases where it is costly for them and yields to neither present nor future material rewards[1].

Moreover, human beings act cooperatively, obey and enforce norms of fairness, even against their self-interest, and the volume of experimental evidence supporting these and other facts that separate us from the selfishness assumption continues to grow (see, for example [12]). The assumption that individuals are self-regarding is in strong conflict with daily observed preferences. *First,* because agents not only care about the outcomes of their economic interactions, but also about the process through which the results are attained. *Second,* because in their decisions agents do not solely consider what they *personally* gain and lose through an interaction. Violating a fairness norm has emotional consequences that enter negatively in the agent's utility function [7]. In addition,we can say that adherence to norms of fairness is underwritten by emotions, and not merely by the expected gain from the repeated interaction [7] (the so-called 'prosocial emotions', such as shame, guilt, empathy or remorse, all of which involve feelings of discomfort at doing something that appears wrong according to one's own values and/or those of other agents whose opinions one values). Furthermore, social scientists (*e.g.*, [6]) have defended the idea that there exist innate moral principles in humans such as fairness which are the product of biological evolution.

In this article, we test the hypothesis that agents have fairness as a moral value, and that the transgression of this moral value of fairness triggers guilt feelings in them. In order to measure the influence of *moral guilt* on the agents' decision-making process, we present an agent-based model of the Prisoner's Dilemma, where the intensity of an agent's guilt feeling plays a determinant role in her course of action. The paper is organized as follows. In Section 2.1 we present an overview of the concept of guilt and the analysis of our game-theoretic model of moral guilt (based on two proposed notions of moral values: Harsanyi's and Rawls') and its influence on strategic decision making. In Section 3 we describe the agent-based model and its implementation. Finally, in Section 3.2 we present some preliminary results and, in Section 4, our ideas for future work.

2 Moral Guilt

While there exist many contrasting theories explain the discrepancy between pure intentional decision and moral behavior, most of them highlight the variability of individual behavior depending on the situational context and group belonging[2].

Furthermore, adherence to moral standards and social norms is underwritten by emotions, the so-called, *prosocial emotions*, such as empathy, shame, guilt, pride or regret.

[1] Cf. J. Mansbridge's monograph [20], where several social scientists from different disciplines argue that individuals have motives for action that go well beyond their egoistic desires and pure rational calculations. People, they suggest, are influenced by feelings of solidarity, altruism and concern for others and their well-being.

[2] A well-integrated model of the ways in which attitudes, norms, and perceived control feed into behavioral intentions and subsequent behavior is proposed by Ajzen's theory of planned behavior [1].

The influence of these type of emotions in the agents' behavior is two-fold [7]: *on the one hand,* they have emotional consequences that affect negatively the agent's preference function; and *on the other hand,* they induce the agent to act in ways that increase the average payoff to other members of the group to whom she belongs.

There are two main trends in guilt literature. *On the one hand,* part of the scholarship considers guilt a belief-based emotion, what is referred to as *'interpersonal guilt'*. The fact that an agent's utility is 'belief-based', in the sense of second-order beliefs (*i.e.* beliefs about other agents' beliefs) is also well-accepted in the literature. The latter has been explained in two ways. *First,* according to the 'social esteem model', where agents care about what others think about them, and thus it represents an element in their utility function (see notably [5]). *Second,* by means of the 'guilt aversion model', were agents care about what others expect of them; that is, agents feel guilty for "hurting their partners [...] and for failing to live up to their expectations, [which motivated them to] alter their behavior [to avoid guilt]."[4] (see also [3] and [9]). *On the other hand,* theories of *'moral guilt'* define guilt as a 'self-conscious' emotion, triggered by the violation of one's moral standards and internalized (social) norms. In this paper, we present a model of guilt in this latter sense.

2.1 The Model

We present a game-theoretic analysis of normative guilt and of its influence on strategic decision making. The intensity of a player's guilt feeling is defined as the difference between the degree of ideality of the *actual* state and the degree of ideality of the *counterfactual* state that could have been achieved had the player chosen a different action. The model assumes a player has two different motivational systems: an endogenous motivational system determined by the player's desires and an exogenous motivational system determined by the player's moral values. Moral values, and more generally moral attitudes (ideals, standards, etc.), originate from an agent's capability of discerning what from his point of view is (morally) *good* from what is (morally) *bad.* If an agent has a certain moral value, then he thinks that its realization ought to be promoted because it is *good* in itself. A similar distinction has also been made by philosophers and by social scientists. For instance, Searle [24] has recently proposed a theory of how an agent may want something without desiring it and on the problem of reasons for acting based on moral values and independent from desires. In his theory of morality [17,16], Harsanyi distinguishes a person's *ethical preferences* from her *personal preferences* and argues that a moral choice is a choice that is based on ethical preferences.

2.2 Guilt-Dependent Utility

Let us first introduce the notion of normal form game.

Definition 1 (Normal form game). *A normal form game is a tuple* $\Gamma = (N, \{S_i\}_{i\in N}, \{U_i\}_{i\in N})$, *where:*

- $N = \{1, \ldots, n\}$ *is a set of players;*
- S_i *is player* i*'s set of strategies;*

– $U_i : \prod_{i \in N} S_i \longrightarrow \mathbb{R}$ *is agent i's personal utility function mapping every strategy profile in* $\prod_{i \in N} S_i$ *to a real number* (i.e., *personal utility of the strategy profile for player i).*

Let $2^{Agt*} = 2^N \setminus \{\emptyset\}$ be the set of all non-empty sets of players (*alias* coalitions). For notational convenience we write $-i$ instead of $N \setminus \{i\}$. For every $J \in 2^{Agt*}$, we define the set of strategies for the coalition J to be $S_J = \prod_{i \in J} S_i$. Elements of S_J are denoted by $s_J, s'_J, \ldots$ For notational convenience, we write S instead of S_N and we denote elements of S by $s, s', \ldots$ Every strategy s_J of coalition J can be seen as a tuple $(s_i)_{i \in J}$ where player i chooses the individual strategy $s_i \in S_i$.

The following definition extends the definition of normal form game with a *moral* component. Namely we assume that players in a game also evaluates outcomes with respect to their ideality degree, *i.e.*, how much a given outcome conforms to the player's moral values.

Definition 2 (Normal form game with moral values). *A normal form game with moral values is a tuple* $\Gamma^+ = (N, \{S_i\}_{i \in N}, \{U_i\}_{i \in N}, \{I_i\}_{i \in N})$ *where:*

– $(N, \{S_i\}_{i \in N}, \{U_i\}_{i \in N})$ *is a normal form game;*
– $I_i : \prod_{i \in N} S_i \longrightarrow \mathbb{R}$ *is agent i's ideality function mapping every strategy profile in* $\prod_{i \in N} S_i$ *to a real number* (i.e., *the ideality of the strategy profile for player i).*

The preceding notion of ideality corresponds to Harsanyi's notion of ethical preference.

We define guilt as the emotion which arises from the comparison between the ideality of the current situation and the ideality of a counterfactual situation that could have been achieved had the player chosen a different action. In particular, intensity of guilt feeling is defined as the difference between the ideality of the current state and the ideality of the best alternative state that could have been achieved had the player chosen a different action.

Definition 3 (Guilt). *Given a normal form game with moral values* $\Gamma^+ = (N, \{S_i\}_{i \in N}, \{U_i\}_{i \in N}, \{I_i\}_{i \in N})$ *the guilt player i will experience after the strategy profile s is played, denoted by Guilt(i,s), is defined as follows:*

$$Guilt(i,s) = I_i(s) - \max_{a_i \in S_i} I_i(a_i, s_{-i}) \tag{1}$$

We assume that guilt affects the utility function of a certain player depending on the player's degree of guilt aversion. More precisely, the higher the influence of guilt on the utility of a given decision option, the more guilt averse the player. The extent to which a player's utility is affected by his guilt feeling is called *degree of guilt aversion*. The following definition describes how a player's utility function is transformed depending on the player's guilt and on the player's degree of guilt aversion.

Definition 4 (Guilt-dependent utility). *Given a normal form game with moral values* $\Gamma^+ = (N, \{S_i\}_{i \in N}, \{U_i\}_{i \in N}, \{I_i\}_{i \in N})$ *the guilt-dependent utility of the strategy profile s for agent i is defined as follows:*

$$U_i^*(s) = U_i(s) + \delta_i(Guilt(i,s)) \tag{2}$$

where δ_i *is a nondecreasing function* $\delta_i : \mathbb{R} \longrightarrow \mathbb{R}$ *such that* $\delta_i(0) = 0$.

The previous definition of guilt-dependent utility is related with the definition of regret-dependent utility proposed in regret theory [18,19,15]. Specifically, similarly to Loomes & Sugden's regret theory, we assume that computation of emotion-dependent utility consists in adding to player i's personal utility the value $\delta_i(Emotion(i,s))$ which measures the intensity of player i's current emotion.[3] There are several possible instantiations of the function $\delta_i(Guilt(i,s))$. For example, it might be defined as follows:

$$\delta_i(Guilt(i,s)) = c_i \times Guilt(i,s) \tag{3}$$

where $c_i \in \mathbb{R}^+ = \{x \in \mathbb{R} | x \geq 0\}$ is a constant measuring player i's degree of guilt aversion.

2.3 Grounding Moral Values on Personal Utilities

In the preceding definition of normal form game with moral values a player i's utility function U_i and ideality function I_i are taken as independent. Harsanyi's theory of morality provides support for an utilitarian interpretation of moral motivation which allows us to reduce a player i's ideality function I_i to the utility functions of all players [17,16]. Specifically, Harsanyi argues that an agent's moral motivation coincides with the goal of maximizing the collective utility represented by the weighted sum of the individual utilities.

Definition 5 (Normal form game with moral values based on Harsanyi's view). *A normal form game with moral values $\Gamma^+ = (N, \{S_i\}_{i\in N}, \{U_i\}_{i\in N}, \{I_i\}_{i\in N})$ is based on Harsanyi's view of morality if and only if for all $i \in N$:*

$$I_i(s) = \sum_{j\in N} k_{i,j} \times U_j(s) \tag{4}$$

for some $k_{i,1}, \ldots, k_{i,n} \in [0, 1]$.

The parameter $k_{i,j}$ in the previous equation can be conceived as the agent i's *degree of empathy* towards agent j. This means that the higher the degree of empathy of agent i towards agent j, the higher the influence of agent j's personal utility on the degree of ideality of a given alternative for agent i. In certain situations, it is reasonable to suppose that an agent has a maximal degree of empathy towards all agents, *i.e.*, $k_{i,j} = 1$ for all $i, j \in N$. Under this assumption, the previous equation can be simplified as follows:

$$I_i(s) = \sum_{j\in N} U_j(s) \tag{5}$$

An alternative to Harsanyi's utilitarian view of morality is Rawls' view [22]. In response to Harsanyi, Rawls proposed the *maximin* criterion of making the least happy agent as happy as possible: for all alternatives s and s', if the level of well-being in the worst-off position is strictly higher in s than in s', then s is better than s'. According to this

[3] On the ground of empirical evidence, Loomes & Sugden also suppose that the function δ_i should be convex. To keep our model simpler, we do not make this assumption here.

well-known criterion of distributive justice, a fair society should be organized so as to admit economic inequalities to the extent that they are beneficial to the less advantaged agents.[4] Following Rawls' interpretation, an agent's moral motivation should coincide with the goal of maximizing the collective utility represented by the individual utility of the less advantaged agent.

Definition 6 (Normal form game with moral values based on Rawls' view). *A normal form game with moral values $\Gamma^+ = (N, \{S_i\}_{i\in N}, \{U_i\}_{i\in N}, \{I_i\}_{i\in N})$ is based on Rawls' view of morality if and only if for all $i \in N$:*

$$I_i(s) = \min_{j\in N} U_j(s) \tag{6}$$

3 Simulation

We have developed an agent-base model in order to test the various hypothesis on the model. In this article, we mainly investigate the influence of the way to compute ideality and the influence of the guilt aversion on agent behaviors. To this purpose, we have chosen the standard Prisoner's Dilemna [21] as frame of the interactions between agents.

The model is implemented using the GAMA platform[5] [26], an open-source generic agent-based modeling and simulation platform. It provides an intuitive modeling language with high-level primitives to define agents and their environment. GAMA has been used successfully to develop a large spectrum of models, from simple and abstract models (as the one presented here) to large-scale models (including lot of different kinds of agents and needing a huge amount of data).

3.1 Model Description

The following paragraphs describe in greater details the computational implementation of the model described in section 2.1.

Global Variables and Parameters

The game that agents will play is the same for all agents, it (and in particular the payoff matrix) is thus a global variable. It is a standard Prisoner's Dilemma whose payoffs are the following: R = 2, T = 3, S = 0 and P = 1[6]. Being a standard Prisoner's Dilemma type of game, each agent has two possible strategies: cooperate (C) or defect (D). In addition, as we will see, we shall modify the ideality computation method (*i.e.* Harsanyi's and

[4] It has to be noted that Rawls' theory of justice is specified in terms of justice over primary goods. Rawls' list of primary goods includes for instance basic liberties and rights, freedom of movement and free choice of occupation, income and wealth, the social bases of self-respect. This difference is however beyond the scope of the present article. See [25] for a discussion on this issue.

[5] `http://code.google.com/p/gama-platform/`

[6] Although for this first version of the article, we employ the standard payoffs mentioned, the four payoff values might be modified, as they are parameters of the simulation.

Rawls' measures) in order to test their influence on the simulation results. The number of time-steps of the simulation might also be changed, allowing us to analyze the influence of the learning process on the results. Finally, the maximum guilt aversion level and the discretization step of the guilt aversion may also be altered.

Agents

The model is composed of one unique kind (or *species*) of agents named 'people'. Each agent is characterized by a guilt aversion level (*guiltAversion*), a positive float number lower or equal to the global parameter (*guiltAversionInitMax*), and an *history* of previous interactions. Each agent *i*'s *history* is a complex structure (a mapping function) associating each other agent *j* already met to a list containing: (1) the number of interactions between both agents, (2) the number of interactions in which *j* chose to cooperate with *i*, and (3) the overall payoff won by *i* from such interactions with agent *j*. As we will see in the following paragraph, the two first elements of the list are taken into account in the computation of the expected utility, whereas the last one is only an indicator of the 'quality' of the interaction between both agents. Derived from the expected utility obtained from the combination of these two first elements, each agent will compute a guilt dependent utility matrix (containing a modified utility $U^{*}value$) from the game utility matrix. It is also important to note that agents are not aware of the guilt aversion level of their interaction partner. They thus make their decision only depending on other agents behavior (their moves).

Learning: Fictitious Play

In order to explain Nash equilibrium (and selection among various Nash equilibria), game theorist have traditionally used different kinds of adjustment models (cf. for example [27] or [14]); mainly *replicator dynamics* (*i.e.* the relative prevalence of any strategy has a growth rate proportional to its payoff relative to the average payoff) and *simple belief learning* (*i.e.* players adjust their beliefs as they accumulate experience, and that current beliefs influence the current choice of strategy). As shown empirically by [10], in both symmetric (single population) and two-type population games, "the learning model is slightly better at explaining the single population data and much better at explaining the two population data."

Thus, in our simulation, we use a simple belief learning process known as *'fictitious play'*, or as the 'Brown-Robinson learning process'. The algorithm was introduced by Brown [8] as an algorithm for finding the value of a zero-sum game, and first studied by Robinson [23]. It assumes that players noiselessly best respond to the belief that other players' current actions will be equal to the average of their actions in all earlier periods. Informally, we can describe it as follows. Let us assume two players playing a finite game repeatedly. After arbitrary initial moves in the first round, where each player chooses a single pure strategy; then both players construct sequences of strategies according to the following rule: at each step, a player considers the sequence chosen by the other player, she supposes that the other player will randomize uniformly over that sequence, and she chooses a best response to that mixed strategy. That is, in every round each player plays a myopic pure best response against the empirical strategy distribution of her opponent (a player's sequence is treated as a multi-set of strategies, one of which is selected uniformly at random). At each time-step, the chosen best-response is added to a player's sequence, and her strategy sequence get extended.

Initialization

The initialization is limited to the creation of the agents and the initialization of the game from simulation parameters. The history, as it has been implemented right now, makes interactions with one agent totally independent from interactions with others: an agent's expected utilities will be computed taking into account only the ratio of cooperation on total interactions with each other agents. We can thus create only one agent per possible guilt aversion level in order to explore all the space of possible interactions given all other parameters. We thus create one agent per guilt aversion level from 0 to a maximal chosen value of the guilt aversion parameter ('*guiltAversionInitMAx*'), for each discretization step (concretely each 0.1, in this case). Although establishing an upper threshold for the guilt aversion parameter might seem arbitrary, in the following section 3.2 we will see that, from a given degree of guilt aversion, results in terms of agents' payoffs do not vary.

Model Dynamics

At each simulation time-step we randomly pair agents. For each pair, each agent will first compute the expected utilities associated to each of the possible strategies (C and D), depending on the probability distribution (computed from the history) of their mate's strategies. Note that agents choose blindly the first time the interact and then select, according to the recorded history, the pure best response against the empirical strategy distribution of their opponent (cf. paragraph 3.1); that is, the strategy that maximizes her expected utility. In case both strategies have the same expected utility, agents choose randomly. After choosing, each agent is informed of the strategy the other agent played, and computes her payoffs. Agents update their history: (1) they number of interactions with that given opponent; (2) the number of interactions in which j cooperated, if it is the case; and (3) they payoff won (cf. paragraph 3.1). This payoff represents the real payoff of the game, without taking into account the guilt element; that is, independently of the agent's ideality notion. The simulation is iterated until the limit time-step chosen. The history is then registered to be analyzed afterwards.

3.2 Results: Harsanyi's and Rawls' Idealities Comparison

In Figures 1 and 2, we illustrate behaviors emerging from the interactions during the game being played. Both axes represent the guilt aversion values from 0 to 10 (with a step of 0.1). Each guilt aversion value represents also one single agent, as we have chosen to create 1 agent per each guilt aversion value. Figures represent thus the behavior that is emerging from the interactions of the agent on the vertical axe with the one on the horizontal axe. Note that we stopped simulations after 5000 steps[7]. In both cases, we launched 20 replications (without observing any variability in the results despite the randomness of the first move).

[7] This number should be big compared to the number of agents, to allow a high ehough number of interactions with all other agents.

Red (resp. green) areas in Figures represent the fact that agents in these interval always play the strategy profile D,D (resp. C,C).

Harsanyi's Ideality

We first use Harsanyi's algorithm of ideality in order to compute the agent's modified utility function U^*. For this first simulation, we keep all the experiment parameters at their default values. We can observe results in the following Figure 1, it represents the convergent behavior of agents on vertical axe (i) interactions with agents on the horizontal one (j).

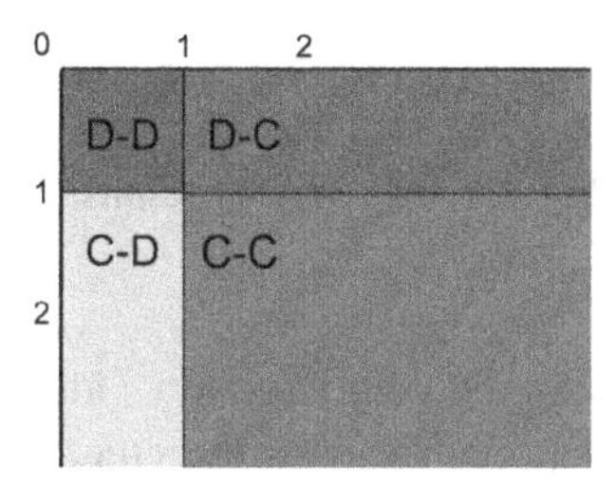

Fig. 1. Harsanyi's ideality results

For Harsanyi's case, there are not very remarkable results. However, we shall note that high guilt aversion does not 'pay off'. On the contrary, less guilt averse agents have a higher average payoff (indeed, guilt averse agents interacting with guilt seeking, get 'cheated on' and their average payoff is the 'sucker' one, that is, zero; whilst their opponent benefits from their defection and obtains the maximum payoff, three).

Furthermore, when implementing Harsanyi's ideality algorithm, the learning process implemented in the agent is not involved into the decision-making process. Indeed, it is easy to see that if an agent has a degree of guilt aversion lower than one, in the game with transformed utility U^* the agent's strategy D strongly dominates the agent's strategy C. Therefore, for every possible probability distribution over the opponent's strategy, D is the strategy which maximizes the agent's expected utility. On the contrary, if an agent has a degree of guilt aversion higher than one, in the game with transformed utility U^* the agent's strategy C strongly dominates the agent's strategy D. Therefore, for every possible probability distribution over the opponent's strategy, C is the strategy which maximizes the agent's expected utility. Hence, an agent with degree of guilt aversion lower than one will always play D, whereas an agent with degree of guilt aversion higher than one will always play C. Agents with degree of guilt aversion equal to one are exactly in the co-joint point where the expected utilities of the two strategies C and D are always equal (i.e., the expected utilities of C and D for an agent with degree of guilt aversion equal to one are equal, for every possible probability distribution over the opponent's strategy). Thus, an agent with degree of guilt aversion equal to one will always play in a random way.

Rawls' Ideality

Analogously, we now use Rawls' algorithm of ideality the agent's U^*, keeping all the experiment parameters at their default values. Figure 2 shows a schematic representation of the results, organized similarly to previous figure 1.

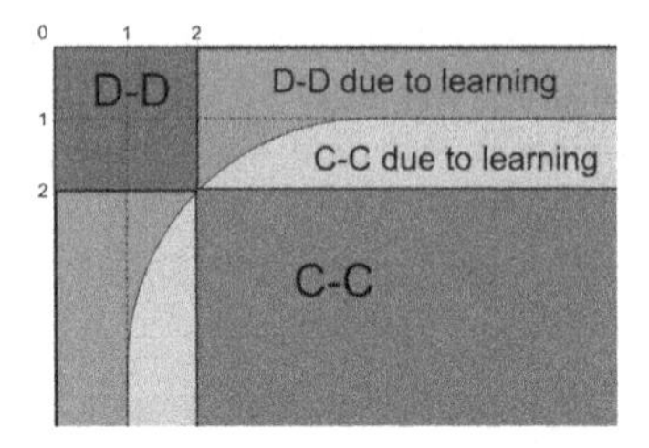

Fig. 2. Rawl's ideality results

Unsimilarly to the results obtained with Harsanyi's ideality, we observe that in this case guilt aversion plays an important role in the emergence of fairness as a moral value. Indeed, agents with a higher degree of risk aversion have a higher average payoff than those who do not present guilt aversion (that is, those who do not experience guilt feelings triggered by the transgression of the fairness value).

In addition, in Rawls' case we have some nuances that we did not have in Harsanyi's case. In particular, the strategy chosen by the agents not only depends on their degree of guilt aversion, but also on the number of interactions they have been part of. This highlights the influence of the learning curve on the decision-making process.

4 Conclusions and Future Work

Hence we can finish by concluding that, from an evolutionary point of view, it is the moral values *à la* Rawls that emerge and that guilt aversion does play an important role in the suitability over time of fairness. As R. Frank stated, "[t]he fact that it might sometimes be best to ignore moral emotions does not imply that it is always, or even usually, best to ignore them. If we are to think clearly about the role of moral emotions in moral choice, we must consider the problems that these emotions were molded by natural selection to solve. Most interesting moral questions concern actions the individual would prefer to take except for the possibility of causing undue harm to others. Unbridled pursuit of self-interest often results in worse outcomes for everyone. In such situations, an effective moral system curbs self-interest for the common good. [Furthermore,] moral systems must not only identify which action is right, they must also provide motives for taking that action." [13, pp. 7-8]

This article presents an on-going piece of research that is much to be completed. Although many, we shall note here some of the possible research questions to investigate further. Firstly, we would like to test the model with different population distributions (according to their degree of guilt aversion), in order to explore which is the minimum percentage of guilt averse agents that is necessary for sustaining cooperation. Secondly, in the present model, an agents' learning algorithm is specific for every agent with whom they have interacted; that is, there is no 'global' learning, in the sense of an intuition of the global trend to cooperate or to defect of the rest of the population. Furthermore, it would also be interesting to analyze the results for a case-scenario were the degree of guilt aversion of the agents is perceivable by their opponents (cf. [13]).

Acknowledgement. This work is part of the EmoTES ("Emotions in strategic interaction : theory, experiments, logical and computational studies") research project. The EmoTES project is funded by the French National Research Agency (ANR).

References

1. Ajzen, I.: The theory of planned behavior. Organizational Behavior and Human Decision Processes 50(2), 179–211 (1991)
2. Axelrod, R., Hamilton, W.D.: The evolution of cooperation. Science 211(4489), 1390–1396 (1981)
3. Battigalli, P., Dufwenberg, M.: Guilt in games. The American Economic Review 97(2), 170–176 (2007)
4. Baumeister, R.F., Stillwell, A.M., Heatherton, T.F.: Guilt: An interpersonal approach. Psychological Bulletin 115, 243–267 (1994)
5. Benabou, R., Tirole, J.: Incentives and prosocial behavior. The American Economic Review 96(5), 1652–1678 (2006)
6. Binmore, K.: Natural justice. Oxford University Press, Oxford (2005)
7. Bowles, S., Gintis, H.: Prosocial emotions. Working Paper 02-07-028, Santa Fe Institute (2003)
8. Brown, G.W.: Iterative solutions of games by fictitious play. In: Koopmans, T.C. (ed.) Activity Analysis of Production and Allocation, pp. 374–376. Wiley, New York (1951)
9. Charness, G., Dufwenberg, M.: Promises and partnership. Econometrica 74, 1579–1601 (2006)
10. Cheung, Y., Friedman, D.: A comparison of learning and replicator dynamics using experimental data. Journal of Economic Behavior and Organization 35, 263–280 (1998)
11. Dawkins, R.: The God Delusion. Houghton Mifflin Co., Boston (2006)
12. Fehr, E., Gaechter, S.: Fairness and retaliation: The economics of reciprocity. Journal of Economic Perspectives 14(3), 159–181 (2000)
13. Frank, R.H.: The Status of Moral Emotions in Consequentialist Moral Reasoning. In: Zak, P.J. (ed.) Moral Markets: The Critical Role of Values in the Economy. Princeton University Press (2007)
14. Fudenberg, D., Levine, D.: Learning in Games. MIT Press, Cambridge (1998)
15. Halpern, J., Pass, R.: Iterated regret minimization: a new solution concept. Games and Economic Behavior 74(1), 194–207 (2012)
16. Harsanyi, J.C.: Cardinal welfare, individualistic ethics, and interpersonal comparisons of utility. Journal of Political Economy 63, 309–321 (1955)
17. Harsanyi, J.C.: Morality and the theory of rational behaviour. In: Sen, A.K., Williams, B. (eds.) Utilitarianism and Beyond, pp. 39–62. Cambridge University Press, Cambridge (1982)
18. Loomes, G., Sugden, R.: Regret theory: An alternative theory of rational choice under uncertainty. The Economic Journal 92(368), 805–824 (1982)

19. Loomes, G., Sugden, R.: Testing for regret and disappointment in choice under uncertainty. The Economic Journal 97, 118–129 (1987)
20. Mansbridge, J.J.: Beyond self-interest. University of Chicago Press (1990)
21. Poundstone, W.: Prisoner's Dilemma. Doubleday, NY (1992)
22. Rawls, J.: A theory of Justice. Harvard University Press, Cambridge (1971)
23. Robinson, J.: An iterative method of solving a game. Annals of Mathematics 54(2), 296–301 (1951)
24. Searle, J.: Rationality in Action. MIT Press, Cambridge (2001)
25. Sen, A.: Collective choice and social welfare. Holden-Day, San Francisco (1970)
26. Taillandier, P., Vo, D.-A., Amouroux, E., Drogoul, A.: GAMA: A simulation platform that integrates geographical information data, agent-based modeling and multi-scale control. In: Desai, N., Liu, A., Winikoff, M. (eds.) PRIMA 2010. LNCS, vol. 7057, pp. 242–258. Springer, Heidelberg (2012)
27. Weibull, J.: Evolutionary Game Theory. MIT Press, Cambridge (1995)

Punishment and Gossip: Sustaining Cooperation in a Public Goods Game

Francesca Giardini[1], Mario Paolucci[1], Daniel Villatoro[2], and Rosaria Conte[1]

[1] ISTC-CNR, Roma, Italy
{francesca.giardini,mario.paolucci,rosaria.conte}@istc.cnr.it
[2] IIIA-CSIC, Barcelona, Spain
dvillatoro@iiia.csic.es

Abstract. In an environment in which free-riders are better off than cooperators, social control is required to foster and maintain cooperation. There are two main paths through which social control can be applied: punishment and reputation. Our experiments explore the efficacy of punishment and reputation on cooperation rates, both in isolation and in combination. Using a Public Goods Game, we are interested in assessing how cooperation rates change when agents can play one of two different reactive strategies, i.e., they can pay a cost in order to reduce the payoff of free-riders, or they can know others' reputation and then either play defect with free-riders, or refuse to interact with them. Cooperation is maintained at a high level through punishment, but also reputation-based partner selection proves effective in maintaining cooperation. However, when agents are informed about free-riders' reputation and play Defect, cooperation decreases. Finally, a combination of punishment and reputation-based partner selection leads to higher cooperation rates.

Keywords: Reputation, Punishment, Agent-based simulation, Social Control, Public Goods Game.

1 Introduction

Social control is an emergent social phenomenon, which allows the costs of prosocial behavior to be redistributed over a population in which cooperators live side-by-side with non-cooperators. Some specific phenomena are usually subsumed under the large heading of social control, including ostracism [3,15] and altruistic punishment. The latter is defined as a costly aggression inflicted to cheaters by members of the group who did not necessarily undergo attacks from the punished, nor get direct benefits out of the sanction applied [9]. According to *strong reciprocity* theory, the presence of individuals who altruistically reward cooperative acts and punish norm violating behavior at a cost to themselves sustains cooperation and promotes social order [11]. However, the act of punishment results in an immediate reduction of welfare both for the punisher and for the punished individual, thus posing several problems, like efficiency [27,7], and the risk of counter-aggression [19].

B. Kamiński and G.Koloch (eds.), *Advances in Social Simulation*,
Advances in Intelligent Systems and Computing 229,
DOI: 10.1007/978-3-642-39829-2_10, © Springer-Verlag Berlin Heidelberg 2014

An alternative solution can be found in weak reciprocity supported by reputation. Knowing about others' past behaviors is crucial to avoid cheaters and select good partners, especially when the group is large and it is not possible to directly witness all the interactions. Moreover, reputation allows the costs of social control to be reduced and distributed among individuals [12]. The importance of reputation in supporting cooperation has been proven in laboratory experiments [28], evolutionary models [20], and simulation studies [8]. Reputation is a signal that conveys socially relevant information about one's peers, and plays a fundamental role in identyfing cheaters and isolating them.

Notwithstanding their importance in supporting cooperation, costly monetary punishment and reputation spreading have never been directly compared. Moving from the work of Carpenter [4], in which the provision of public goods is not negatively affected by the size of the group but by the ability of mutual monitoring among agents, we enriched the original model by designing agents who spread reputation about their previous partners. Extending previous research on cooperation and reputation [8]and on cooperation and punishment [27], we explore the performance of punishment and reputation as mechanisms for social control, and we test their effects on cooperation rates both in isolation and in combination.

Here we present a simulation platform to compare the effectiveness of costly punishment and reputation spreading in maintaining cooperation in a population in which defectors have a selective advantage because they exploit others' contributions, without paying the costs of cooperation. Using a Public Goods Game [4], we measure cooperation rates in mixed populations in which there are pure cooperators, pure cheaters and agents who play reactive strategies. Our contribution adds to existing literature in three ways:

1. we introduce a systematic exploration of two different social control mechanisms, and we test them also in combination with other parameters, like the group size and the costs of punishing;
2. we specify two different mechanisms for reputation: *Refuse* and *Defect* (they will be extensively explained in Section 3). *Refuse* is a partner choice mechanism which permits gossipers to avoid free-riders, whereas *Defect* is a social control mechanism that leads gossipers to defect against non-cooperators. Both these mechanisms are present in human societies, in which we use reputational information to avoid cheaters (when this is possible), or to treat them as we expect them to treat us. This difference between these reactions makes it important to compare them and to understand the conditions that make one mechanism more effective than the other.
3. we assess the extent to which reputation spreading and punishment are comparable mechanisms for social control, by comparing directly agents' average contributions when costly monetary punishment and reputation spreading are available.

In Section 2 we will introduce related work, in Section 3 we describe the simulation model, and Section 4 will present the simulation results. In Section 5 we will draw some conclusions and we will also sketch some ideas for future work.

2 Related Work

In their evolutionary history, humans have developed several mechanisms for the emergence and establishment of social norms [2]. Punishment and reputation are among the most widespread and effective mechanisms to sustain cooperation and they are specially interesting for virtual societies in which the efficacy of enforcing mechanisms is limited by a combination of factors (like their massive size, their spontaneity of creation and destruction, and dynamics). There is a large body of evidence showing that humans are willing to punish non-cooperators, even when this implies a reduction in their payoffs [10], and this is true also in simulation settings. Villatoro et al. [27] have analyzed the effect of sanctioning on the emergence of the norm of cooperation, showing that a monetary punishment accompanied with a norm elicitation, that is, sanctioning, allowed the system to reach higher cooperation levels at lower costs, when compared with other punishment strategies.

Reputation is, along with punishment, the other strategy used to support cooperation, even if it works in a completely different way. If punishing means paying a cost in order to make the other pay an even higher cost for his defection, reputation implies that the information about agents' past behavior becomes known, and this allows agents to avoid ill-reputed individuals. In Axelrod's words [1]: *"Knowing people's reputation allows you to know something about what strategy they use even before you have to make your first choice"* (p.151). The importance of reputation for promoting and sustaining social control is uncontroversial and it has been demonstrated both in lab experiments [23] and in simulation settings, in which reputation has proven to be a cheap and effective means to avoid cheaters and increase cooperators' payoffs [21].

When partner selection is available, reputation becomes essential for discriminating between good and bad partners, and then to be protected against exploitation. Giardini and Conte [13] presented ethnographic data from different traditional societies along with simulation data, showing how reputation spreading evolved as a solution to the problem of adaptation posed by social control, and highlighting the importance of gossip as a means to reduce the costs of cheaters' identification. The effect of partner selection has been studied also by Perrau and others [6], who have analyzed the effect of ostracism in virtual societies, obtaining high levels of tolerance against free-riders. However, in the work of Perrau, agents do not explicitly transmit information about other agents, they only reason about the interactions.

In the multi-agent field, several attempts have been made to model and use reputation, especially in two sub-fields of information technologies, i.e., computerized interaction (with a special reference to electronic marketplaces), and agent-mediated interaction (for a review, see [18]). Models of reputation for multi-agent systems applications [29,24,16] clearly show the positive effects of reputation, and there are also interesting cases in which trust is paired with reputation (for a couple of exhaustive reviews, see [22,26]).

More specifically, Sabater and colleagues [25,5] developed a computational system called REPAGE in which different kinds of reputational information were taken into account and the role of information reliability in a market-like simulation scenario was addressed. Analogously, Giardini and colleagues [8] showed that reputation was a means to punish untruthful informers *without bearing the costs of further retaliation*, at the same time protecting the system from collapsing. In addition, for certain percentages of cheating rates, in both studies, the authors showed that reputation played a relevant role in enhancing the quality of production in an artificial cluster of interacting firms.

3 The Model

Moving from the simulation framework developed in [4], we designed a simulation platform in NetLogo in order to compare the performance of costly punishment and reputation spreading in mixed populations in which different types of agents play a Public Good Game (PGG), the classical experimental model used to investigate social dilemmas [17]. In this game, agents decide whether to free-ride or to contribute[1] a fixed amount (a contribution of 1 unit) to a public pool. The sum of all the contributions is multiplied by a benefit factor (set to 3 in the current model[2]) and the resulting quantity is divided amongst all the participants in the group, without considering their contributions. This is a classic public good where free-riding would be the utility maximizing strategy at the individual level; however, if all agents adopted that strategy, this would result in the overexploitation of resources and in a worse outcome at the group level (the so-called *Tragedy of the Commons* [14]).

Algorithm 1. Description of punisher's behaviors

```
for Number of Timesteps do
    Random group formation of the population;
    Agents take First Stage decision;
    Gather and Distribution of the Public good in each group;
    First Stage Decisions are made public within the group;
    Agents make Second Stage decision;
    Punishment Execution;
end for
```

Agents are either non-reactive or reactive types. In the former category we find *Cooperators* (C), who always contribute to the common pool, and *Free-riders* (FR), who never contribute to the common pool. Reactive agents change their

[1] Note that the decision in this framework is binary whether to cooperate or not. In other works, specially in those of experimental economics, this decision has to be taken in a continuum, deciding how much to contribute from a total amount of money common to all agents.

[2] According to the game design, in order for contribution to be irrational for a utility-maximizer individual, the tokens in the pot must be multiplied by an amount smaller than the number of players and greater than 1.

behaviors in response to the percentage of detected free-riders in their group: when the number of known defectors in a group is too high, i.e., it exceeds a certain threshold, agents become active and their strategy changes (as described in Algorithm 1). Each and every agent is endowed with an initial amount of 50 points that can be used to cooperate or to punish others; regardless of the strategy, agents are culled from the game when their payoff goes to zero and they are not replaced. The other strategies are:

- *Tit-for-Tat* (TFT): They start as cooperators but if active, they start free-riding until exiting from the active state.
- *Nice Punishers* (NP): They contribute in the passive state; once active, they punish free-riders at a cost to themselves and making *FR*s pay a cost, but they continue to cooperate in the PGG. This behavior will continue until the agent exits the active state (by being assigned in a group with a number of cheaters below the threshold).
- *Mean Punishers* (MP): They contribute in the passive state; once active, they punish free-riders and free ride themselves in the PGG until they exit the active state.

To the above types, developed by Carpenter [4], we add two more types, in order to compare punishment and reputation. Active Gossipers transmit and receive information and they integrate their personal experience and the reputational information received from other Gossipers in order to react against Free-Riders. Gossipers' behavior is described in Algorithm 2 (2). Gossipers can be:

- *Nice Gossipers* (NG): Agents contribute in the passive state; once active, they start spreading information about free-riders, and cooperate in the PGG.
- *Mean Gossipers* (MG): Agents contribute in the passive state; once active, they start spreading information about free-riders, but they always defect in the PGG when active.

Algorithm 2. Description of gossipers' behaviors

```
for Number of Timesteps do
    Random group formation of the population;
    if Group has bad reputation then
        Apply reputation strategy
    else
        Take First stage decision according to active/passive status.
    end if
    Gather and Distribution of the Public good in each group;
    First Stage Decisions are made public within the group;
    Agents make Second Stage decision;
    Reputation diffusion;
end for
```

In this work, we explore the efficacy of two different ways of using reputation. Reputation allows agents to be informed about other members of their group before interacting with them, hence this information can be used in two ways, either to retaliate against free-riders or to refrain from interacting with them. The former modality, called "Defect", consists in Gossipers not contributing to the PGG when playing in a group containing too many cheaters. The latter modality, "Refuse" allows agents to refuse the interaction by skipping a turn, taking their contribution away from the pools, and paying the price of not receiving any dividend, if the number of known defectors in their current group assignment would make them active. Refusal cannot be performed twice in a row. In both modalities, Gossipers transmit information about cheaters, informing their peers of the identity of non-cooperators and using the information they receive from them.

4 Experiments and Results

We run three different sets of simulations, each one lasting for 100 time steps. Each experiment was repeated 20 times for each combination of the selected parameters. The cost of contributing to the Public Good was set to 1, and the sum of all the contributions was multiplied by a benefit factor set to 3. The public good, i.e., the resulting quantity, was divided among all the group members, without considering whether they contributed or not. Simulations started with equal proportions of each strategy. Variables of interest are summarized in Table 1.

Table 1. Parameters of the simulation

PARAMETERS				
COST (cost of punishing)	0.2	1		
PUNISHMENT (cost of being punished)	2	5		
GROUP SIZE	5	10	25	50
INFORMATION TRANSMITTED	0	1	10	

In our first experiment we tested the effect of punishment as a partner control mechanism on cooperation rates measured as the total number of agents playing C divided by the total number of active agents per time step. Nice and Mean Punishers became active when they detected more than 20% of defections in their group. Punishing costed the punisher x and the punished agent y, with $y \geq x$. We identified 4 different combinations of Punishment and Cost: LpLc (low punishment, low cost), HpLc (high punishment, low cost), LpHc (low punishment, High cost) and HpHc (high punishment and high cost). The cooperation rates are affected by group size, and they change according to the different combinations of punishment and cost (Figure 1). When punishment is low and the cost is high, cooperation rates are the lowest for every group size.

Both HpHc and HpLc allows cooperation to reach 100%, no matter the group size and in a quite short amount of time. This is probably due to the fact that high punishment leads Free-Riders to use up all their resources, while protecting the other agents. Moreover, once the percentage of tolerated Free-riders goes below the fixed threshold, all the other populations contribute, thus maintaining cooperation stable and complete.

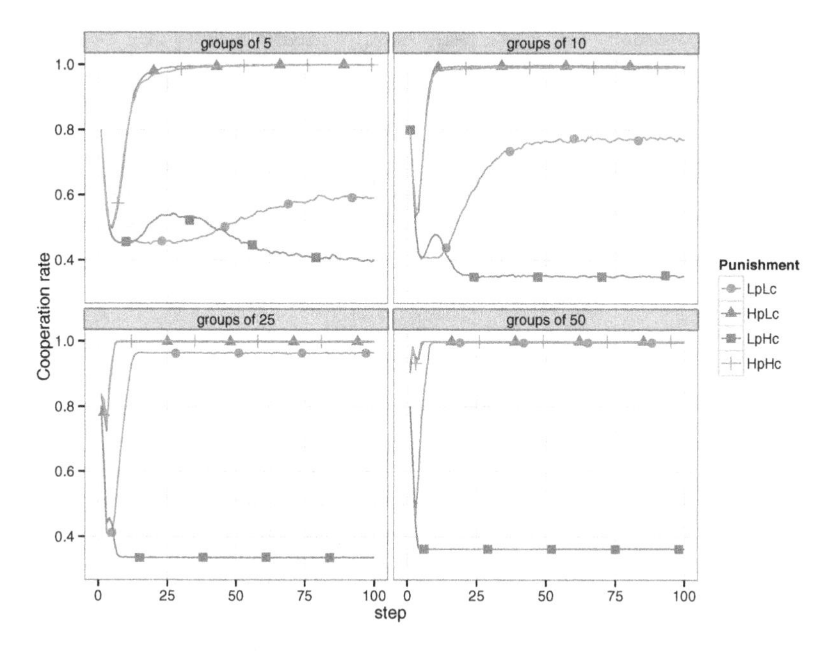

Fig. 1. Cooperation rates over time for different group sizes in mixed populations with C, FR, TFT, NP and MP. When punishment is costly (HpLc, HpHc), Free-riders are easily controlled and cooperation can be maintained. When the cost is high and the punishment is not (LpHc) the cooperation rate is always lower than in the other cases.

Our second experiment was designed to test the effects of information spreading in isolating Free-Riders and maintaining cooperation. We were also interested in comparing the efficacy of the two modalities, and to assess which one worked better. We tested the effectiveness of "Refuse" and "Defect" for different amounts of information (i) available. Manipulating the number of gossips transmitted $\{i = 0; 1; 10\}$ for each time-step, we wanted to test whether the amount of information had an effect on the ability of gossipers to react directly (playing "Defect"), or indirectly (playing "Refuse") against Free-riders.

When agents used the "Defect" modality, cooperation rates declined for every group size. In the "Defect" modality (upper part of Figure 2), cooperation rates were higher in the first 20 periods, but they showed a rapid decrease in the last 20 periods. Increasing the amount of information about other agents made this decrease steeper, even if this pattern remained stable for different group sizes. This is due to the fact that playing "Defect" triggered even more defections, thus reducing the overall cooperation rate and making the decrease even steeper when the amount of available information was higher.

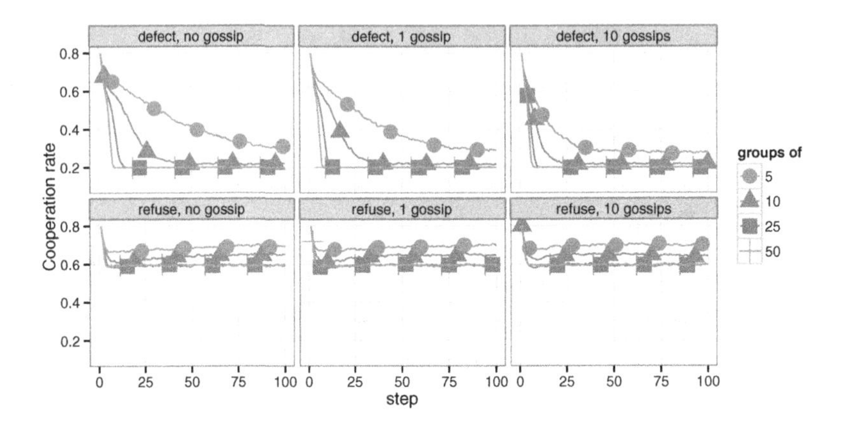

Fig. 2. Cooperation rates over time for Defect (top) and Refuse (bottom) strategies for Gossipers. In larger groups (25 and 50 agents) the amount of information has a negative effect on cooperation rates when the reaction is based on defection, as compared to the Refuse strategy. In this latter case, cooperation rates are stable and quite high.

On the contrary, when gossipers played "Refuse", cooperation levels were quite high for all group sizes. Small groups showed higher cooperation rates (Figure 2, bottom panel), because refusing dangerous interactions in small groups was effective in isolating cheaters and in making cooperators interact with other cooperators. Taking into account the first periods of the simulation experiments, we saw that there are interesting differences in the cooperation rates, not only between groups of different sizes, but also among situations with zero, one or ten items of information (3).

In a scenario in which all strategies are loaded, we compared the performance of populations in which all the strategies were present (C, FR, TFT, NP, MP, NG and MG) for different combinations of costs (HpHc, HpLc, LcHp, LcLp), for different group size (5 and 50). Figure 3 (4) shows that punishment and reputation can have a combined effect that boosts cooperation to 1 in large groups in every situation but in the LpHc. The negative effect of Defect is confirmed also when there are both Punishers and Gossipers, and it is also made more relevant by the higher amount of information.

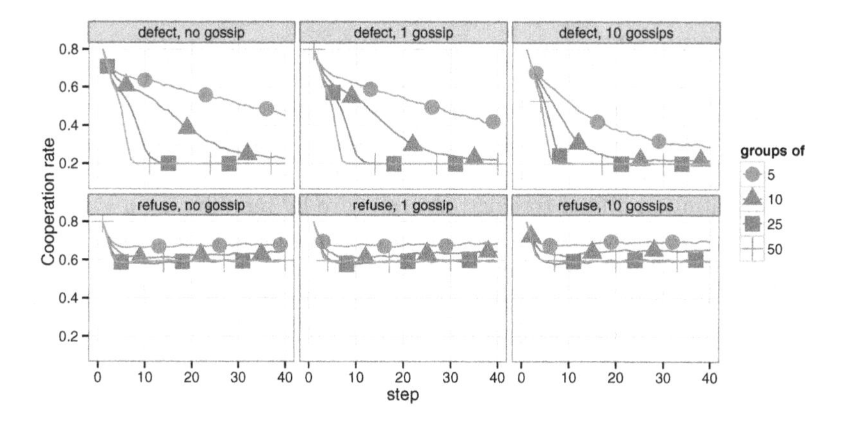

Fig. 3. Cooperation rates in the first 40 ticks of the simulation experiment. When Gossipers can refuse the interaction, receiving more information (10 gossips) increases the cooperation rates in the initial phases of the simulation, especially for groups of 5 and 10 agents.

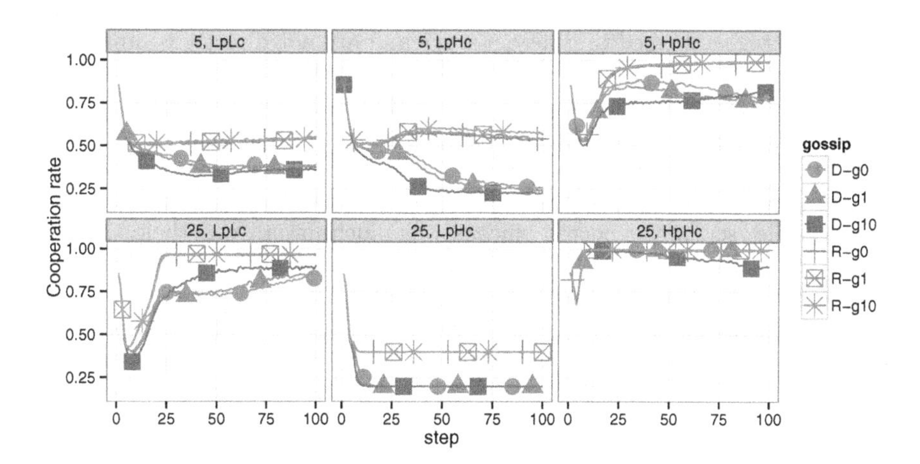

Fig. 4. Cooperation rates for all populations for different combinations of costs of punishment and number of gossips for small (5 agents) and large groups (25 agents). Both small and large groups achieve the highest percentages of cooperation when punishment is costly and being punished is costly as well (HpHc), even if this happens mostly in the larger group, where cooperation rate is 1 for all the populations except for the one including agents playing "Defect" and transmitting 10 gossips. This population has a good performance only when the group is big (25 agents) and the costs of punishment are low (LpLc). "Refuse" is the best strategy and its efficacy is increased by the amount of information exchanged.

5 Discussion and Future Work

In this work we proposed a direct comparison between reputation spreading and costly monetary punishment for controlling free-riding in a Public Goods Game. These strategies, punishment and reputation spreading, were tested in order to investigate their effects on cooperation rates in mixed populations, and to assess how effective they are in preventing cooperation from extinction in different settings. Our contribution added to existing literature by introducing a systematic exploration of different elements, in isolation and in combination. The effects of these strategies in isolation have been tested in a variety of settings and environments, but it is also interesting to compare them directly. Costly punishment is an effective strategy in promoting cooperation, and here it works quite well in selecting out the free-riders, especially when the cost of being punished is much higher than the cost of punishing, as expected. Even more interesting is our finding that in larger groups the effect of punishing, even when it is not costly, can support cooperation much better than in smaller groups.

We also introduced two different strategies based on reputation with the goal of evaluating two alternative ways of using information about cheaters. The *Refuse* strategy can be considered pro-active and it proved to be the best strategy in terms of cooperation rates, for every population composition. On the other hand, the *Defect* strategy was designed as a direct retaliation against cheaters, with negative consequences on the average cooperation rates in our populations. These two strategies worked in different ways and are not directly comparable, but their usage may shed new light on the conditions that make one mechanism more effective than the other, also considering that in humans societies the two options are usually available at the same time. It is worth noting that, on average, cooperation rates with Refuse were close to 70%, which is a very high percentage, especially because Gossipers were not able to affect Free-Riders' payoffs directly, as Punishers did, so their expected efficacy was much lower. Nonetheless, being able to use reputation to identify Free-Riders before interacting with them was really effecting in preventing exploitation and fostering cooperation.

This work represents a first step in a process of exploring different combinations of direct, i.e., punishment, and indirect, i.e., reputation, mechanisms for promoting cooperation in social dilemmas. Our platform will allow us to explore single parameters and different combinations of them, with the aim of understanding how they work, and in what way their combination may determine the success of a given strategy. We do not claim that reputation is a better mechanism, but our data show that it is worth exploring the possibility that, for given combinations of parameters, reputation would be more effective in sustaining social control.

Regarding future works, with the advent of social networking platforms like Facebook or Twitter, and motivated by the Living Labs philosophy, a possible extension would be to perform experiments on social networks sites with real subjects. Because of the public access to the messages shared by the users and the underlying social network that connects then, Twitter could be the best platform where to perform our experiment. We plan to identify active communities within the social platform and made them play with bots developed by us. Our bots will allow us to generate controlled situations, and study the different treatments discussed in this paper. Measures like the virtual earnings of the players, the messages sent amongst users and the changes on the underlying social network will be observed and used as indicators to extract our results.

Acknowledgements. The authors wish to thank two anonymous referees for advice. This work has been partially supported by the PRISMA project (PON04a2 Aunder), within the Italian National Program for Research and Innovation (Programma Operativo Nazionale Ricerca e Competitivitá 2007-2013).

References

1. Axelrod, R.: The evolution of cooperation. Basic Books (1984)
2. Axelrod, R.: An evolutionary approach to norms. The American Political Science Review 80(4), 1095–1111 (1986)
3. Bowles, S., Gintis, H.: The evolution of reciprocal preferences. Working papers. Santa Fe Institute (2000)
4. Carpenter, J.P.: Punishing free-riders: How group size affects mutual monitoring and the provision of public goods. Games and Economic Behavior 60(1), 31–51 (2007)
5. Conte, R., Paolucci, M., Sabater Mir, J.: Reputation for innovating social networks. Advances in Complex Systems 11(2), 303–320 (2008)
6. de Pinninck Bas, A.P., Sierra, C., Schorlemmer, M.: A multiagent network for peer norm enforcement. Autonomous Agents and Multi Agent Systems 21, 397–424 (2010)
7. Dreber, A., Rand, D., Fudenberg, D., Nowak, M.: Winners don't punish. Nature 452, 348–351 (2008)
8. Giardini, F., Di Tosto, G., Conte, R.: A model for simulating reputation dynamics in industrial districts. Simulation Modelling Practice and Theory (SIMPAT) 16(2), 231–241 (2008)
9. Fehr, E., Gachter, S.: Altruistic punishment in humans. Nature 415(6868), 137–140 (2002)
10. Fehr, E., Gachter, S.: Altruistic punishment in humans. Nature 415(6868), 137–140 (2002)
11. Fehr, E., Henrich, J.: Is strong reciprocity a maladaptation? on the evolutionary foundations of human altruism. IZA Discussion Papers 712. Institute for the Study of Labor, IZA (February 2003)
12. Giardini, F., Conte, R.: Gossip for social control in natural and artificial societies. Simulation: Transactions of the Society for Modeling and Simulation International (2011)
13. Giardini, F., Conte, R.: Gossip for social control in natural and artificial societies. Simulation 88(1), 18–32 (2012)
14. Hardin, G.: The Tragedy of the Commons. Science 162(3859) (1968)
15. Hirshleifer, D., Rasmusen, E.: Cooperation in a repeated prisoners' dilemma with ostracism (1989)

16. Huynh, T.D., Jennings, N.R., Shadbolt, N.R.: Certified reputation: how an agent can trust a stranger. In: Proceedings of the Fifth International Joint Conference on Autonomous Agents and Multiagent Systems, AAMAS 2006, pp. 1217–1224. ACM, New York (2006)
17. Ledyard, J.O.: Public goods: A survey of experimental research. In: Handbook of Experimental Economics, pp. 111–194. Princeton University Press (1995)
18. Mui, L., Mohtashemi, M., Halberstadt, A.: Notions of reputation in multi-agents systems: a review. In: AAMAS, pp. 280–287 (2002)
19. Nikiforakis, N.: Punishment and counter-punishment in public good games: Can we really govern ourselves? Journal of Public Economics 92, 91–112 (2008)
20. Nowak, M.A., Sigmund, K.: Evolution of indirect reciprocity by image scoring. Nature 393(6685), 573–577 (1998)
21. Pinyol, I., Paolucci, M., Sabater-Mir, J., Conte, R.: Beyond accuracy. reputation for partner selection with lies and retaliation. In: Antunes, L., Paolucci, M., Norling, E. (eds.) MABS 2007. LNCS (LNAI), vol. 5003, pp. 128–140. Springer, Heidelberg (2008)
22. Ramchurn, Sarvapali, Sierra, C., Godo, L., Jennings, N.R.: Devising a trust model for multi-agent interactions using confidence and reputation. Applied Artificial Intelligence 18(9-10), 833–852 (2004)
23. Rockenbach, B., Milinski, M.: The efficient interaction of indirect reciprocity and costly punishment. Nature 444(7120), 718–723
24. Sabater, J., Sierra, C.: Reputation and social network analysis in multi-agent systems. In: Proceedings of the First International Joint Conference on Autonomous Agents and Multiagent Systems (AAMAS 2002), Bologna, Italy, pp. 475–482 (2002)
25. Sabater, J., Paolucci, M., Conte, R.: Repage: Reputation and image among limited autonomous partners. Journal of Artificial Societies and Social Simulation 9(2) (2006)
26. Sabater, J., Sierra, C.: Review on computational trust and reputation models. Artif. Intell. Rev. 24, 33–60 (2005)
27. Villatoro, D., Andrighetto, G., Conte, R., Sabater-Mir, J.: Dynamic sanctioning for robust and cost-efficient norm compliance. In: Proceedings of the Twenty-Second International Joint Conference on Artificial Intelligence, IJCAI (2011)
28. Wedekind, C., Milinski, M.: Cooperation Through Image Scoring in Humans. Science 288(5467), 850–852 (2000)
29. Yu, B., Singh, M.P.: Distributed reputation management for electronic commerce. Computational Intelligence 18(4), 535–549 (2002)

How Much Rationality Tolerates the Shadow Economy? – An Agent-Based Econophysics Approach

Sascha Hokamp[1] and Götz Seibold[2]

[1] Institute of Economics, BTU Cottbus, P.O. Box 101344, 03013 Cottbus, Germany
Sascha.Hokamp@tu-cottbus.de
[2] Institute of Physics, BTU Cottbus, P.O. Box 101344, 03013 Cottbus, Germany
goetz@physik.tu-cottbus.de

Abstract. We calculate the size of the shadow economy within a multi-agent econophysics model previously developed for the study of tax evasion. In particular, we analyze deviating behavior depending on the fraction of rational agents which aim to pursue their self interest. Two audit mechanisms are considered within our model, that are, (i) a constant compliance period which is enforced after black market activities of an agent have been detected and (ii) a backauditing method which determines the compliance period according to the particpation rate in the shadow economy within a previously preassigned time interval. We calibrate our simulation with respect to experimental evidence of tax compliance in France and Germany and give estimates for the percentage of selfish agents in these countries. This implies different policy recommendations that may work to fight the shadow economy, tax evasion, and the like.

Keywords: shadow economy; econophysics; multi-agent model.

1 Introduction

Theoretical approaches to account for shadow economy and tax compliance are often based on the seminal work of Allingham and Sandmo (1972) which incorporates potential penalties, tax rates and audit probabilities as basic parameters in order to evaluate the expected utility of tax payers. Obviously one of the reasons why people participate in the shadow economy is to circumvent the tax system. Vice versa the shadow economy is often taken as a proxy for the amount of tax evasion (Alm et al., 2012) although the former is naturally driven by additional factors. Buehn and Schneider (2012) define shadow economy "as all market-based legal production of goods and services that are deliberately concealed from public authorities ....". The size of this production is of course depending on efficient audit mechanisms and penalties which provides the link for the application of theories originally developed for the modelling of tax compliance.

In this regard agent-based models have been set up as a comparatively new tool for analyzing tax compliance issues. In fact, an essential feature of any agent-based model is the direct non-market based interaction of agents, which is combined with some process that allows for changes in individual behavior patterns[1]. Therefore, agent-based tax

[1] An exception is the work by Szabó et al. (2009, 2010).

B. Kamiński and G.Koloch (eds.), *Advances in Social Simulation*,

Advances in Intelligent Systems and Computing 229,
DOI: 10.1007/978-3-642-39829-2_11,

evasion models may be categorized according to the features of this individual interaction process. In fact, in econophysics models this process is driven by statistical mechanics using the Ising model (Ising, 1925) that is well known in physics and describes objects which can be in one of two states and interact on a given lattice structure. Examples include Zaklan et al. (2008, 2009), Lima and Zaklan (2008), and Lima (2010) which have identified the Ising states with compliant and non-compliant tax payers. In contrast, if the interacting process is driven by parameter changes that induce behavioral changes via a utility function and (or) by stochastic processes that do not have physical roots, these models belong to the economics domain. Examples include Mittone and Patelli (2000); Davis et al. (2003); Bloomquist (2004, 2008); Korobow et al. (2007); Antunes et al. (2007); Szabó et al. (2009, 2010); Hokamp and Pickhardt (2010); Méder et al. (2012); Nordblom and Žamac (2012); Andrei et al. (2013); Hokamp (2013); Pellizari and Rizzi (2013) of which some are summarized by Bloomquist (2006) and Pickhardt and Seibold (2013).

In agent-based tax evasion models of the econophysics type the Ising model is used to mimic conditional cooperation among agents (Zaklan et al., 2009). Yet, the actual patterns and levels of tax evasion in these models depend on two additional factors: the network structure of society and the tax enforcement mechanism. The network structure is implemented by alternative lattice types and tax enforcement consists of the two economic standard parameters audit probability and penalty rate. To this extent, rational behavior patterns are essentially reconstructed by means of statistical mechanics.

In previous work Pickhardt and Seibold (2013) and more recently Seibold and Pickhardt (2013) have extended the Ising-based econophysics approach to tax evasion toward the implementation of different agent types. This theory is able to reproduce results from agent-based economics models (Hokamp and Pickhardt, 2010) so that it should be also appropriate for a quantitative analyis. Following this idea we aim in the present contribution to apply the model to an analysis of the shadow economy in France and Germany with respect to the percentage of rational agents in both countries. Note that in the present paper we use the term 'shadow economy' synonymous with the participation in black market services.

In Sec. 2 we outline the basic ingredients of our econophysics model and exemplify the approach for a black market with homogeneous agents. We apply our model to the analysis of the shadow economies in Germany and France in Sec. 3 where we deduce the essential parameters entering the simulations from previous experimental and agent-based investigations in the literature. Finally, we discuss our results for different enforcement schemes and give policy recommendations to combat the shadow economy in Sec. 4.

2 The Agent-Based Econophysics Approach

Our considerations are based on the Ising model hamiltonian

$$H = -J\sum_{\langle ij\rangle} S_i S_j - \sum_i B_i S_i \tag{1}$$

where J describes the coupling of Ising variables (spins) $S_i = +1, -1$ between adjacent lattice sites denoted by $\langle ij\rangle$. In the present context $S_i = +1(-1)$ is interpreted as a participant in the white (black) market. The following results are not sensitive to the specific lattice geometry and we implement the model on a two-dimensional square lattice with dimension 1000×1000. Eq. (1) contains also the coupling of the spins to a local magnetic field B_i which can be associated with the morale attitude of the agents and corresponds to the parameter γ_i in the theory of Nordblom and Žamac (2012). In addition, our model contains a local temperature T_i which measures the susceptibility of agents to external perturbations (either influence of neighbors of magnetic field). We then use the heat-bath algorithm [cf. Krauth (2006)] in order to evaluate statistical averages of the model. The probability for a spin at lattice site i to take the values $S_i = \pm 1$ is given by

$$p_i(S_i) = \frac{1}{1+\exp\{-[E(-S_i)-E(S_i)]/T_i\}} \tag{2}$$

and $E(-S_i) - E(S_i)$ is the energy change for a spin-flip at site i. Upon picking a random number $0 \leq r \leq 1$ the spin takes the value $S_i = 1$ when $r < p_i(S_i = 1)$ and $S_i = -1$ otherwise. One time step then corresponds to a complete sweep through the lattice.

Following Hokamp and Pickhardt (2010) we consider societies which are composed of the following four types of agents: (i) *selfish a-type agents*, which take advantage from black market activities ($S_i = -1$) and, thus, are characterized by $B_i/T_i < 0$ and $|B_i| > J$; (ii) *copying b-type agents*, which conform to the norm of their social network and thus copy the behavior with respect to black or white market participation from their neighborhood. This can be modelled by $B_i << J$ and $J_i/T_i \gtrsim 1$; (iii) *ethical c-type agents*, which have large moral doubts about participating in black market services and thus are parametrized by $B_i/T_i > 0$ and $|B_i| > J$; (iv) *random d-type agents*, which act by chance, within a certain range, due to some confusion about the attribution of services to the black or white market. We implement this behavior by $B_i << J$ and $J/T_i << 1$. The parameters distinguishing the different agent types are taken from Pickhardt and Seibold (2013) and Seibold and Pickhardt (2013).

Furtheron we implement different enforcement schemes into our model. Here we first consider the case where the detection of a black market participating agent enforces its compliance over the following h time steps. This is the procedure which has been invoked in Zaklan et al. (2008, 2009); Lima (2010); Pickhardt and Seibold (2013) and also implemented in a randomized variant in Lima and Zaklan (2008). Second, we also study lapse of time effects, i.e. the situation where a detected agent is also screened over several years in the past by the (tax) authorities (i.e. backaudit). This variant has been studied within an econophysics tax compliance model in Seibold and Pickhardt (2013). If tax evasion is detected in the current time period, the backaudit comprises also an inspection of the preceding b_p time steps. Denote with n_e the number of time steps over which the agent was evading within the backaudit plus current period. Then the period k over which the agent is reinforced to be compliant is set to $k = n_e * h$. For example, for a convicted agent in the current time step, inspection of the preceding $b_p = 5$ time steps reveals three periods where he was evading. Setting $h = 2$ yields a number of $(3+1)*2 = 8$ periods where he is forced to be compliant. Thus the above limit of fixed compliance period h is recovered in the limit of zero backaudit $b_p = 0$ since then $k = (0+1)*h = h$.

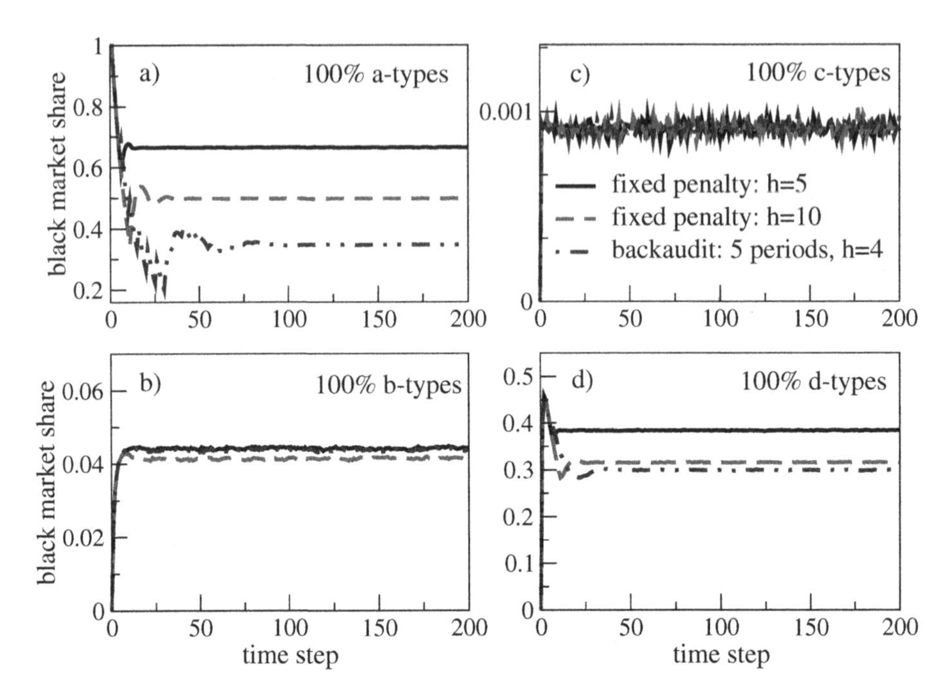

Fig. 1. Time evolution for the participation in black market services for a society consisting of 100% a-type (panel a), 100% b-type (panel b), 100% c-type (panel c), and 100% d-type (panel d) agents. Results are reported for different enforcement mechanisms: fixed compliance period $h = 5$ (solid, black), $h = 10$ (dashed, red), and backaudit (dashed-dotted, blue). Audit probability is $p_a = 10\%$ in each case.

Before analyzing the heterogeneous agent model it is instructive to consider first the case of a black market with all agents being of the same type. The resulting percentage of black market share as a function of time is shown in Fig. 1 where we also compare the different enforcement schemes.

The first case of endogenous non-compliant selfish agents is shown in Fig. 1a. At time step zero we correspondingly set the share of black market participation to $p_{bm} = 1$. Due to the enforcement mechanisms the black market share is significantly reduced because at each time step a certain percentage of the remaining non-compliant agents are forced to become compliant. Before reaching a stationary value small oscillations are observed since after h time steps the first detected agents can become non-compliant again. Notably, the black market share is reduced strongest for the backaudit mechanism

Fig. 1b reports the result for copying b-type agents. As initial condition all agents are set to 'compliant'. Since b-types tend to copy the behavior of their social network only few of them change their behavior and the equilibrium value for the black market share for all enforcement mechanisms approaches a rather small value between 4% and 5%. It should be noted that the equilibrium value is independent of the initial condition. If we would have set all agents to non-compliant at time step zero, the audits would have reduced the black market share to the same equilibrium value than shown in Fig. 1.

The time evolution for ethical agents is reported in Fig. 1c. Here the initial black market participation is set to $p_{bm} = 0$ and there is only a very small probability that one of the agents becomes non-compliant. Since ethical agents avoid black market participation the results are also almost independent of the audit probability. Hence, any positive audit probability would be inefficient in this case.

Finally, Fig. 1d shows the black market share for d-type agents. Since these agents act by chance their participation in black market activities would be of the order of $\sim 50\%$ without any audit. The different enforcement mechanisms then lead to a further reduction of this equilibrium value where similar to the a-types the backaudit mechanism is most effective.

3 Exploring the Shadow Economy

We now turn to the analysis of the shadow economies in France and Germany within our model which requires the estimation of the corresponding specific agent compositions. In general, tax experiments provide average data and do not provide individual compliance data of tax payers. The work by Bazart and Pickhardt (2011) is a notable exception and allows to extract the percentage of fully compliant individuals (i.e. essentially c-type agents in our terminology). Subject to the small group sizes of 5 subjects Bazart and Pickhardt (2011) obtain a full compliance ratio of 5% and 20% for Germany and France, respectively[2].

Unfortunately no data are available on the percentage of d-type agents. However, Andreoni et al. (1998) report that about seven percent of U.S. households overpaid their taxes in 1988. If we anticipate that about the same amount of people underpays their taxes we arrive at a percentage of $\sim 15\%$ of d-type agents [cf. also Hokamp and Pickhardt (2010)]. For simplicity, this percentage is adopted in equal measure for France and Germany[3].

In the following we will denote the parameters derived for France (i.e. 20% c-types and 15% d-types) as parameter set 'F' and the parameters derived for Germany (i.e. 5% c-types and 15% d-types) as parameter set 'G'. The aim is then to determine the fraction of a- and b-type agents which participate in the shadow economy in these countries. Calculations are performed for two different enforcement schemes which both are based on the probability p_a for an audit at a given lattice site (agent). For both, France and Germany we set $p_a = 0.1$. Concerning the backaudit enforcement we consider a backaudit period of $b_p = 5$ time steps which is compatible with the limitation period of 5 years in 2006/2007 where the experiments by Bazart and Pickhardt (2011) have been conducted.

In order to obtain values on the shadow economies we adopt the values from Buehn and Schneider (2012) which have estimated the shadow economy of 162 countries.

[2] Note that the compliance ratio only specifies the percentage of subjects which behaved fully compliant in each of the rounds of the experiment. The average compliance rate which determines tax evasion is larger.

[3] One may of course expect that due to different tax pressure this number differs between France, Germany and the U.S.. Future experiments can help to resolve this element of uncertainty for the percentage of d-type agents.

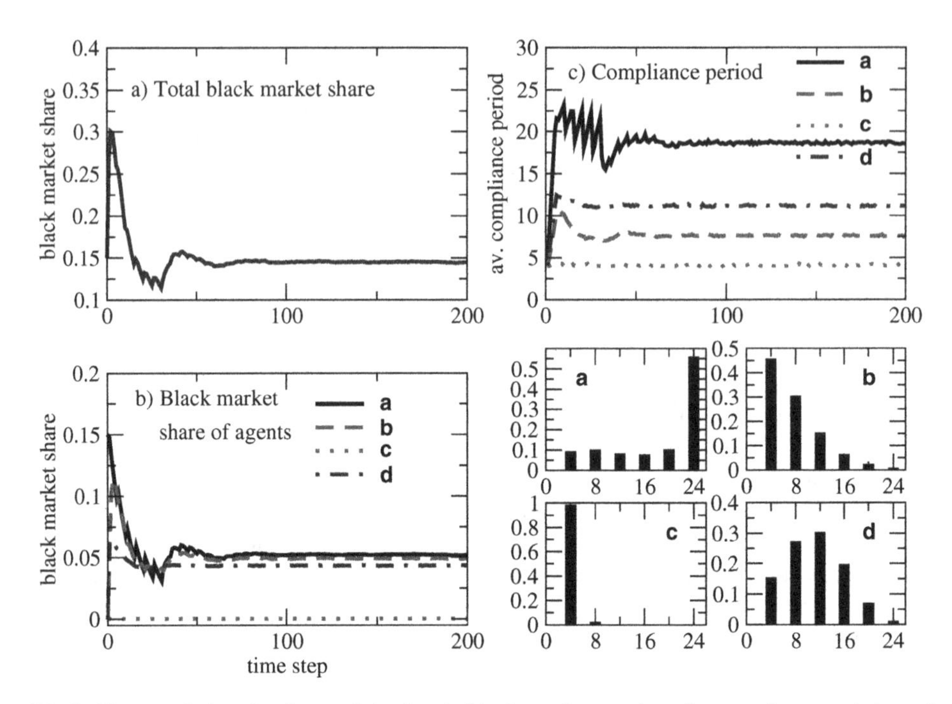

Fig. 2. Time evolution for the participation in black market services for a society consisting of 15% a-type, 50% b-type, 20% c-type, and 15% d-type agents. The upper left panel compares the extent and dynamics of black market participation for backaudit periods $b_p = 5$. The lower left panel breaks down the participation probability to the individual agent types. The upper right panel displays the average forced compliance period (or penalty) for the individual agent types and the lower right panels show the corresponding distribution. Audit probability is $p_a = 10\%$ in each case.

According to this analysis the size of the shadow economy in France and Germany is 15% and 16% of the official GDP, respectively, averaged over the period 1999 - 2007. These numbers are supported by the data of Elgin and Öztunali (2012) which report 16.53% for OECD EU-countries in the period from 2001-2009.

Fig. 2a shows the dynamics of black market participation for the parameter set 'F' appropriate to France and 15(50)% a(b)-type agents. Initial conditions are chosen such that all agents but the a-types are set to white market participants so that the black market share at time step 'zero' just reflects the percentage of selfish a-types. Note, however, that the equilibrium result for large time steps does not depend on these initial conditions. As can be seen from Fig. 2b the initial increase in the first time step is due to the b-types which copy the black market participation from the a-type agents. In the following periods black market participation is reduced to the (back)audit and approaches an equilibrium value after passing a transient regime. For this parameter set a-, b- and d-type agends contribute equally with $\approx 5\%$ to the total black market share. It is also instructive to monitor the average compliance period of the convicted agents as shown in Fig. 2c and the actual distributions which are displayed in the lower right panels of

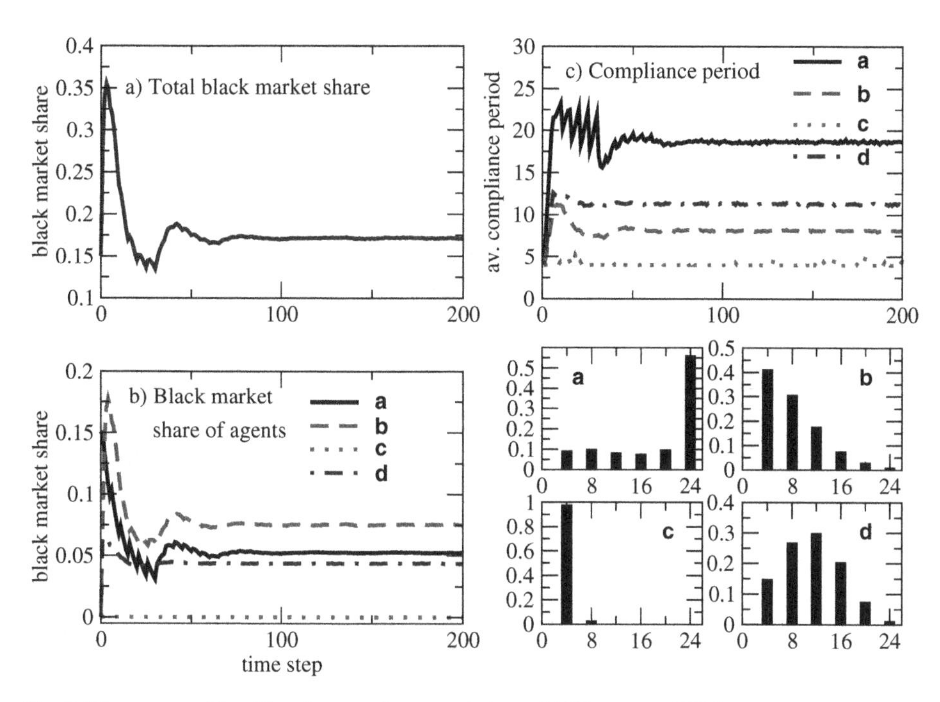

Fig. 3. The same as Fig. 2 but for a society consisting of 15% a-type, 65% b-type, 5% c-type, and 15% d-type agents

Fig. 2. Naturally most a-types are penalized with the maximum compliance period of $k = 6 * 4 = 24$ time steps. Only those which have been convicted also in previous time steps (and therefore have to stay compliant) are penalized with a reduced compliance period. On the other hand, the (few) convicted c-type agents are penalized with a compliance period of $k = 4$ since their probability of repeatedly being non-compliant within the backaudit period is vanishingly small.

Fig. 3 shows the analogous results for the parameter set 'G' derived for Germany and also 15% of a-type agents. The reduced percentage of (compliant) c-types as compared to France is compensated by the increased number of b-type agents which enhance the black market participation as becomes apparent from Fig. 3a. In fact, the largest contribution is now from the b-types (cf. Fig. 3b) in contrast the previous parameter set. Note especially that the percentage of convicted b-types with a compliance period of 4 time steps is reduced from $\sim 45\%$ in Fig. 2 to $\sim 40\%$ in Fig. 3 due to the concomitant reduction of c-types which define the social norm of adjacent b-type agents.

We are now in the position to evaluate the black market share as a function of a-type agents for the two parameter sets 'F' and 'G' derived above. Fig. 4 reports the corresponding results for different audit schemes, i.e. fixed compliance period with $h = 5, 10$ and backauditing with backwards auditing period of $b_p = 5$ and scaling factor $h = 4$ (cf. above). Within all auditing schemes we find a larger black market share for the parameter set as deduced for Germany due to the smaller (larger) percentage of c-(b-) type agents. The horizontal lines in Fig. 4 indicate the size of the shadow

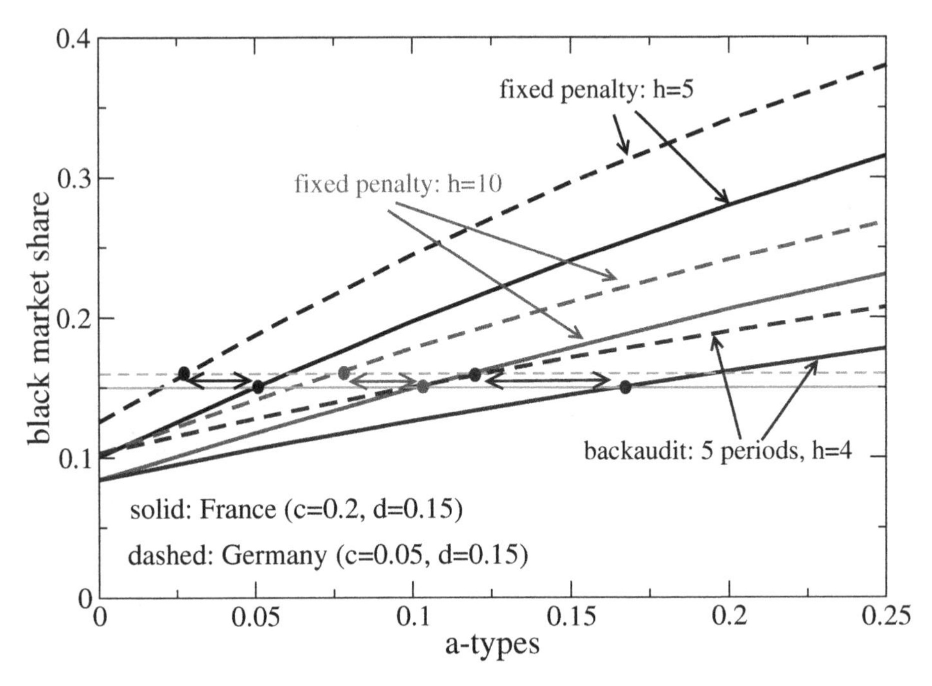

Fig. 4. Black market share for the two parameter sets derived for France (solid lines) and Germany (dashed lines) as a function of a-type agents. The horizontal lines indicate the size of the shadow economy in both countries and the intersections (indicated by dots) fix the respective percentage of a-types. Results are reported for different enforcement mechanisms: fixed compliance period $h = 5$ (black), $h = 10$ (red), and backaudit (blue).

economy for France and Germany as reported by Buehn and Schneider (2012) and the intersections with the tax evasion curves are indicated by dots for each audit mechanism. We find that the difference of a-types between both countries which is compatible with these values (as illustrated by arrows) is between 3% and 5% with the large difference obtained for the backaudit scheme. Since backauditing is more efficient in reducing black market services it is compatible with a larger percentage of a-type agents than audit schemes with fixed compliance period.

4 Policy Recommendations and Discussion

The larger percentage of rational a-types resulting for the parameter set 'F' suggests that audit mechanisms to combat the shadow economy are more efficient in France than in Germany. It is also interesting that the difference between rational a-types in both countries can be traced back to the much larger percentage of c-types in the parameter set 'F'. Inspection of the experiment by Bazart and Pickhardt (2011) reveals that this difference is due to the larger full compliance ratio for french female subjects (41%) than for german females (10%) whereas the full compliance ratio between male (France: 4.8%, Germany: 3.7%) does not differ significantly. Note, however, that the group sizes in these investigations were rather small so that further experiments are required in order to substantiate the corresponding data.

In our econophysics model we have assumed a constant (over time) distribution of agent types. On the other hand, Nordblom and Žamac (2012) have set up a economic model which describes how social beings update their personal norms. Implementing these mechanisms in our model would allow for the transformation between different agent types and is an interesting perspective for future research.

References

Allingham, M.G., Sandmo, A.: Income Tax Evasion: A Theoretical Analysis. Journal of Public Economics 1, 323–338 (1972)

Alm, J.: Measuring, explaining, and controlling tax evasion: lessons from theory, experiments, and field studies. International Tax and Public Finance 19(1), 54–77 (2012)

Andrei, A., Comer, K., Koehler, M.: An agent-based model of network effects on tax compliance and evasion. Journal of Economic Psychology (in press, 2013), `http://dx.doi.org/10.1016/j.joep.2013.01.002`

Andreoni, J., Erard, B., Feinstein, J.: Tax Compliance. Journal of Economic Literature 36(2), 818–860 (1998)

Antunes, L., Balsa, J., Respício, A., Coelho, H.: Tactical exploration of tax compliance decisions in multi-agent based simulation. In: Antunes, L., Takadama, K. (eds.) MABS 2006. LNCS (LNAI), vol. 4442, pp. 80–95. Springer, Heidelberg (2007)

Bazart, C., Pickhardt, M.: Fighting Income Tax Evasion with Positive Rewards. Public Finance Review 39(1), 124–149 (2011)

Bloomquist, K.M.: Modeling Taxpayers' Response to Compliance Improvement Alternatives. Paper presented at the Annual Conference of the North American Association for Computational Social and Organizational Sciences, Pittsburgh, PA (2004)

Bloomquist, K.M.: A Comparison of Agent-based Models of Income Tax Evasion. Social Science Computer Review 24(4), 411–425 (2006)

Bloomquist, K.M.: Taxpayer Compliance Simulation: A Multi-Agent Based Approach. In: Edmonds, B., Hernández, C., Troitzsch, K.G. (eds.) Social Simulation: Technologies, Advances and New Discoveries. Premier References Series, ch. 2, pp. 13–25 (2008)

Buehn, A., Schneider, F.: Shadow economies around the world: novel insights, accepted knowledge, and new estimates. International Tax and Public Finance 19(1), 139–171 (2012)

Davis, J.S., Hecht, G., Perkins, J.D.: Social Behaviors, Enforcement, and Tax Compliance Dynamics. The Accounting Review 78(1), 39–69 (2003)

Elgin, C., Öztunali, O.: Shadow Economies around the World: Model Based Estimates. Working Papers 2012/05. Bogazici University, Department of Economics (2012)

Hokamp, S., Pickhardt, M.: Income Tax Evasion in a Society of Heterogeneous Agents - Evidence from an Agent-based Model. International Economic Journal 24(4), 541–553 (2010)

Hokamp, S.: Dynamics of tax evasion with back auditing, social norm updating and public goods provision - An agent-based simulation. Journal of Economic Psychology (in press, 2013), `http://dx.doi.org/10.1016/j.joep.2013.01.006`

Ising, E.: Beitrag zur Theorie des Ferromagnetismus. Zeitschrift für Physik 31(1), 253–258 (1925)

Korobow, A., Johnson, C., Axtell, R.: An Agent-based Model of Tax Compliance with Social Networks. National Tax Journal 60(3), 589–610 (2007)

Krauth, W.: Statistical Mechanics; Algorithms and Computations. Oxford University Press (2006)

Lima, F.W.S., Zaklan, G.: A Multi-agent-based Approach to Tax Morale. International Journal of Modern Physics C: Computational Physics and Physical Computation 19(12), 1797–1808 (2008)

Lima, F.W.S.: Analysing and Controlling the Tax Evasion Dynamics via Majority-Vote Model. Journal of Physics: Conference Series 246, 1–12 (2010)

Méder, Z.Z., Simonovits, A., Vincze, J.: Tax Morale and Tax Evasion: Social Preferences and Bounded Rationality. Economic Analysis and Policy 42(2), 171–188 (2012)

Mittone, L., Patelli, P.: Imitative Behaviour in Tax Evasion. In: Luna, F., Stefansson, B. (eds.) Economic Simulations in Swarm: Agent-based Modelling and Object Oriented Programming, pp. 133–158. Kluwer Academic Publishers, Dordrecht (2000)

Nordblom, K., Žamac, J.: Endogenous Norm Formation Over the Life Cycle - The Case of Tax Morale. Economic Analyis & Policy 42(2), 153–170 (2012)

Pellizari, P., Rizzi, D.: Citizenship and power in an agent-based model of tax compliance with public expenditure. Journal of Economic Psychology (in press, 2013), `http://dx.doi.org/10.1016/j.joep.2012.12.006`

Pickhardt, M., Seibold, G.: Income Tax Evasion Dynamcis: Evidence from an Agent-based Econophysics Model. Journal of Economic Psychology (in press, 2013), `http://dx.doi.org/10.1016/j.joep.2013.01.011`

Seibold, G., Pickhardt, M.: Lapse of time effects on tax evasion in an agent-based Econophysics model. Physica A 392(9), 2079–2087 (2013)

Szabó, A., Gulyás, L., Tóth, I.J.: Sensitivity analysis of a tax evasion model applying automated design of experiments. In: Lopes, L.S., Lau, N., Mariano, P., Rocha, L.M. (eds.) EPIA 2009. LNCS, vol. 5816, pp. 572–583. Springer, Heidelberg (2009)

Szabó, A., Gulyás, L., Tóth, I.J.: Simulating Tax Evasion with Utilitarian Agents and Social Feedback. International Journal of Agent Technologies and Systems 2(1), 16–30 (2010)

Zaklan, G., Lima, F.W.S., Westerhoff, F.: Controlling Tax Evasion Fluctuations. Physica A: Statistical Mechanics and its Applications 387(23), 5857–5861 (2008)

Zaklan, G., Westerhoff, F., Stauffer, D.: Analysing Tax Evasion Dynamics via the Ising Model. Journal of Economic Interaction and Coordination 4, 1–14 (2009)

Studying Possible Outcomes in a Model of Sexually Transmitted Virus (HPV) Causing Cervical Cancer for Poland

Andrzej Jarynowski[1,2,4] and Ana Serafimovic[3]

[1] Smoluchowski Institute, Jagiellonian University, Cracow, Poland
[2] Department of Sociology, Stockholm University, Sweden
[3] Department of Mathematics, Stockholm University, Sweden
[4] Central Institute for Labor Protection, National Research Institute, Warsaw, Poland
andrzej.jarynowski@uj.edu.pl

Abstract. The aim of this paper is to supplement knowledge about the spread of sexually transmitted diseases through computer simulations. The model has aggregated the most important properties of HPV infections and development of cervical cancer in demographically changing Polish society. The main goal is the authoritative analysis of the potential epidemiological control strategies and their impact on situation in Poland in the future 25 years. Constructed model shows indication that vaccination with screening organized alongside would be effective measures against cervical cancer. It also alarms authorities that processes like aging of society and increase of sexual activity (which are taking place in Poland at the moment) could recall epidemic, if prevention would not act properly.

Keywords: epidemic modeling, social model simulation, system research, STI, HPV, cervical cancer.

1 Introduction

Human papillomavirus, or HPV, is a sexually transmitted virus infection, which is not only the main, but also necessary risk factor for developing cervical cancer [2], [4], [1] - second most common type of cancer in women. Accordingly, these infections are quite widespread - in 1990s it has been shown that about 70 percent of the sexually active population have acquired a virus of this type at some point of their lives [7]. Out of 30 types of HPV virus that are known to infect genital areas, 15 are high risk or oncogenic, although it might pass as long as twenty years for cancer to develop from the time one gets infected by such a virus [3]. Among the oncogenic HPVs, the most severe one is type 16, present in about half of all cervical cancer cases [4]. Recent studies have shown that the main safety precaution with respect to cervical cancer is going to be a combination of vaccination and screening - since only type specific vaccines are available and there are as many as 15 high risk HPVs. As well as that, screening

B. Kamiński and G.Koloch (eds.), *Advances in Social Simulation*, Advances in Intelligent Systems and Computing 229,
DOI: 10.1007/978-3-642-39829-2_12,

alone, although a very reliable method for all age groups [2], has decreased cancer incidences in developed countries, but globally, licensing proper vaccines would be of great importance, especially since the morbidity of cervical cancer is high in developing countries [3]. However, even when a proper vaccine is introduced, it will take time for the decrease in the incidences (and fatality) of cervical cancer to become visible, due to the slow progression of such cancers; this also confirms that screening should be organized along with the vaccination [2].

Given that the infection is transmittable via sexual contact, vaccinating a certain amount of females alone would be a better strategy than vaccinating the same amount of both sexes, and there have already been studies that confirmed this assumption [3]. Another reason to decide for women particularly is that the possible consequences of infection are much more severe in females (cervical cancer).

Table 1. Proportions of activity groups [4]

Age (years)	Highest activity	Moderately high activity	Moderate activity	Lowest activity
15 – 19	0.015	0.03	0.135	0.82
20 – 24	0.015	0.025	0.34	0.62
25 – 34	0.01	0.02	0.21	0.76
35 – 64	0.005	0.01	0.09	0.895

Table 2. Mean rates of sexual partner change (new partners per year) for activity groups [4]

Age (years)	Highest activity	Moderately high activity	Moderate activity	Lowest activity
15 – 19	15	3.50	1.34	0.48
20 – 24	17.5	0.96	0.38	0.14
25 – 34	15	0.67	0.21	0.08
35 – 64*	7.5	0.45	0.08	0.04

Investigation of Sexuality. Investigation of sexuality is not an easy task, especially in Poland (this middle-income country is classified by WHO in the field of sexually transmitted diseases at the level of countries of the third world [8]). There are no recorded data, perfect research methods or fully verifiable results anywhere in the world and more so in Poland. In this study we decide to use results of a Finish survey (Table 1) and adjust them to the Polish reality. Adjustsing was done by fitting real cancer cases (Figure 2) and scaling parameter was around 0.35 in 90th. This means, that values from Finish suveys must be divided by around 3. None of the surveys from other coutries have as high respondent rate as in Finland. Moreover, all other surveys show bias. For example in south- and east-european studies males report much more sexual partners than females.

Mathematical Modeling. The area of epidemic modeling is explored by researchers from different academic backgrounds: physicians, physicists, mathematicians, statisticians, computer scientists, geographers (spacial epidemiology) and sociologists. All of them have something to add and this project is based on different approaches from mathematical to computational approach (simulations) and it also comprises concepts of physics, sociological and statistical analysis [11]. This work is specious, because it is interdisciplinary and shedding light on methods which have not been used very much in epidemiology until this moment. Epidemiological models based on human-to-human transmission from differential equation point of view do exist in older literature [12], but in more recent work in the area, agent-based models appear more often. Mathematical models and computer simulation start to play significant role as quantity of social interactions [13] is enormous, but more important than simulations are real data especially register-based.

Environment of Study. The core of this research are computer simulations. First, a model that represents sexually active population of Poles was developed. The model operates on the level of the whole country (action and state of agents are simulated at the level of the whole country [9], [10] - above 20 million people for Poland). In this study, however, we limit ourselves to the standard epidemic models, where categories of population are subgroups based on the model of SIR (Susceptible-Infected-Removed) [10]. We built the model

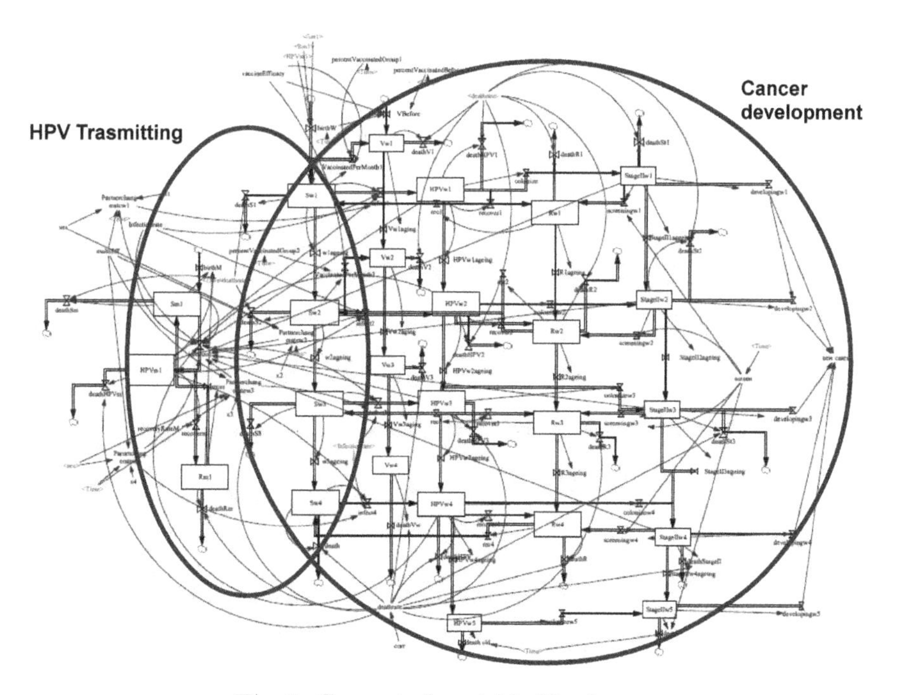

Fig. 1. Concept of model in Vensim

describing the epidemiology and consequences of HPV for Poland in the environment of Vensim (Figure 1) based on our previous model for Sweden [14]. We adapted it to the demographic structure of Poles (unfortunately most of the parameters concerning sexuality and medical properties were estimated based on data for other populations - mostly Nordic - Table 3). The results obtained for vaccination strategies agreed with the work of scientists sponsored by the Merc AND Co [15] and GalaxySmith [16] (vaccine producers). Model contains age structure, because sexual behavior, as well as other medical parameters, is age depended. The existing model consists only of the main fraction of the population (heterosexual [17]) and takes into account the demographic changes over form last 25 years and predicts for next 25. Sexuality (number of new sexual partners in a given time interval) was randomly chosen in every step of simulation from empirical distribution for subgroup from given age category. Unfortunately, we do not have Polish data with such resolution. Only few of the Polish studies about sexuality consider aspects interesting for us and none of them was performed on big enough, representative sample (according Izdebski's research [18]). We observe the change of sexual behavior (average number of partners is increasing over time [19]), and the model will replicate that.

2 Problem Statement

The time between getting infected by HPV and developing cancer can be twenty years or more, therefore a dynamic model of human behavior would be very useful, so that simulations can be made and different scenarios compared [3].

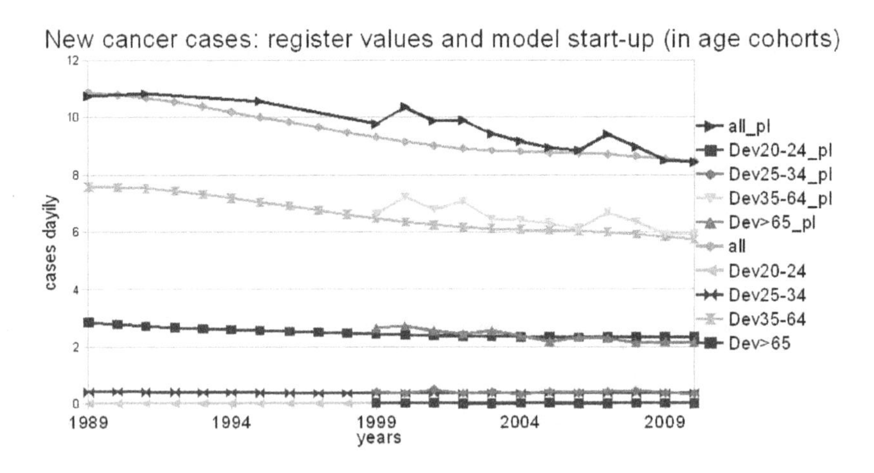

Fig. 2. Fitting area for calibaring model parameters. Number of new cancer cases for different age cohorts ($20-24$, $25-34$, $35-64, \geq 65$) from historical registry (pl) are compared with simulated values.

Our model is based on HPV transition between people due to heterosexual contact. We assume that women infected by HPV could recover and acquire temporal immunity or could be long-term colonized by the virus and could then develop cancer. The model helps us understand the process and predict possible realizations and allows us to give suggestions on how to cope with an outbreak. We test the following scenarios:

a) no obligatory vaccination or vaccination of 14 years old girls (before sexual debut - the optimal srategy according to [4], [4],[3], [1], [16]);
b) with 50% or 75% of sexuality (linear) increase;
c) for different health care progress in effective screening frequencies of testing for cell pathology until frequency up to every 2.5 or 7.5 years.

3 Methods and Model Design (Code Is Available [20])

We model part of Polish population which is sexually active ($15 - 64$ years old). We assume a temporal naturally acquired immunity and lifelong vaccine acquired immunity, with given efficacy. Male population is divided into 3 stages: Susceptible (Sm), Infectious (HPVm), Recovered (Rm) and population is dynamic due to natural birth and death rates. For women, apart from Sw, HPVw, Rw, additional stages Long-term colonized (StageIIw), Vaccinated (Vw) and Having cancer (Cancer), are allowed and all except Cancer are divided into different age groups. We also introduced changing of society by aging, birth and death (which was set to affect older groups much more than younger ones). We assume that around the age of 65 people are not changing partners, so death rate means not natural death, but rather removing from sexually active society. We are still tracing women above 65, who were already infected (in stage HPV or StageII) and can develop the cancer. We also allow dying in sexually active lifespan due to some other cause, with age-dependent death rate. Birth, which is actually to be understood as turning 15 years and potentially beginning a sexual life is interpolated from register data for time up until 2013 (now) and extrapolated for near future. In our model, Polish society is slowly aging in waves (as it seems to be happening in reality, due to specific demographic structure (Figure 4(a),4(b)). Medical properties of cancer developing are known to be age-dependent as well as sexual activity, so we decided to choose age groups with respect to data form reports. We used reports of Finish sexuality, in which the society is divided in groups of 5 years intervals. Cancer development properties were set up seperatly for all age-cohorts. As a trade-off for having the smallest numbers of groups, but caching main differences in behavior, we choose to devide women into the following age cohorts.

1) $15 - 19$ (initiation of sexual live)
2) $20 - 24$ (most active sexual group)
3) $25 - 34$ (stabilization of sexual live)
4) $35 - 64$ (sexual stagnation and stronger susceptibility to cancer)
5) ≥ 65 (no sexuality and cancer development)

Table 3. Model parameters

Parameter name	Description	Value	Reference
inflow	daily inflow of youth rate in Poland	Figure 4(b)	aprox. [5]
deathrate	daily death rate in Poland	sex, age and population depended	aprox. [5], [23]
Infectionrate	Infectivity per partnership	0.5	[15],[3]
*Partner-changratew*1 − 4	Partner change rates in women	Stochastic, from Table 1-2	[4]
*colonizew*1 − 5	HPV progression rate to Stage II (per months)	Age $15-24$: $(0.1*(0.1/72+0.2/36)$ $25-34$: $0.5*(0.1/72+0.2/36)$ $35-64$: $0.35/72+0.2/36$ ≥ 65: $1.2*(0.35/72+0.2/36)$	[4] and [23]
*recover*1 − 5	HPV regression rate to immune (per month)	Age $15-34$: $0.6/18$ ≥ 35: $0.15/18$	[4]
*screening*1 − 5	Natural rate from stageII to immune (per month)	Age $15-34$: $0.65/72$ ≥ 35: $0.4/72$	[4]
*developing*2 − 5	Rate from stageII to cancer (per month)	$15-24$: $(0.1*(0.13/120)$ $25-34$: $0.5*(0.13/120)$ $35-64$: $0.13/120$ ≥ 65: $1.2*(0.13/120)$	[4], [23]
PercentVaccinated-Before and *Group*1 − 2	vaccinated before corresponding cohorts	90% 5%	own assump.
vaccineEfficacy	Percent of women for whom the vaccine works	95%	[1]
survive	5-year survival rate	0.41	[22]
screen	Frequency of screening up to	2.5 or 7.5 years	own assump.
sexuallity	Linear increase of sexuallity	50% or 75%	own assump.
recoveryRateM	Recovery rate from HPV for men (per month)	0.4/18	[4]
immunityLost	Rate from R to S for all (per month)	7.5/12	[4], [3]
multistrain	Sum effect of HPV 16 and 18	2	own assump.

Time

Simulations cover 50 years, from 1989 until 2039 (around 25 before the present year - 2013, to fit model parameters and around 25 years afterwards, to predict the future situation). Starting year - 1989 was chosen because of political transformation in Poland. This is when a collapse of social norms began, like those about sexuality. Moreover, before that date, there is no full registry on new cancer cases and until 1999 only part of the data are available (since 1999 National Cancer Registry [21] has been publishing very detailed data on-line). Another important change happened around 2006, when Poland introduced popular screening program. At present (2013), it is recommended to women between 25 and 60 to screen every 3 years, but only little above 20% of them do it [24] (effective screening frequency is around every 15 years). To capture this in the model, we differentiate screening values into 3 periods (Figure 3):

- until 2005: there was no regular screening, and tests are done when physicians ask for that,
- 2006 – 2015: when regular screening procedures are introduced,
- 2016 – 2039: where we predict that screening effectiveness will be continuously improving (starting from every 10 years, which means that 1/3 of women will follow 'every 3 years rule').

We choose only 25 years timespan for prediction, because of the rapid change in medical treatments. For example, screening test in our model is allowed to detect only women "permanently" colonized, in StageII, but there already exist PCR-tests (where even genomic sequence can be obtained), which detect presence of virus in every stage. However, they are too expensive for common use now. Demographic prediction also cannot look too much ahead, because we did not take into account factors such as migration. Time step of simulation is 0.1 day (it has to be very small to avoid averaging because of stochastic character of simulation).

Demographics. Population in this model consists of sexually active Poles, who are leaving the system after reaching 65 years old. Exception are women from

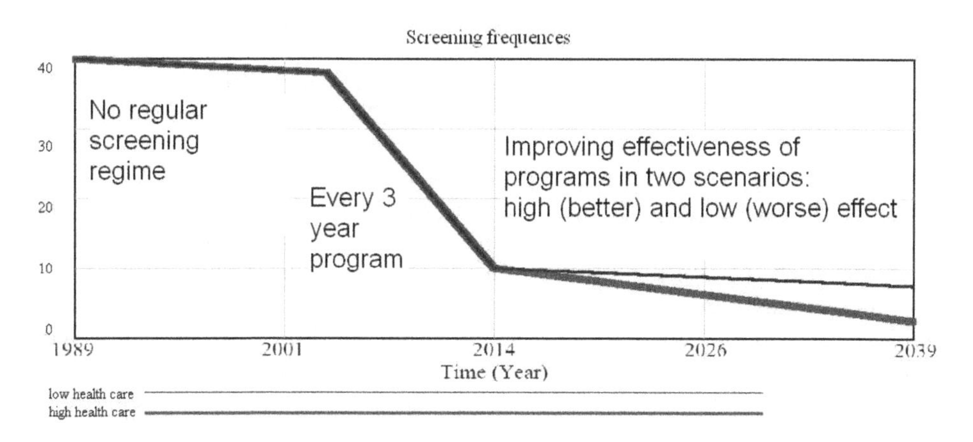

Fig. 3. Screening frequencies (every x years)

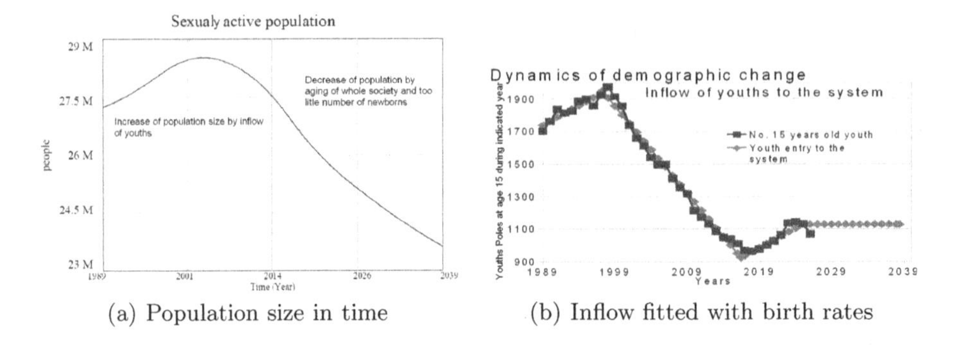

(a) Population size in time

(b) Inflow fitted with birth rates

Fig. 4. Demographics

subgroups HPVw and StageII, who can still develop cancer until their death caused by any other factors (natural death). Life expectancy of those women is growing according to Statistics Polland - GUS [5] and it is modeled as linear increase for the whole time horizon of simulation. All agents could leave system due to natural death rate which is a function of population size, age and sex. Inflow to the system are the young of age 15 (respectively more boys than girls according to GUS). We can extrapolate number of the young coming into the system for the next 15 years of simulation (until 2027) by scaling number of young Poles who will grow up and start their sexual life in future. We can observe two characteristics of Poland - wave structure, and birth rate is mainly decreasing (Figure 4(b)). In effect population is aging and dying out. After the pick year in the number of the young coming into the system, which is around 2035, this number would probably decrease (as it can be observed in first years after pick). We decided not to follow this prediction, and following the rule to be as much conservative as possible, the rates of coming in later years (after 2035) are assumed constant. In effect, Polish sexually active population (corresponding to workers age span) will decrease from 27.5 in 2013 to 23.5 million in 2039 (Figure 4(a)). This value alone should alarm the authorities, and probably it is overestimated, since emigration of Polish citizens has not been taken into account.

Transition. Transition of disease can take place due to sexual contact, with given probability. It has been estimated in other paper between 0.4 to 0.6, so we choose it as 0.5. With respect to Finish data set, individuals in our model have a randomly chosen number of new contacts each day, which follows the empirical distribution for given age groups from the Finish surveys (rescaled to Poland). Stochasticity was introduced not at the level of individuals, but at the level of subgroup (limitation of Vensim). One can understand that all people

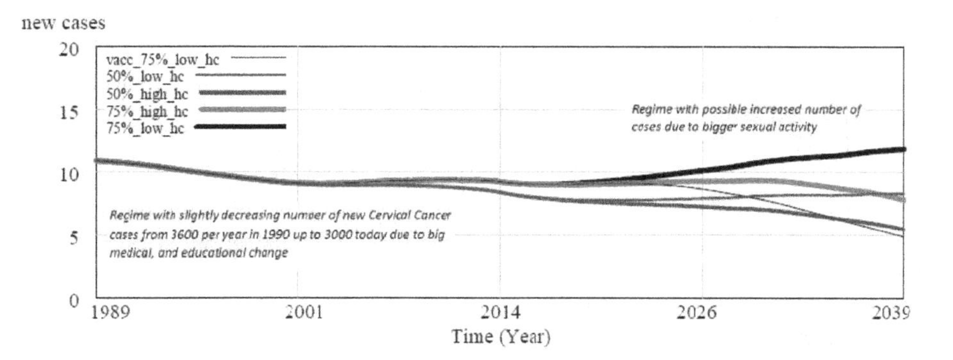

Fig. 5. New cancer cases per day, for different scenarios

in a given subgroup can be very active or passive, at different iterations. We introduce change of sexuality as a linear increase of newborn's sexuality (the young at age 15 coming into the system) during 50 years of simulation. In the older age categories, sexuality increase is delayed, and agent bring their 'sexual liberality' by aging (coming from younger to the older subgroup).

Cancer Developing. Once a woman is infected, she can either recover or be "permanently" colonized. From colonized state, she can be screened according to the program, or for other reasons, and find out about the infection. We assume that disease is curable in 100% of the cases if treated at this stage. Those women at the stage StageII who haven't been screened have a risk of developing cancer. Once a woman has acquired cancer, she can either survive or die, with given probabilities.

Immunity. Agents acquire temporal immunity due to natural recovery (and are moved from HPV or SrageII to R), but after some time they lose immunity and became Susceptible (S). Moreover, recovering from HPV does not protect against new infections forever, even for the same strain, and other modelers assume immunity period of 5 or 10 years [1,4].

Vaccination. We decided to vaccinate obligatory (in one scenario) only 14 years old girls (vaccination takes place before sexual debut) This is represented as a flow of the fraction covered by vaccination straight from birth flow to stage vaccinated (Vw) then they stay in this set of stages until death (life-long vaccine). We allow vaccinating older cohort of girls (20 – 24) up to 5% of subgroup population (for all scenarios). It represents voluntary decision of some girls or their parents (it's recommended by vaccine producers to vaccinate up to 26 years old girls). In the model it is implemented as the flow from the Susceptible (Sw) to Vaccinated (Vw).

4 Results and Validation

Performed simulations agreed in historical area (Figure 2) for all new cancer cases [21] registered (since 1989) and also for given age-cohorts (registry made since 1999). Possible regimes of future states show increase or decrease of new cancer case (Figure 5). Only vaccination will allow Poles to continue process of decreasing cancer prevalence in Poland. For other scenarios for different health care level or sexuality increase it could be better of worse, but even most pessimistic assumptions (75% of sexulatity increase and effective screening frequency at the level of every 7.5 year) are still not the worst possible as we can imagine.

We show trajectories of different stocks and flow of model (Figure 7, 6) to check if possible outcomes seem to be realistic and validate model this way. We can observe huge increase of HPV (in both stages) and cancer prevalence in cohort of oldest woman (Figure 7, 6 StageIIw5, HPVw5 and development5), which was already predicted by other cancer researchers [21], [24].

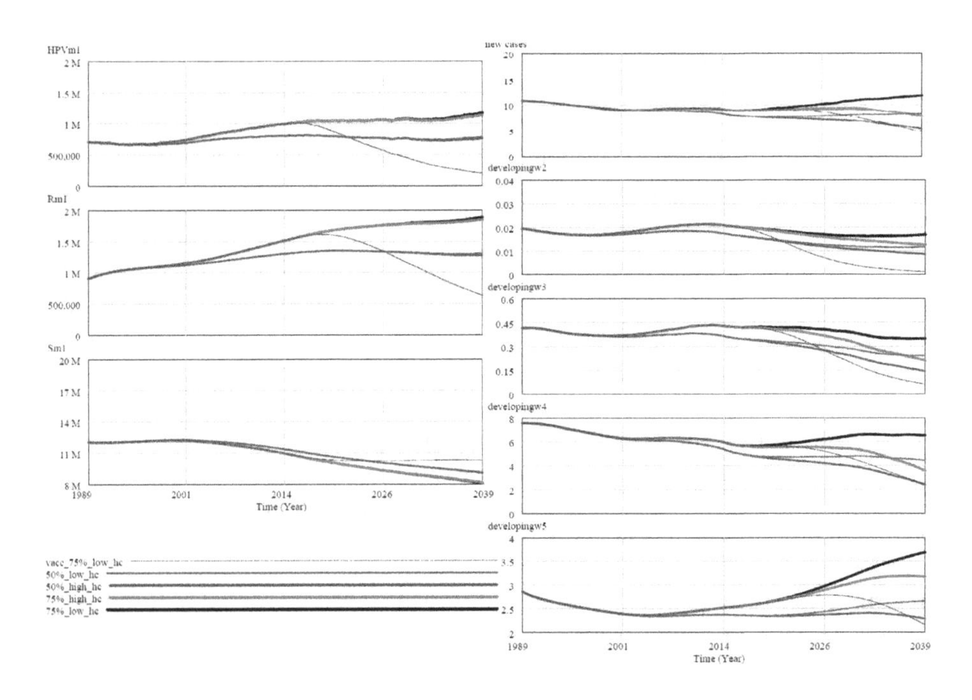

Fig. 6. Chosen stocks and flows in time. Flows show number of women who developed cancer per day for given age cohorts: development2: 20-24, development3: 25-34, development4: 35-64, development5: over 65. Stocks show number of men in stages: Sm, HPVm, Rm (subpopulation size in time).

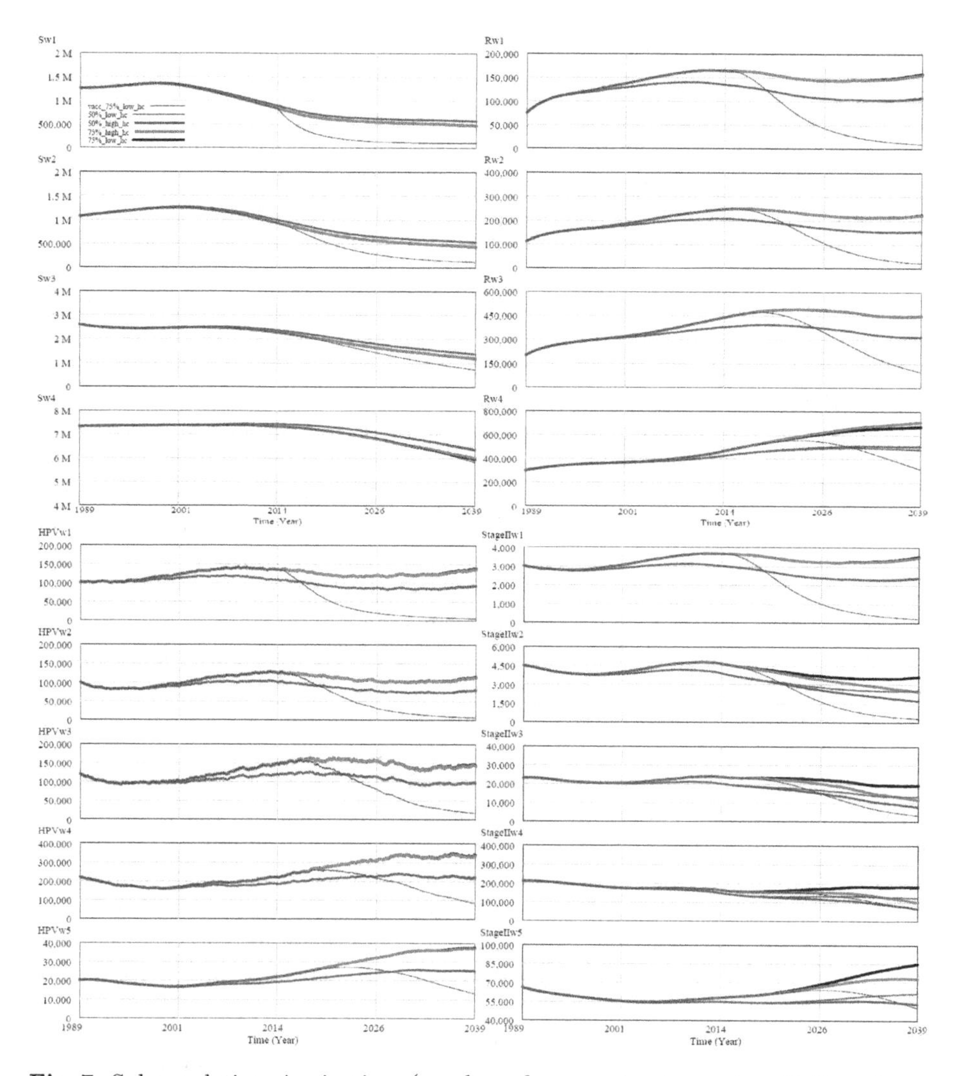

Fig. 7. Subpopulation size in time (number of women in stages: Sw, HPVw, Rw or StageIIw for given age cohorts. 1: 15-19, 2: 20-24, 3: 25-34, 4: 35-64, 5: over 65.

5 Conclusion and Discussion

We compared trajectories of new cancer cases detection for all scenarios (Figure 5). Fitted plateau (with small decay - Figure 2) in the first 25 years of simulation (1989-2014) corresponds to the situation in Poland (around $10-9$ deaths per day), so this setting of the model can be used to test alternative scenarios. The best strategy, in the long run and with respect to the total number of cervical cancer cases, is to vaccinate girls before starting sexual life (we even chose to test these scenarios at the worst values of other control parameters), and this could be expected taking into account corresponding studies from other countries. For scenarios without vaccination, we also found that it is necessary to stay with short screening intervals, because it also has a big impact on results. If sexuality would increase about 75% or more, we could observe a change in trend, and growing number of new cancer cases in near future. Our feeling might be that such an increase of sexuality is not too much, because it was estimated that at this point Poles have less than 2 times lifelong sexual partners than Swedes or Finns. Moreover aging of Polish society is also involved in growing oncological problems. The intervention seems to be needed, because we tried to be as much conservative as possible (model parameters mostly approach expected value from lower side).

Our approach could be extended to be more realistic. This model takes into account HPV just as one aggressive strain. There are at least two pathogens: type 16 and type 18 [4], which can cause cancer and existing vaccines are protecting from both of them. A multistrain model would of course be more realistic, but at least two times more parameters would be needed. In order to imitate multistrain reality, we needed to increase the amount of infection, so we multiplied the partner change rates by 2. By that we assume that properties of both strains are the same, and that infection or cancer development have to be independent processes. This is not true in general, but epidemiology of type 16 and type 18 is similar. Also, evolution helps us justify the independence assumption, because those strains are fighting against each other, and it is very rare to be colonized by both of them (after the secondary infection only one will survive). We assume that vaccine acquired immunity is lifelong, but it is already known not to be true [4]. The trials of this vaccine have not been running for more than 20 years and no longitudinal studies have been done. However, some of the women have lost their immunity after only 10 years. As suggested by geographers working in spatial epidemiology, aspects such as education and whether one lives in urban or rural area also have a great impact on parameters such as sexual activity, and it would be more realistic if these aspects were taken into account as well. However, it is difficult to find data on activity with respect to such factors, and further disaggregation of the population would lead to a hardly tractable model. Moreover, the model is already very complex (with more than 100 parameters), but we did our best to validate it by sensitivity analyses of parameters and functional of control variables (Figure 7, 6).

Acknowledgments. We would like to thank to Anna Franzen or Fredrik Liljeros from Stockholm University, Lisa Brouwers and Sharon Kuhlmann form Swedish Institute of Disease Control and Andrzej Grabowski from National Research Institute in Warsaw for cooperation. AJ thanks to Swedish Institute for invitation to Sweden.

References

1. Garnett, G.P., Kimb, J.J., French, K., Goldie, S.J.: Modelling the impact of HPV vaccines on cervical cancer and screening programmes. Vaccine 24S3, S3/178–S3/186 (2006)

2. Andrae, B., Kemetli, L., Sparen, P., Silfverdal, L., Strander, B., Ryd, W., Dillner, J., Tornberg, S.: Screening-Preventable Cervical Cancer Risks: Evidence From a Nationwide Audit in Sweden. J. Natl. Cancer Inst. 100, 622–629 (2008)
3. Kim, J.J.: Mathematical Model of HPV Provides Insight into Impacts of Risk Factors and Vaccine. PLoS Med. 3(5), e164 (2006)
4. Barnabas, R.V., Laukkanen, P., Koskela, P., Kontula, O., Lehtinen, M., et al.: Epidemiology of HPV 16 and cervical cancer in Finland and the potential impact of vaccination: Mathematical modelling analyses. PLoS Med 3(5), e138 (2006)
5. GUS-Glowny Urzad Statystyczny Statistics Poland, `http://www.gus.gov.pl`
6. Moller, T., Anderson, H., Aareleid, T., et al.: Cancer prevalence in Northern Europe: the EUROPREVAL study. Annals of Oncology 14, 946–957 (2003)
7. Syrjanen, K., Hakama, M., Saarikoski, S., Vayrynen, M., Yliskoski, M., Syrjanen, S., Kataja, V., Castreen, O.: Prevalence, incidence, and estimated life-time risk of cervical human papillomavirus infections in a nonselected Finnish female population. Sexually Transmitted Diseases 17(1), 15–19 (1990)
8. WHO reorts, WHO (2012), `http://www.euro.who.int/en/what-we-do/health-topics/Life-stages/sexual-and-reproductive-health/country-work`
9. Brouwers, L., et al.: MicroSim: Modeling the Swedish Population (2003)
10. Buda, A., Jarynowski, A.: Life time of correlation, vol. 1. Wydawnicwto Niezalezne, Wroclaw (2010)
11. Grabowski, A.: The relationship between human behavior and the process of epidemic spreading in a real social network. EPJ. B 85, 248 (2012)
12. Bernoulli, D.: An attempt at a new analysis of the mortality caused by smallpox and of the advantages of inoculation to prevent it. Mem. Math. Phy. Acad. Roy. Sci. Paris (1766)
13. Liljeros, F.: Preferential attachment in sexual networks. PNAS 104, 26 (2007)
14. Franzen, A., Jarynowski, A., Serafimovic, A.: HPV in Sweden, presentation (2012), `http://th.if.uj.edu.pl/~gulakov/hpv_newest.pdf`
15. Isinga, P.: Model for assessing HPV vaccination strategies. Emerging Infectious Diseases 13(1) (2007)
16. Ryding, J.: Seroepidemiology as basis fordesign of HPV vaccination program. Human Papillomavirus vaccination: Immunological and epidemiological studies, Lund (2008)
17. Dosekan, O.: An overview of the relative risks of different sexual behaviours on HIV transmission. Current Opinion in HIV and AIDS 5(4), 291 (2010)
18. Izdebski, Z.: Sexualluty of Polse in beginning od XXI century. Jagiellonian University Press, Krakow (2012)
19. Lew-Starowicz, Z.: Introduction to Sexuology. PZWL, Warszawa (2010)
20. Jarynowski, A.: code in Vensim, `http://th.if.uj.edu.pl/~gulakov/hpv_model.mdl`
21. Reports based on data of National Cancer Registry. Centre of Oncology in Poland, `http://85.128.14.124/krn`
22. Manji, M.: Cervical cancer screening program in Saudi Arabia: Action is overdue. Ann. Saudi Med. (2000)
23. Young Australians: their health and wellbeing 2011. Australian Institute of Health and Welfare (June 2011)
24. Institute of Oncology, IO, `http://www.io.gliwice.pl`

The Interaction of Social Conformity and Social orientation in Public Good Dilemmas

Friedrich Krebs

Centre for Environmental Systems Research, Kassel, Germany
krebs@usf.uni-kassel.de

Abstract. This paper reports on an agent-based model that represents social mobilisation as the collective provision of a public good. The model represents processes of social influence, temporal dynamics, and collectives of agents each behaviourally grounded in the HAPPenInGS theory (**H**eterogeneous **A**gents **P**roviding **P**ublic **G**oods) of preference-guided action selection of individuals in public good dilemmas [1]. We introduce the HAPPenInGS theory and present results of a series of simulation exercises scrutinising the interplay of individual-level preferences, perceptions and behaviours, and macro-level outcomes.

Keywords: Public good dilemma, heterogeneous preferences, social orientations, social conformity.

1 Introduction and Motivation

In many real world contexts individuals find themselves in situations where they have to decide between options of behaviour that serve a collective purpose or behaviours which satisfy one's private interests, ignoring the collective. In some cases the underlying social dilemma [2] is solved and we observe collective action [3]. In others social mobilisation is unsuccessful. To assess this phenomenon the explicit consideration of heterogeneous individual preferences and social structure plays a key role.

This paper reports on an agent-based social simulation (ABSS) that represents social mobilisation as the collective provision of a public good. In large populations such public good provision occurs on the level of spatially confined population subgroups. In the model we capture this spatial heterogeneity by assuming that diverse public goods exist or emerge on fixed locations in a common environment. It is further assumed that each public good has a fixed spatial extent defining the area where contributions are accumulated and where benefits of the public good may emerge. In addition to their spatial location, agents are embedded in a social structure allowing them to observe the behaviour of other agents within their acting group and to exchange information through social networks which may span groups. The behavioural target variable of an agent is its contribution to the public good in the next time step of the model. Possible interference with the generally multi-facetted decision-making in contexts other than the public good provision is not considered. Decision-making is modelled by an implementation of the HAPPenInGS theory (**H**eterogeneous **A**gents **P**roviding **P**ublic **G**oods) of preference-guided action selection of individuals in public good dilemmas [1]. HAPPenInGS is grounded in theoretic concepts and empirical

B. Kamiński and G. Koloch (eds.), *Advances in Social Simulation*,
Advances in Intelligent Systems and Computing 229,
DOI: 10.1007/978-3-642-39829-2_13, © Springer-Verlag Berlin Heidelberg 2014

results on social dilemmas and public goods, as well as in psychological theory of decision-making.

The HAPPenInGS theory was previously applied in in an agent-based model of a case study in the context of climate change adaptation [1, 4, 5]. The case study investigates the role of local neighbourhood support as a supplement to public health care during heat waves. Neighbourhood support is a public good which is locally provided by volunteering activities of neighbourhood residents. Likewise, the benefits of the public good emerge locally in the respective neighbourhood districts. The model represents the collective dynamics of the provision of neighbourhood support in a population of agents each behaviourally grounded in the HAPPenInGS theory and initialised from large-scale socio-empirical data.

The work reported in the following sections of this paper abstracts from a particular case study context and presents a typical middle-range ABSS [6, 7] which represents the most important properties and dynamics of the problem domain while remaining sufficiently abstract to allow for a systematic analysis of the collective behavioural dynamics and macro-level patterns implied by the underlying theory and model assumptions. With the model we aim to demonstrate what effect agent-level balancing between possibly conflicting outcome preferences has on individual behaviours and perceptions, and on the level of the collectively provided public good. Therefore, among the "Sixteen Reasons Other Than Prediction to Build Models" [8] our goal is to "Illuminate Core Dynamics" [8].

The paper proceeds as follows: Section 2 gives a brief overview of the HAPPenInGS theory. Section 3 reports on the setup of the abstract agent-based model HAPPenInGS-A. The results of the dynamical analysis of the abstract ABSS are reported in section 4. Section 5 concludes.

2 Actor Perspective: The HAPPenInGS theory

HAPPenInGS represents actor heterogeneity as subjective preferences that guide individual decision-making. HAPPenInGS states that an individual's decision about his behaviour in the public good dilemma is based on a subjective appraisal of each of his investment options with respect to his preference dimensions. The psychological theoretical foundation of decision-making in HAPPenInGS is the Theory of Planned Behaviour (TPB, [9]) with the extension of social value orientations, the latter being a well introduced concept in experimental social dilemma research. The TPB concerns the explanation and prediction of individual deliberative behaviour and is the most widely applied social-psychological theory of human decision-making [10]. The theory is unique in its empirical foundations and its numerical tractability. Furthermore, the calculus involved is based on multi-attribute utility on the level of individual actors which offers a simple and straightforward means of implementation in ABSS exercises.

HAPPenInGS covers three preference dimensions (stated in italics). Firstly, the attitude towards an investment option is represented in terms of the individual advantage the decision-maker associates with the respective behaviour, i.e. the expected subjective benefit of the *public good*. As a second dimension, the expected *social conformity* of an investment option with existing social norms is represented. This dimension covers the influence of "injunctive social norms" [11]. In general, "thick network(s) of social interactions" [12] can yield social coordination [13, 14] which in turn helps to overcome social dilemmas. The essential understanding is that individuals possess and utilise social knowledge in social dilemma situations [15]. Finally, we represent social value orientations as a preference for a desired ratio between the decision-maker's contribution to the public good and the contributions of the other members of the providing group. This representation of an expected conformity of an investment option with a desired *distribution of investments* accounts for the concept of social value orientations or social motives [16, 17], which are essential individual factor influencing cooperation in social dilemmas [18].

In HAPPenInGS the evaluation of the preference dimensions is based on an individual's perception of its situational context. The environmental sub context provides the decision-maker with local and subjective perceptions of the present state of the public good and of other (case-specific) non-social external factors relevant to the decision process. Behaviours that are put into action change the state of the environment. The actor perceives the consequences of his and collective behaviours as well as the generally varying temporal/spatial salience of the public good. This perception process constructs the actor's internal representation of the expected benefit of the public good. In its social context an individual perceives the investment behaviours of its social network peers and actively constructs an internal representation of the (present) social norm. Furthermore, the individual perceives the investment behaviours of other members of its providing group. This perception is not evaluated in a behavioural normative sense. Instead the individual compares the perception to its own investment behaviour and evaluates the situation with respect to its social value orientations.

3 Embedding the Individual: Agent-Based Model Setup

The agent's environment context has a grid topology. Some of the patches are populated by agent groups of a fixed size n. In each simulation step the level of the public good on a group's patch is calculated from the contributions of the group members. An agent's investment state variable x is one of 11 distinct behavioural options that reflect investments of 0.0, 0.1 up to 1.0 in steps of 0.1. Based on the contributions of the n agents we determine the level of the generated public good from a logistic curve describing production function of the public good. For low individual investments x_i the value is close to 0. For high values of x_i (for all i) it approaches 1.0. As for the success of the public good provision only the sum of the investments of all group members is crucial, unequally distributed contributions may yield identical success. This allows cases of free riding to occur e.g. if substantial levels of the public good were provided by a majority of group members and a minority refuses to invest. Furthermore, patches vary in the success of the public good provision by the local agent group. Other spatial relations or dynamics are not considered in the abstract model.

The social context is represented as a directed social network where a network link from agent A to agent B means that B is influenced by A. In the model, the investment behaviour of A is perceivable by B and contributes to B's evaluation of its social conformity goal along with the behaviours of all other agents having outgoing social network links to B. In principle, an agent's social network peers may be located on arbitrary patches of the environmental context, i.e. network links may span agent groups.

HAPPenInGS forms the theoretical framework of agent decision-making in HAPPenInGS-A. Accordingly, an agent's selection of a behavioural option is guided by the three preference dimensions stated by the theory which are shown in the first column of Table 1. An agent's knowledge about the effectiveness of his behavioural options is represented by a utility calculation which relates the perceptions of an agent and his expectations about the outcome of executing a behaviour to the agent's subjective preferences.

Table 1. Preferences according to HAPPenInGS, the respective parameters, and assessment criteria along with an example preference set. An agent's preferences are weighting factors for the criteria. The last column displays the formulas used to calculate an agent's estimation of the satisfaction of a given preference. Here x is the investment level of the behavioural option evaluated, c(x) stands for the expected level of the public good, abs() calculates the absolute value.

HAPPenInGS preference	**Preference parameter (example value)**	**Criterion**
Public good	*publicGoodPreference* (1.0)	c(x)
Distribution of investments	*selfPreference* (0.1)	1.0 – x
	othersPreference (0.3)	1.0 - mean x of other group members
Social conformity	*socialConformityPreference* (0.1)	1.0 - abs(x - mean x of social network peers)

In HAPPenInGS-A, agents use this subjective utility to assess, compare and rank their behavioural options based on their preferences and perceptions. To do so, each agent perceives the present success of the collective action, i.e. the level of the public good provided by its respective group, and supposes that the n-1 other agents of its group keep to their previous investment decisions in the next time step. Thus, each agent can estimate the success of the collective action associated with each of its possible next investment decisions. Moreover, an agent perceives the average level of contributions within its group which enables the assessment of the expected conformity of an investment behaviour with a given preference for the distribution of investments in the providing group. Finally, agents refer to their social network and determine the average level of contributions by their peers as a proxy for a perceived social norm. Table 1 gives an overview of the preference structure along with an example preference set. Based on their preference set, agents can determine the expected utility of each investment option x. The expected subjective overall utility of investment option x is calculated by adding up the criteria values for a given x weighted by the respective preference value. The presented utility approach allows for the representation of agent heterogeneity in various aspects: Firstly, different preference sets may be defined representing agent types that differ in their basic orientations (understood as persistent personality traits). Secondly, agents are heterogeneously embedded in their local neighbourhood group and include the observed behaviours of other group members in their subjective utility estimation. Thirdly, subjective utility reflects the embedding of an agent in its network of social peers and opinion dynamics based on observed behaviours.

In HAPPenInGS-A the final selection of an investment option is represented by a probabilistic choice model (see e.g. [19]) based on the expected overall utilities of each of the behavioural options. This decision process is triggered if there is at least one behavioural option with higher expected utility than the utility achieved by the agent in the previous time step. Else the previous investment decision is kept. Exploration is triggered with a probability of 1% and modelled by having agents select a random investment level (uniformly distributed).

Simulations are initialised as follows: The environmental context consists of 20x20 equally sized, square patches. During initialisation 20 patches are randomly selected and populated by agents. On each selected patch we initialise 20 agents forming a neighbourhood agent group. Agent types are defined by preference profiles according to HAPPenInGS which are represented in the agents' persistent state variables. For heterogeneous populations the relative ratio between agent types is preserved on the level of neighbourhood groups. For all agents the investment state variable is initialised to 0.0. After setting up the agents in the environmental context, the social network is initialised. We use a stylised version of the network initialisation process proposed in 20: To account for baseline homophily we link each agent to 5 randomly selected agents from its neighbourhood group. To account for inbreeding homophily each agent is in addition linked to 5 agents of the same agent type selected at random from the full population. Agent locations and the social network remain fixed throughout the simulation.

4 Dynamical Analysis

The main goal of this section is to investigate the macro-level patterns generated by HAPPenInGS-A. Such patterns constitute and describe the collective as well as the temporal-dynamic implications of the HAPPenInGS theory.

4.1 Basic Agent Types

We first summarise previously reported results [1, 21] which investigate the influence of agents' social orientation (i.e. the weighting between *selfPreference* and *othersPreference*) on agent behaviours and on the success of the collective action in terms of a sensitivity analysis of the respective preference parameters of HAPPenInGS-A. For this sensitivity analysis the preference for social conformity was set to 0 (*socialConformityPreference*=0.0), i.e. agents disregard social influences. A particular outcome of the sensitivity analysis is the identification of prototype parameter combinations representing basic types of agents with regard to social orientation (cf. Fig. 2. in [21]):

- **Altruistic agents** stress altruistic tendencies relative to egoistic tendencies (*publicGoodPreference*=1.0, *selfPreference*=0.1, *othersPreference*=0.3*)*. In average, these agents show contributions up to 0.42 and achieve high PG levels between 0.8 and 0.9.
- Vice versa **egoistic agents** stress egoistic tendencies (*publicGoodPreference*=1.0, *selfPreference*=0.3, *othersPreference*=0.1*)*. For this type, average contribution levels do not exceed 0.07 yielding negligible PG levels between 0.0 and 0.1.

The basic agent types are to be understood as discrete representatives of ranges of parameter combinations in HAPPenInGS-A which lead to qualitatively similar macro-level outcomes. Altruistic and egoistic types are located at the two extreme ends of the spectrum of social orientations that are covered by HAPPenInGS-A and can be seen as HAPPenInGS counterparts of prosocials and proselfs as found in experimental work on PG dilemmas.

4.2 Social influence in Altruistic Populations

This section analyses the effect of increasing the weight put on the social conformity preference in populations that are homogeneously composed of representatives of the altruistic basic agent type (for a more detailed discussion including other agent types refer to [1]). The preference sets considered in this sensitivity analysis combine the prototype social orientation preference with different settings for the social conformity preference. For each combination of parameter settings 20 independent runs over 400 steps are performed. Simulation results are reported with respect to three different indicators:

- Agents' behaviours in terms of their mean contribution (black).
- The standard deviation of agent behaviours within a population (vertical dashed, black).
- The temporal stability (convergence) of agent behaviours as the total decrease or increase of contribution relative to the previous time step on the individual level (delta contribution, blue).
- Agents' achieved social conformity as the mean value of their perceived social conformity (red). For each agent the social conformity perception is calculated according to the criterion formula in Table 1. It reflects an agent's absolute attainment of its social conformity preference.
- The success of the collective action as the mean level of the PG of all groups (green).

We report simulation results in terms of mean values of the respective 20 runs performed per simulation tick. We pick three distinct settings for the weight of the social conformity preference during decision-making: In the first diagram of Fig. 1 we display simulation results for *socialConformityPreference*=0 which reflects the extreme case where social networks have no influence in the decision process. In the aggregated simulation results discussed above we find that for a medium setting for the importance of social conformity (*socialConformityPreference*=0.2) success of the PG provision and the achieved social conformity have their maxima. Therefore, we pick this setting for a second in-depth analysis that is shown in the second diagram of Fig. 1. Finally, we show results for the extreme case of *socialConformityPreference*=0.5.

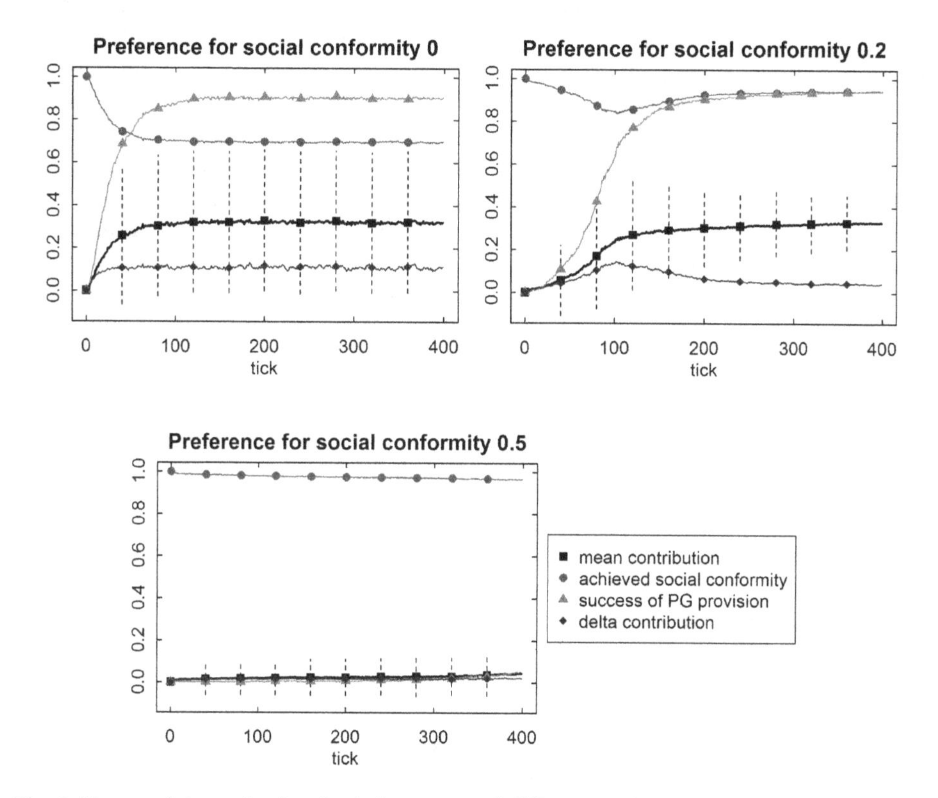

Fig. 1. Temporal dynamics for altruistic agents and different preferences for social conformity. We show mean values of 20 runs performed per simulation tick. See text for further explanation. Adapted from [1].

In all three diagrams of Fig. 1, due to the homogeneous initialisation of the agents with zero contributions, the social conformity perceived by the agents starts on the maximum level during the initial simulation steps. If the preference for social conformity is high this passive social norm is sustained throughout the simulation and social mobilisation does not take place. For more moderate preferences for social conformity we observe successful social mobilisation. In both cases the achieved social conformity drops during the initial phase of social mobilisation. For the first case (preference for social conformity 0) the maximum success of PG provision is reached after 150 ticks and sustained until the end of the simulation. In contrast, for the moderate social conformity preference mobilisation proceeds slower but continuously improves towards the end of the simulation. This slow-down of social mobilisation may be attributed to an on-going process of social adjustment: Clearly, compared to the first diagram of Fig. 1, the achieved social conformity does not drop as steeply and instead continues to increase during the simulation after tick 100. Likewise, for the medium social conformity setting we observe a decrease of the standard deviation of the contributions (dashed black) which does not occur for the low social conformity setting. An additional difference between simulations with low and medium settings for the social conformity preference lies in the development of the behavioural stability indicator (delta contribution in blue) that reflects the individual change in behaviour between two consecutive simulation ticks: While under both settings contributions settle around a mean of 0.3, in the absence of social conformity preferences individual behaviours appear to fluctuate significantly (around 0.1) from one tick to the next throughout the simulation, i.e. agents continuously adjust their behaviour. With medium social conformity preference this effect is dampened and delta contribution approaches zero over time. Apparently, the striving for social conformity not only decreases the spread of behaviours within the population; it also decreases the spread of behaviours selected by one agent over time. The following section goes a step further and reports on simulation results for agent populations with heterogeneous preference profiles.

4.3 Social Influence in Heterogeneous Populations

This section investigates the interplay between social conformity and social orientations in heterogeneous agent populations. We set up populations with different ratios of altruistic agents and egoistic agents, and different (population) global settings of the social conformity preference. The investigated parameter combinations are documented in Table 2. Again, for each population composition and parameter settings 20 independent runs over 400 steps are performed.

Table 2. Agent types, respective preference sets and population composition used in the sensitivity analysis. Two basic types of agents regarding social orientation are investigated. For each basic type three different settings for the social conformity preference are considered. For each preference set a total of 9 population compositions is investigated.

socialConformityPreference	Proportion of agents in population	
0.0, 0.1, and 0.2	Altruists	0.1, 0.2,…, 0.9
	Egoists	1 – proportion of altruists

Results are reported with respect to the following extension of three of the performance indicators introduced in the previous section:

- Agents' behaviours in terms of their mean contribution (black), mean contributions of altruistic agents (dashed black), and mean contributions of egoistic agents (dotted black).

- Agents' achieved social conformity as the mean value of their perceived social conformity (red) and the respective means for altruistic and egoistic agents (dashed respectively dotted red).
- The success of the collective action as the mean value of all groups (green).

In the diagrams, we compare the performance of heterogeneous populations with increasing social conformity preference. To assess performance we calculate means of the performance indicators during the last 200 steps for each run. In the diagrams we show the mean values of these aggregations and standard deviations of the 20 initialisations used.

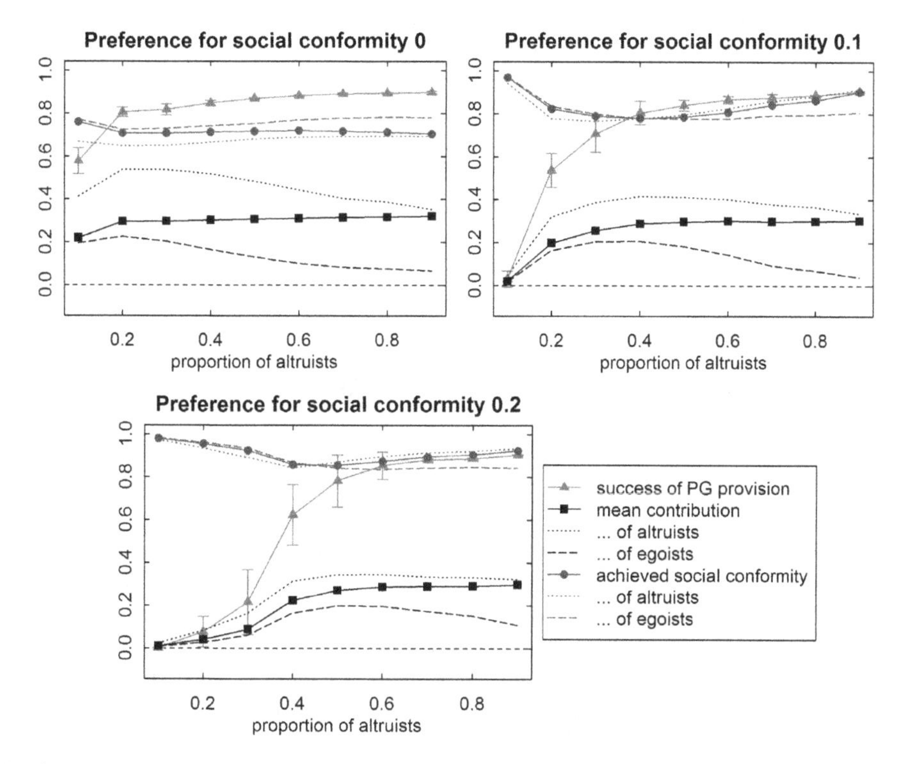

Fig. 2. Aggregated results for different population compositions and different preferences for social conformity. For each run performed we calculate mean values of the performance indicators during the last 200 simulation steps. The error bars show the standard deviation of the mean values of the 20 runs performed for a distinct parameter setting. See text for further explanation. Adapted from [1].

Unsurprisingly, in all three diagrams of Fig. 2 we observe that the success of the collective action generally increases with increasing proportion of altruists. However, significant differences exist depending on preference weight of social conformity. For the lowest setting for social conformity we observe that when the proportion of altruists is at least 0.2 the success of the PG provision reaches high levels. Even if 90% of the population are egoists, a moderate PG level of around 0.6 is provided. When the preference weight of social conformity increases, the proportion of altruists required for achieving high levels of the PG increases: In the first diagram of Fig. 2 high levels of the PG are provided for populations with at least 40% of altruistic agents. Likewise, in the second diagram more than 50% of altruists are required. Apparently, social coherence effects inhibit mobilisation due to a lock-in of passive behaviour with increasing preference for social conformity and with decreasing proportion of altruists. Especially for the highest preference for social conformity we observe that altruists show investment behaviours that obviously contradict their social orientations: For the social conformity preference 0.2, if the size of the altruistic subpopulation does not exceed 30% of the population, altruists contribute less than 0.15. Yet, such behaviour yields high social conformity in a predominantly egoistic population which apparently overrides individual social orientations during agent decision-making.

Finally, it may be seen that the inequality of behaviours between altruists and egoists decreases when preference for social conformity increases. As a result, altruists tend to contribute less and egoists tend to contribute more with increasing social conformity preference. Clearly, when the preference for social conformity gains importance during agent decision-making, the topology of the social network increasingly influences investment behaviours. Therefore, prevalent behaviours of the respective other subpopulation contribute more to an individual's social conformity perception and in turn trigger a process of inter-population social adjustment.

5 Synthesis of Results and Conclusions

The focus of the HAPPenInGS theory is the micro-level of decision-making in public good dilemmas. The purpose of HAPPenInGS-A is to supplement the bottom-up macro-level perspective on public good provision by applying methods of ABSS. In doing so, HAPPenInGS-A represents processes of social influence, temporal dynamics, and collectives of agents each behaviourally grounded in HAPPenInGS. The key question of this paper was on the interrelation of individual-level preferences and macro-level outcomes. We tackled the question by exploring the range of implications of the HAPPenInGS theory through a series of simulation exercises. In particular we performed a dynamical analysis of the interplay of social conformity preferences and social orientations on the individual level, and the macro-level consequences.

Some results mainly confirmed expectations and therefore demonstrated the consistency of HAPPenInGS: For instance, high preferences for social conformity might prevent social mobilisation even in predominantly altruistic populations. On the other hand it could be shown that medium social conformity preferences foster social mobilisation and in addition decrease inequalities of contributions on the individual level. In addition to the general plausibility of such results, they demonstrate the emergent complexity which arises from a simple and abstract theory like HAPPenInGS in the context of an ABSS exercise. Other results of the analysis of HAPPenInGS-A were less intuitive: For example the assessment of the temporal dimension of social mobilisation showed that if agents strive for social conformity in addition to following altruistic preferences, the speed of social mobilisation is reduced. Clearly, the underlying process of inter-individual social adjustment demonstrates the interplay between individual value orientations and social norms in HAPPenInGS. However, neither the existence of the process as such nor its assessment can readily be guessed from the HAPPenInGS theory beforehand. Instead the results are genuine contributions of the methodical approach of ABSS.

Furthermore, the simulation experiments allowed narrowing down the circumstances under which social conformity preferences can conflict with social value orientations. Under such conditions individual behaviours may emerge to be inconsistent with individual value orientations – egoists contribute more, but in turn altruists contribute less. Again, such implications illustrate well the added value of the method of ABSS. Interestingly, there is empirical work demonstrating the existence of this "Boomerang Effect" for the case of household energy conservation [22]. The study shows that when households are provided with a descriptive social norm [11] indicating average energy consumption behaviour within their residential neighbourhood they appear to adjust their consumption behaviour towards this norm. In other words, the social norm tends to interfere with other factors influencing decision-making. In line with the results of HAPPenInGS-A, the study showed that this adjustment is symmetrical in nature: Households consuming more energy than the average reduce their consumption while those already saving energy increase their consumption towards the mean.

Finally, when the influence of social conformity preferences on individual decision-making increases it has to be expected that likewise properties of the underlying social network topology impact macro-level outcomes. In political sociology research in this direction has been undertaken for the case of the collective action of political participation [23]. Yet, such approaches lack psychological soundness in the representation of individual decision-making. This gap could be filled by scenario simulations based on HAPPenInGS.

Acknowledgements. This research was funded by the German Federal Ministry of Education and Research (BMBF). I wish to thank Andreas Ernst for fruitful inputs and discussions.

References

1. Krebs, F.: Decision-Making in Public Good Dilemmas: Theory and Agent-Based Simulation Kassel (2013), `http://nbn-resolving.de/urn:nbn:de:hebis:34-2013051442711`
2. Dawes, R.M.: Social dilemmas. Annual Review of Psychology 31, 169–193 (1980)
3. Olson, M.: The logic of collective action. Harvard University Press, Cambridge (1965)
4. Krebs, F., Holzhauer, S., Ernst, A.: Modelling the Role of Neighbourhood Support in Regional Climate Change Adaptation. Appl. Spatial Analysis, 1–27 (2013)
5. Krebs, F., Holzhauer, S., Ernst, A.: Public Good Provision in Large Populations: The Case of Neighbourhood Support in Northern Hesse. In: European Social Simulation Association, Proceedings of ESSA 2011, Montpellier (2011)
6. Gilbert, N.: Agent-based models. Sage Publ., Los Angeles (2008)
7. Squazzoni, F.: Agent-based computational sociology. Wiley, Hoboken (2012)

8. Epstein, J.M.: Why Model? Journal of Artificial Societies and Social Simulation 11, 12 (2008)
9. Ajzen, I.: From intentions to actions: A theory of planned behavior. In: Kuhl, J., Beckman, J. (eds.) Action-control: From Cognition to Behavior, pp. 11–39. Springer, Heidelberg (1985)
10. Ajzen, I., Fishbein, M.: The Influence of Attitudes on Behavior. In: Johnson, B.T., Albarracin, D., Zanna, M.P. (eds.) The Handbook of Attitudes, pp. 173–221. Lawrence Erlbaum Associates Publishers, Mahwah (2005)
11. Cialdini, R.B., Kallgren, C.A., Reno, R.R.: A Focus Theory of Normative Conduct: A Theoretical Refinement and Reevaluation of the Role of Norms in Human Behavior. In: Zanna, M.P. (ed.) Advances in Experimental Social Psychology, vol. 24, pp. 201–234. Academic Press (1991)
12. Gardner, G.T., Stern, P.C.: Environmental problems and human behavior. Allyn & Bacon, Boston (2000)
13. Kopelman, S.: The effect of culture and power on cooperation in commons dilemmas: Implications for global resource management. Organizational Behavior and Human Decision Processes 108, 153–163 (2009)
14. van Dijk, E., Wilke, H.A.M.: Coordination Rules in Asymmetric Social Dilemmas: A Comparison between Public Good Dilemmas and Resource Dilemmas. Journal of Experimental Social Psychology 31, 1–27 (1995)
15. Ernst, A.: Ökologisch-soziale Dilemmata. Psychologische Wirkmechanismen des Umweltverhaltens. Psychologische Wirkmechanismen des Umweltverhaltens Beltz Psychologie-Verl.-Union. Beltz, Weinheim (1997)
16. Liebrand, W.B.G., McClintock, C.G.: The ring measure of social values - computerized procedure for assessing individual differences in information processing and social value orientation. European Journal of Personality 2, 217–230 (1988)
17. Messick, D.M., McClintock, C.G.: Motivational bases of choice in experimental games. Journal of Experimental Social Psychology 4, 1–25 (1968)
18. Weber, J.M., Kopelman, S., Messick, D.M.: A conceptual review of decision making in social dilemmas: Applying a logic of appropriateness. Personality and Social Psychology Review 8, 281–307 (2004)
19. Janssen, M.A., Ahn, T.K.: Learning, signaling, and social preferences in public-good games. Ecology and Society 11 (2006)
20. Holzhauer, S., Krebs, F., Ernst, A.: Considering baseline homophily when generating spatial social networks for agent-based modelling. Computational and Mathematical Organization Theory, 1–23 (2012)
21. Krebs, F., Ernst, A.: The interplay of social orientation and social conformity in neighbourhood support. In: Ernst, A., Kuhn, S. (eds.) Proceedings of the 3rd World Congress of Social Simulation (2010)
22. Schultz, P.W., Nolan, J.M., Cialdini, R.B., Goldstein, N.J., Griskevicius, V.: The Constructive, Destructive, and Reconstructive Power of Social Norms. Psychological Science 18, 429–434 (2007)
23. Siegel, D.A.: Social Networks and Collective Action. American Journal of Political Science 53, 122–138 (2009)

Effects of Corporate Tax Rate Cuts on Firms' Investment and Valuation: A Microsimulation Approach

Keiichi Kubota[1] and Hitoshi Takehara[2]

[1] Chuo University, Tokyo, Japan
kekubota@tamacc.chuo-u.ac.jp
[2] Waseda University, Tokyo, Japan
takehara@waseda.jp

Abstract. This paper investigates changes in firm values triggered by a hypothetical corporate tax rate cut for Japan. We use multiplicative production functions and firms' investment decision changes over time via the accumulated process of retained earnings on the simulated path. We find changes in corporate tax rates can enhance the market value of firm equity in most cases, while there are some cases in which the effects are neutral or even detrimental for firm value. We interpret that these mixed results are caused by joint effects of current provisions of tax loss carry-forward and the net balance of tax deferred accounts. We find that past profitability and variability of each firm is crucial to hit exact threshold points at which firms experience value appreciations or not. The results possess important implications to both regulators and corporate financial managers.

Keywords: corporate taxation, tax loss carry-forward, deferred tax assets and liabilities, retained earnings, earnings variability.

1 Introduction

The analyses on changes in firm value from tax rate changes present important implications for policy makers as well as corporate managers. In the current paper we employ a microsimulation approach (Shahnazarian, 2011), in which we assume a multiplicative production function for each individual firm and simulate the evolvement of accounting accounts of firms. Based on hypothetical corporate tax rates we compute fundamental values of firms with the residual income valuation model of Ohlson (1995) and assess effects of tax rate cuts utilizing a simulation method in which each firm's past data at the micro level are used to obtain the necessary parameter values.

In this paper we focus primarily on the corporate tax rate cut case[1], because it is widely known that the corporate tax rate in Japan is much higher than in other

[1] The analysis of effects of tax rate changes on individual financial income is our ongoing work. After the East Japan Big earthquake, the government has considered an increase in overall individual tax rate. Also, even though the corporate tax rate had been cut 5% in fiscal year 2012, an additional recovery special tax was levied and the net tax cut became smaller.

B. Kamiński and G. Koloch (eds.), *Advances in Social Simulation*,
Advances in Intelligent Systems and Computing 229,
DOI: 10.1007/978-3-642-39829-2_14, © Springer-Verlag Berlin Heidelberg 2014

developed countries. Until fiscal year 2011, for Japanese large corporations, the national corporate tax rate was 30%, the prefecture tax rate 5% and the city tax rate 12.3%, adding up to an effective tax rate of 40.78%, which is applicable to our sample of Tokyo Stock Exchange listed firms.

We focus on individual firms' financial statement data and investigate effects of tax rate cuts on firm value by utilizing the accounting residual income valuation model, and identify conditions under which the effects of tax rate changes are favorable, neutral, or rather detrimental. Furthermore, in doing so, we take into consideration both tax loss carry-forward allowances and changes in net deferred tax assets and liabilities. Unlike Kubota and Takehara (2012), where the straightforward method by Graham (1996) is used to compute time series income before tax, we use a microsimulation approach (Shahnazarian, 2011). That is, actual data from each item of firms' assets are used to estimate micro-based firm production functions, and then extrapolate the future net income path before the corporate tax. Next, by discounting the stream of this future income on an after-tax base, we compute the fundamental value of firms in which the cost of equity estimates are based on estimates from the unconditional Fama and French three factor model (Fama and French, 1993).

Section 2 discusses the motivation of our study and raises our research agenda. Section 3 formulates the process by which corporate investment evolves over time based on items from financial statements and the corresponding valuation model. Section 4 explains our data, reports basic observations, and explains the simulation method we employ. Section 5 reports the simulation results and Section 6 concludes.

2 Taxation and Firm Value

2.1 Motivation

Corporate tax shield benefits or tax burdens arise from two sources: from debt or non-debt sources. In this paper we focus on the latter. When Japanese firms incur losses, they can extend their tax-loss shield against future income, like in many other countries. Note that in the U.S., it can be charged to both future and past income. [2]Hence, these allowances from non-debt sources help firms decrease their future tax burden and possibly reduce the cost of capital.

The Japanese financial reporting system and the tax reporting system follow the so-called uniform reporting system like other continental European countries. [3]However, from fiscal year 1999, the tax deferral account in balance sheets was admitted to be recorded in consolidated financial statements on the condition that such an accrued amount is expected to be reversed with a high probability within five years. In spite of this new tax-timing difference allowance, the Japanese system can still be classified

[2] In Japan, the current Corporation Tax Act (Article 57 Paragraph 3) allows for firms with reported losses to deduct their losses against their future profits up to a maximum of nine years. This used to be seven years. Also, the provision of tax loss carry-back allowances is provided, although this has not been implemented since 1992 except for the firm liquidation case.

[3] See Cummins et al. (1994) for these classifications.

as a uniform reporting system in the sense that depreciation methods, inventory costing, or other major accounting choices have to follow uniform reporting both for tax and financial accounting purposes.

2.2 Objectives of the Study

In this paper we try to assess the importance of the provisions mentioned above, when the corporate tax rate is changed, which could affect firm valuation. The effects of tax rate changes on firm valuation becomes complex with state-dependent integral equations for a multi-period case, and so we use a simulation method. Note the tax loss carry-forward allowances are valid only up to a maximum of seven years during our sampling period by the Corporation Tax Act in Japan. The book entry of deferred tax assets is also expected to be reversed within five years in order to be certified by CPAs. However, note that future income streams of firms possess infinite lives as going concerns. Every year, new entry of tax allowances when a firm incurs losses and deferred tax assets or liabilities are recorded will generate accumulated processes with finite lives.

We emphasize that our analysis is quite important from the viewpoint of both regulators and corporate managers. Regulators would want to assess the directions of new investment behavior by firms triggered by the statutory tax rate cut for the purpose of implementing better economic policies, and at the same time investment decisions chosen by corporate managers may change through their rational responses to corporate tax rate changes (MacKie-Mason, 1990). Hence, such changes in firm behavior will significantly change the allocation of scare resources both in industries and in an economy.

With these considerations we try to empirically identify the type of firms which undergo favorable changes, neutral effects, or unfavorable changes, triggered by the enactment of corporate and individual financial income tax rate cuts.

3 Corporate Investment and Valuation

3.1 Evolution of Firms' Retained Earnings and Investment

We construct a corporate investment model with time evolvement equations of the accounting variables as follows.

$$
\begin{aligned}
dca_{j,t}/TA_{j,t-1} &= c_{0,g}^{dca}\cdot(1/TA_{j,t-1}) + c_{1,g}^{dca}\cdot RE_{j,t-1}/TA_{j,t-1} + \varepsilon_{j,t} \\
InvMA_{j,t}/TA_{j,t-1} &= c_{0,g}^{MA}\cdot(1/TA_{j,t-1}) + c_{1,g}^{MA}\cdot RE_{j,t-1}/TA_{j,t-1} + \varepsilon_{j,t} \\
InvBU_{j,t}/TA_{j,t-1} &= c_{0,g}^{BU}\cdot(1/TA_{j,t-1}) + c_{1,g}^{BU}\cdot RE_{j,t-1}/TA_{j,t-1} + \varepsilon_{j,t} \\
InvGW_{j,t}/TA_{j,t-1} &= c_{0,g}^{GW}\cdot(1/TA_{j,t-1}) + c_{1,g}^{GW}\cdot RE_{j,t-1}/TA_{j,t-1} + \varepsilon_{j,t} \\
InvP_{j,t}/TA_{j,t-1} &= c_{0,g}^{P}\cdot(1/TA_{j,t-1}) + c_{1,g}^{P}\cdot RE_{j,t-1}/TA_{j,t-1} + \varepsilon_{j,t}
\end{aligned}
\tag{1}
$$

In the system of equations (1), *dca* denotes changes in current assets. *InvMA*, new investment in machinery and equipment, *InvBU,* new investment in buildings and structures, *InvGW*, change in goodwill, *InvP*, change in patents and copyrights, and *TA*, the total assets of firms.[4] Subscript j,t means firm j and fiscal year t. The subscript g denotes the pooled data for each industry because we estimate industry-wide coefficients to minimize estimation errors from using individual firm data for estimation. We pool five years of individual firm data for each industry and estimate the coefficients for each industry with OLS. We also assume the error terms in (1) are serially uncorrelated.[5] Then, given industry-wide coefficients, we plug in accounting variables to get estimates of the variables for each firm.

Based on the time evolvement of the accounting variables above, we assume each firm is equipped with the following power production function.

$$OIBD_{j,t} = \gamma_{g,0} \cdot (CA_{j,t-1} + dca_{j,t})^{\gamma_{g,1}} (MA_{j,t-1} + InvMA_{j,t})^{\gamma_{g,2}} e^{\sigma_{j,t} \cdot w} \tag{2}$$

In equation (2), *OIBD* denotes operating income before depreciation, *CA*, current assets, and *MA*, the balance of machinery and equipment. For the error term, σ denotes annualized volatility for each firm and w is a standard normal random variable. We estimate the model using the log transformation of this equation and we estimate gammas and sigmas for each firm in each industry with pooled data for five years.

We assume the book and economic depreciation (Samuelson, 1964) coincide with each other and the following (3) holds for depreciation expenses every period. We use a depreciation rate of 25 percent for tax purposes in Japan[6] and depreciation expenses for tax purposes must coincide with depreciation expenses for financial purposes.

Let δ denote the accounting depreciation rate. The accounting depreciation, *DEPR*, is defined in the following equation (3) where *BU* denotes the book value of buildings and structures and *GW* denotes goodwill,

$$DEPR_{j,t} = (MA_{j,t-1} + BU_{j,t-1} + GW_{j,t-1} + P_{j,t-1}) \cdot \delta, \text{ where } \delta = 0.25. \tag{3}$$

With these preparations the time evolvement of firm accounting variables is governed by the following accounting identity equations.

$$\begin{aligned} &CA_{j,t} = CA_{j,t-1} + dca_t,\ MA_{j,t} = \delta \cdot MA_{j,t-1} + InvMA_t, \\ &BU_{j,t} = \delta \cdot BU_{j,t-1} + InvBU_t,\ GW_{j,t} = \delta \cdot GW_{j,t-1} + InvGW_t,\ P_{j,t} = \delta \cdot P_{j,t-1} + InvP_t. \end{aligned} \tag{4}$$

Furthermore, we define the earnings before tax, *EBT*, as

$$EBT_{j,t} = (OIBT_{j,t} - DEPR_{j,t}) + (FI_{j,t} - FE_{j,t}) \tag{5}$$

---

[4] Goodwill is recorded as assets and depreciated at a maximum of 20 years in Japan. Patents and copyrights were depreciated before, but are now fully expensed.

[5] Because we estimate at the industry level, serial correlations will be less serious than the case of estimating at an individual firm level.

[6] The idea is similar to the modified accelerated cost recovery system (MACRS) in the U.S. by the Tax Reform Act of 1986. It was enacted in fiscal year 2009 and from 2011 the rate was reduced to 20 percent.

where *FI* and *FE* denote financial income and financial expenses, which are classified as non-operating expenses in Japan.[7]

Let *DTD* denote the current balance of tax deductible temporary differences which result from deferred tax assets and loss carry-forward balances. Accordingly, net income of the firm, *NI*, and retained earnings, *RE*, are defined as a function of *EBT* and *DTD* as shown in equations (6), (7), and (8).

When the realization of the random variable, *EBT*, is negative, i.e., firm j has a deficit in year t, firm j does not pay taxes and *DTD* increases as in (6) by the amount of additional tax loss carry-forward allowances which are valid for the next seven fiscal years to be expensed against future income.[8] In this case the net income of firm j is negative and is equal to *EBT*. By further assuming that firm j does not pay any dividends to its shareholders when her/his net income is negative, we get the following relations when $EBT<0$.[9]

$$DTD_{j,t} = DTD_{j,t-1} - EBT_{j,t},\ \ NI_{j,t} = EBT_{j,t},\ \ RE_{j,t} = NI_{j,t} \tag{6}$$

On the other hand, when realization of *EBT* is positive but less than or equal to *DTD*, firm j does not pay taxes and at the same time, the asset balance of deductible temporary differences is exactly reduced by the amount of *EBT*,

$$DTD_{j,t} = DTD_{j,t-1} - EBT_{j,t},\ \ NI_{j,t} = (1-\tau_c^*)\cdot EBT_{j,t},\ \ RE_{j,t} = NI_{j,t}\cdot\kappa, \tag{7}$$

where τ_c^* is the effective corporate tax rate and κ is a retention ratio of the firm which we assume to be constant over time.

Finally, when *EBT* is greater than $DTD_{j,t-1}$ as shown in (8), it becomes a normal case in which firms pay full corporate tax rates. Then it coincides with the statutory corporate tax rate when any other investment tax credit cannot be applied to this firm.

$$DTD_{j,t} = 0,\ \ NI_{j,t} = (1-\tau_j^*)\cdot EBT_{j,t},\ \ RE_{j,t} = NI_{j,t}\cdot\kappa \tag{8}$$

Given such a way of computing the evolvement of retained earnings and the balance of tax deductible temporary differences from realizations of net income before-tax for each period, we are ready to proceed to discount future net income after-tax in the next sub-section based on our production functions and evolvement accounting variables.

3.2 Firm Valuation Equation

We compute fundamental value of firms using the Edwards-Bell-Ohlson formula (Ohlson, 1995) and plug in inputs from simulated future income series readjusted on an after-tax basis. The fundamental value is defined in equation (9) where B_t is the

---

[7] In this formulation, we abstract from extraordinary gains and expenses for expository simplicity, but it is included for net income before tax items.

[8] This becomes multiple integral equations over many periods *ex ante* to derive firm valuation, and we resort to a simulation method in this paper.

[9] This assumption is only for expository simplicity here, and in the simulation current dividend payout ratio is used as an input parameter.

owner's equity value measured at its historical cost at the end of the previous period t-1, and the overall numerator denotes the "residual income" as defined by Ohlson (1995). The left hand side variable, V_t^*, thus denotes the computed fundamental value. We define $\tilde{NI}_{t+k}$ on a before tax basis, $\tau^*_{c,t+k}$ is the effective corporate tax rate for period $t+k$. B_t is the owner's equity value measured at its historical cost at the end of the previous period t-1, $\tilde{NI}_{t+k}$ is the future net income stream, $r_{e,t}$ is the discount rate which is the cost of equity, and Φ_{t-1} is the publicly available information set at the end of time t-1,

$$V_t^* = B_t + \sum_{k=1}^{\infty} \frac{E_t(\tilde{NI}_{t+k} \times (1-\tau^*_{c,t+k}) - r_{e,t}\tilde{B}_{t+k-1}) | \Phi_{t-1})}{(1+r_{e,t})^k} \tag{9}$$

In the following, we assume that the cost of equity, $r_{e,t}$, is the rate of return before personal tax, which can vary over time, and we estimate the cost of equity based on publicly available information using the Fama and French model.[10] In order to estimate $r_{e,t}$ we use the unconditional version of the three-factor model (Fama and French, 1993).

Every year we compute the fundamental value of firms on June 30 after all financial statements from the March 31 fiscal year end become publicly available and are approved at shareholder meetings before the end of June, which is a time limit according to Article 24 of Financial Instruments and Exchange Act of Japan. Note we use only the sample of firms whose fiscal year ends March 31, which amounts to more than 90 percent of all the listed firms on the Tokyo Stock Exchange.

Note the valuation equation (11) includes the summation of an infinite time horizon. However, in the Monte Carlo simulation framework, we compute the intrinsic value of firms by assuming that residual income becomes zero after the year t=21. Consequently, after generating the sample paths of residual income from taxable income, we sum up the present value of the residual income for the years t=1,...,20.

4 Basic Data and the Simulation Method

4.1 Data and Basic Observations

Our financial data is based on consolidated financial statements of all Japanese firms listed on the First and Second Section of the Tokyo Stock Exchange starting from calendar year 2000 (fiscal year 1999) through 2009 (2008), excluding financial firms and other firms in industries without data on fixed assets.[11] We exclude sample firms

---

[10] See Brennan (1970) for CAPM specifications where personal tax parameters are included. However, in this paper we do not consider that Miller (1977) equilibrium holds equivocally in an economy and we discount after-tax residual income by the rate of return on stock before the corporate tax basis.

[11] Naturally some firms in a particular industry do not have material fixed assets and do not disclose. We also exclude these firms from our sample.

with negative book values in equity and also exclude firms with less than 1% and higher than 99% of the estimated drift parameters and variance parameters of net income after-tax. Thus, the numbers of firm observations are a maximum of 1,374 firms for calendar year 2007, and a minimum of 989 firms for calendar year 2000.[12] The data source for financial variables is the Nikkei NEEDS Database, and for stock returns, the Nikkei Portfolio Master Database.

4.2 The Simulation Method

We compute the present value of the future net income stream after-tax, using a simulation method proposed by Shahnazarian (2011). He uses the accounting identity equation and its first difference derivative to generate a simulation path of future cash flow and discount this cash flow, in which firms are assumed to react optimally in a dynamic optimization context. However, our approach is different in the sense that we estimate production functions at the industry level and use the residual income model. The definition of net income before-tax in our study is the sum of earnings before-tax plus the net deferred tax balance, divided by statutory tax rates. Note that Japanese accounting standards require firms to report the deferred tax balance on an after-tax basis[13]

The simulations are based on our estimated production function (2) where the argument is based on estimates obtained from the systems of equations (1), and original estimates are obtained from industry-wide OLS regressions. Into this equation we plug in random generators of white noise and generate 10,000 different paths.[14] For each path we extrapolate future net income after-tax, and the stream of this is used as an input to the fundamental valuation equation (9) along with the estimates of each firm's cost of equity. When we compute the net income after-tax as well as the balance of deductible temporary tax differences in equations from (6) to (8), we take into account both the outstanding tax carry-forward balances and future tax-loss possible allowances. This happens when a firm incurs losses on any one of the 10,000 simulation paths 20 years into the future.[15] We use the expected statutory tax rate to readjust the future tax deferral balance by using Japanese accounting standards.[16] In the simulation we assume that the remaining balance on deferred tax

---

[12] The Japanese financial reporting system changed from individual to consolidated filings in fiscal year 1999. This is why we start the sample observations in 2000. The estimation using individual financial statements for a longer observation period is a subject for our future research.

[13] Note again Japanese financial and tax reporting follow the uniform reporting system. However, since fiscal year 1999, the tax deferral account in balance sheets was recorded in financial statements on the condition that such a deferred amount had a high probability of being reversed.

[14] This may be an arguable assumption and the bootstrapping method using past data is our future work. We thank an anonymous referee of the ESSA 2013 Conference for pointing this out.

[15] For firms which have experienced losses less than 7 years ago, the tax loss carry-forward benefits will accrue. We extend our simulations for 20 consecutive years to fully account for future cumulative effects.

[16] The recommendation to include tax deferral accounts in Japanese accounting standards, Accounting Standards for Tax Effect Accounting, was released October 30, 1998 and enacted in April 1999. Some firms began voluntary disclosure in April 1998.

assets is resolved in the first next period when firms incur enough profits, and if not, in consequent next periods. We also assume that firms keep the payout ratio constant at the same value in the past five years if net income after-tax is positive.

5 Effects on Firm Valuation of a Corporate Tax Rate Cut

In Table 1 we report the year-by-year results of our simulations. We compare using a ratio of V_0, the valuation based on the current effective tax rate of 40.87%, to V_1, the hypothetical valuation under the effective tax rate of 35%, which is equivalent to an approximate 5 percent cut in the corporate tax rate in equation (9).[17] Note V_1 is the average value for each firm computed from 10,000 simulation paths. We report empirical firm distributions of the ratio V_1/V_0, called VR in Table 1, which is a hypothetical value under the tax rate cut to the value when the current tax rate is kept intact. All the calibrated values are computed each year for June 30. For example, for calendar year 2000, balance sheet data from March 2000 is used, which is from financial statements of firms for fiscal year 1999. With these computed ratios we contrast firms which are expected to increase values by a tax cut (lower half of Table 1) with firms which are expected to decrease values (upper half of Table 1).

Table 1. Year-by-Year Frequency Distribution of Intrinsic Value Ratio

Intrinsic Value is defined as (*V* at hypothetical tax rate of 35%)/
(*V* at the ongoing effective tax rate of 40.87%)

	2000	2001	2002	2003	2004	2005	2006	2007	2008	2009
VR≦0.8	7.48	7.61	9.24	10.26	8.53	4.94	4.88	3.71	2.83	6.06
0.80<VR≦0.85	1.52	1.24	1.93	1.26	1.81	1.33	0.73	1.16	0.29	0.61
0.85<VR≦0.90	2.33	2.19	2.69	2.69	1.89	2.21	1.60	1.16	0.80	1.07
0.90<VR≦0.95	2.63	2.28	3.03	3.20	2.87	2.14	1.68	1.31	1.23	1.61
0.95<VR≦1.00	3.24	3.14	3.36	4.12	3.77	2.06	2.55	1.46	1.89	2.00
1.00<VR≦1.05	0.00	0.10	0.17	0.25	0.30	0.00	0.07	0.07	0.00	0.00
1.05<VR≦1.10	7.08	9.32	9.66	12.20	7.02	12.61	9.18	9.97	12.91	12.13
1.10<VR≦1.15	38.52	39.30	33.53	33.22	38.79	45.72	50.91	53.64	55.77	46.35
1.15<VR≦1.20	15.67	14.18	11.01	11.86	13.74	13.57	14.20	16.08	15.01	15.66
1.20<VR	21.54	20.65	25.38	20.94	21.28	15.41	14.20	11.43	9.28	14.50

In the table we find that firms with ranges of VR values from 1.10 to 1.15 are within the modal range. Thus, with an effective 5.87% tax rate cut, the value of the firm increases from 5% to 10% for firms at the highest frequency, owing both to pure tax effects and productivity changes due to increased investment. We find that the most effective case for a VR value exceeding 1.20 displays the largest percentage in

[17] In fact, the new bill for reducing the corporate tax rate by 5% passed the Diet in December 2010, but because of the March 11earthquake, the enactment of the bill was delayed until fiscal year 2012.

2002 at 25.38%. However, in the bottom row of the table, the smallest number is 9.28% in 2008 and 11.43% in 2007. In fact, in the past four years, lower percentage numbers for this highest value-ratio group with a VR value above 1.20 are consecutively reported. Accordingly, the result reveals that during the past four years, remarkable firm value appreciations were not expected by the corporate tax cut. The frequency for VR values between 1.10 and 1.15 dramatically decreased in 2009 at 46.35%. There are not conspicuous changes for VR values between 1.05 and 1.10 and 1.15 and 1.20 in the past four years. Hence, the result again suggests, unlike previous years, that 2008 and 2009 may not have been the best time to conduct government tax rate cuts, although a majority of firms could still incur value appreciation due to the cuts.

From the table, the best timing that the corporate tax rate cut should have been enacted was in 2007 when 81.15% (=53.64+16.08+11.43) of the firms belonged to VR values higher than 1.10. The reason for this might be due to the persistent downturn of the Japanese economy during our observation period because future profits were expected to decline, and thus corporate tax reduction effects became smaller. In other words, the decrease in effective tax rates for particular firms might have been smaller than the decrease in statutory tax rates which canceled out (added to) any gain (loss) from deferred tax liabilities (assets).

Noteworthy is the fact that the value decrease among a larger number of firms is observed in 2002 and 2000. For the range of VR within 0.85 and 1.00, 8.2% (=2.33+2.63+3.24) of the sample firms are in this category in 2000 and the corresponding number is 9.08% (=2.69+3.03+3.36) for 2002. Future profits were expected to keep declining (because of the state of the Japanese economy) and corporate tax reduction effects became smaller for non-profitable firms with tax loss carry-forward than those for profitable firms who continued paying taxes at the highest marginal tax rate bracket of 40.87 %. In other words, the decrease in effective tax rates for particular firms might have been smaller than the decrease in statutory tax rates *per se,* and canceled (added to) any gain (loss) from deferred tax liabilities (assets).

In Kubota and Takehara (2012), where they do not consider the secondary investment effect of the corporate tax rate as considered in the current paper, comparable numbers for appreciation case is smaller and the ones for depreciation case is larger. When investment effects are not considered, losing firms cannot enjoy tax loss carry-forward allowances by further increasing investment expenses, and in this case, retained earnings and consequent investment will not increase. For profitable firms, the converse holds and by increasing investment and profits the firms can increase accumulate retained earnings for future investment expansions. For the regulators, special funding with a lower cost for this group of firms and/ or other tax subsidiary instruments as additional measures might have had to be augmented from a social welfare viewpoint to neutralize these positive and negative effects of the tax rate cut.[18]

Table 2 reports an industry-wide distribution of VR ratios as of June 2009. Industry classification is based on 33 classifications of the Tokyo Stock Exchange.

---

[18] Samuelson (1964) argues against accelerated depreciation in the sense of subsidizing large firms at the expense of small firms. Here, the augment goes in a similar fashion where the tax rate cut subsidizes profitable firms, but works against losing and small firms (see Table 5). If losing firms are inefficient ones, it is a natural consequence to exit the market, but for small firms, the implications are not easy to assess. However, this is outside the scope of the current paper.

We impose the condition that there are at least 15 firms in an industry. We find possible positive effects of corporate tax cuts in the food industry, pharmaceuticals, land transportation, and warehousing by counting the percentage of firms whose VR values are larger than 1.05. On the other hand, negative effects are observed among the chemical and transportation industries. A value depreciation of more than 20 % is observed for these two industries (22.22% and 21.95%, respectively). The metal products industry is the next largest with 16.67 % of firms incurring a firm value loss over 20%. Note that the firms in these industries are heavily equipped. Accordingly, when the corporate tax rate cut *per se* is not as effective as shown here, as an alternative governmental corporate tax cut policy, the policy-maker better arrange investment tax credits, faster accelerated depreciation allowances and/or shortened estimated life of fixed assets in order to argument policy effects like the case of the U.S Tax Reform Act of 1986 and investment tax credits in 1962.[19] Other special tax codes can also be used as simultaneous auxiliary devices.

Table 2. Industry-wide Frequency Distribution of Intrrinsice Value Ration at the end of June 2009

Intrinsic Value is defined as (*V* at hypothetical tax rate of 35%)/
(*V* at the ongoing effective tax rate of 40.87%)

	#Firms	VR≦0.8	0.80<VR ≦0.85	0.85<VR ≦0.90	0.90<VR ≦0.95	0.95<VR ≦1.00	1.00<VR ≦1.05	1.05<VR ≦1.10	1.10<VR ≦1.15	1.15<VR ≦1.20	1.20<VR
Construction	35	8.57	0.00	0.00	0.00	0.00	0.00	11.43	48.57	8.57	22.86
Foods	66	1.52	0.00	0.00	1.52	0.00	0.00	22.73	60.61	7.58	6.06
Textiles&Apparels	44	9.09	0.00	4.55	4.55	11.36	0.00	11.36	20.45	18.18	20.45
Chemicals	18	22.22	0.00	5.56	22.22	0.00	0.00	0.00	5.56	5.56	38.89
Pharmaceutical	115	1.74	0.00	0.00	0.00	0.00	0.00	15.65	57.39	13.91	11.30
Glass&Ceramics Products	32	6.25	0.00	0.00	0.00	3.13	0.00	9.38	40.63	28.13	12.50
Iron&Steel	16	0.00	0.00	0.00	6.25	0.00	0.00	6.25	31.25	12.50	43.75
Metal Products	36	16.67	0.00	0.00	0.00	0.00	0.00	5.56	36.11	19.44	22.22
Machinery	44	2.27	0.00	0.00	0.00	2.27	0.00	0.00	40.91	29.55	25.00
Electric Appliances	25	4.00	0.00	0.00	4.00	0.00	0.00	0.00	48.00	20.00	24.00
Transportation Equipment	41	21.95	0.00	12.20	0.00	2.44	0.00	2.44	26.83	17.07	17.07
Precision Instruments	144	4.86	1.39	0.69	0.00	0.69	0.00	8.33	55.56	18.75	9.72
Other Products	159	8.18	0.63	1.26	4.40	1.89	0.00	9.43	33.96	18.87	21.38
Electric Power&Gas	79	1.27	0.00	0.00	0.00	2.53	0.00	17.72	69.62	2.53	6.33
Land Transportation	30	0.00	0.00	0.00	0.00	3.33	0.00	10.00	60.00	6.67	20.00
Warehousing	62	9.68	0.00	0.00	1.61	1.61	0.00	9.68	48.39	12.90	16.13
Wholesale Trade	49	0.00	2.04	6.12	2.04	6.12	0.00	53.06	14.29	4.08	12.24
Retail Trade	20	5.00	0.00	0.00	0.00	5.00	0.00	65.00	5.00	20.00	0.00
Services	26	11.54	0.00	0.00	3.85	3.85	0.00	0.00	42.31	23.08	15.38

[19] Note Samuelson (1964) is the first study which theoretically analyzed the effects of tax depreciation methods on firm value as well as social equity between large and small firms. Note the higher accelerated depreciation rate and smaller salvage values were permitted for tax purposes since 2007 in Japan.

In sum, we empirically and successfully demonstrate that estimated stochastic processes of net income after taking into consideration the balance of both DTA and LCA can reveal the net effect of a corporate tax rate cut on firm values. The result is a new contribution to the literature, in particular, for countries with uniform tax reporting systems.

6 Conclusion

This paper addressed a fundamental query to assess changes in firm values triggered by hypothetical corporate tax rate changes. The study is based on a microsimulation of all listed firms in Japan. We utilized the fundamental residual income valuation model by Ohlson (1995). In disentangling the effects on firm value by corporate tax rate changes, we paid particular attention to the net changes on deferred tax liabilities in the equity account and the contra account of deferred tax assets. Future paths of taxable net income for all individual firms were then computed with industry-wise multiplicative production functions and firm specific accounting variables. By discounting the future income of firms with the equity cost of capital computed with the three-factor asset pricing model (Fama and French, 1993) for each simulation path, we obtain new hypothetical values for all firms after all corporate tax effects were fully taken care of in future periods.

We find firm values may increase, do not change, or decrease even if the corporate tax rate is cut. Hence, for regulators, the timing decision of a tax rate cut is a sensitive decision based on past deferred tax states of firms, past profitability and variability, and business cycles. Similarly, for corporate financial managers, it is important to counteract tax rate cuts in changing investment decisions, again based on deferred tax states, past firm profitability and variability, and business cycles of each industry.

References

1. Brennan, M.: Taxes, Market Valuation, and Corporate Financial Policy. National Tax Journal 23, 417–427 (1970)
2. Cummins, J.G., Harris, T.S., Hassett, K.A.: Accounting Standards, Information Flow, and Firm Investment Behavior. NBER Working Paper No. 4685 (1994)
3. Fama, E.F., French, K.R.: Common Risk Factors in the Returns on Stock and Bonds. Journal of Finance 48, 65–427 (1993)
4. Graham, J.R.: Debt and the Marginal Tax Rate. Journal of Financial Economics 41, 41–73 (1996)
5. Kubota, K., Takehara, H.: Effects of Tax Rate Cut on Equity Valuation: Impact of Firm's Profitability and Variability. Japan Journal of Finance 32, 23–39 (2012)
6. MacKie-Mason, J.K.: Do Taxes Affect Corporate Decisions? Journal of Finance 45, 1471–1493 (1990)
7. Miller, M.: Debt and Taxes. Journal of Finance 32, 261–275 (1977)
8. Ohlson, J.: Earnings, Book Values, and Dividends in Equity Valuation. Contemporary Accounting Research 11, 661–687 (1995)
9. Samuelson, P.A.: Tax Deductibility of Depreciation to Insure Invariant Valuations. Journal of Political Economy 72, 604–606 (1964)
10. Shahnazarian, H.: A Dynamic Micro-Econometric Simulation Model for Firms. International Journal of Microsimulation 4, 2–22 (2011)

Growing Inequality, Financial Fragility, and Macroeconomic Dynamics: An Agent Based Model

Alberto Russo[1], Luca Riccetti[2], and Mauro Gallegati[1]

[1] Università Politecnica delle Marche, Ancona, Italy
alberto.russo@univpm.it
http://sites.google.com/site/albrusso76/
[2] Università "La Sapienza", Roma, Italy

Abstract. Our aim is to analyse the interplay between growing inequality and financial fragility in a complex macroeconomic system. In order to do this, we propose a macroeconomic microfounded framework with heterogeneous agents in which households, firms, and banks interact according to decentralised matching processes. The main result is that growing inequality leads to more macroeconomic volatility, increasing the likelihood of observing large unemployment crises.

Keywords: agent-based macroeconomics, business cycle, crisis, consumption, inequality.

1 Introduction

We aim at investigating the interplay between growing inequality and financial fragility in a complex macroeconomic system. In last decades, many advanced economies experimented a rise of income and wealth inequality. For instance, a decline of the labour share has occurred in advanced economies (about 10% in Europe and Japan and 3-4% in Anglo-Saxon countries since 1980), especially in unskilled sectors ([7]). The decrease of the labour share may cause a lack of effective demand in a context of growing inequality. Moreover, income inequality was relatively low and roughly stable before the 1980s, then it has drastically increased. Actually, consumer credit and other forms of indebtedness have prevented this to happen for a while, but at the cost of an increasing financial instability and a following large crisis.

Regarding the rise of inequality, economists proposed alternative interpretations: from high vs. low skilled workers (and then the economic impact of technological progress on labour market dynamics), to the role of labour flexibility and globalisation, and so on. In this paper we focus on the macroeconomic consequences of raising inequality, and in particular on the possible lack of aggregate demand due to an "excessive" saving of rich and a "too-low" consumption of poor. "Excessive" saving could be due to a precautionary motive, as showed in heterogeneous-agent models with incomplete markets and liquidity constraints

B. Kamiński and G.Koloch (eds.), *Advances in Social Simulation*,
Advances in Intelligent Systems and Computing 229,
DOI: 10.1007/978-3-642-39829-2_15,

(see, for instance, [6] and [1]). However, in a monetary production economy, such as the capitalist system, this situation may lead to a fall of the profit rate that, in turn, may result in lower production. The consequent rise of unemployment may further deteriorates macroeconomic conditions.

In order to analyse this complex scenario, we propose a macroeconomic micro-founded framework with heterogeneous agents in which households, firms, and banks interact according to a decentralized matching process presenting common features across four markets: goods, labour, credit and deposits. For a comprehensive description of the modelling framework see [9]. In general, the idea is to start from simple (adaptive) behavioural rules at the individual level in order to reproduce the emergence of aggregate regularities ([13]). In other words, we build the macroeconomy from the *bottom up* ([4]).

In our setting, agents are boundedly rational and follow (relatively) simple rules of behaviour in an incomplete and asymmetric information context: households try to buy consumption goods from the cheapest supplier, they also try to work in the firm offering the highest wage; firms try to accumulate profits by selling their products to households (they set the price according to their individual excess demand) and hiring cheapest workers; workers update the asked wage according to their occupational status (upward if employed, downward if unemployed); households' saving goes into bank deposits; given the Basilea-like regulatory constraints, banks extend credit to finance firms' production; firms choose the banks offering lowest interest rates, while households deposit money in the banks offering the highest interest rates.

In this agent based macroeconomic setting, we assume that firms' financial structure is based on the Dynamic Trade-Off theory ([5]). According to this theory, firms have a "target leverage", that is a desired ratio between debt and net worth, and they try to reach it by following an adaptive rule governing credit demand. This capital structure has a relevant role in influencing the leverage cycle, with important consequences on macroeconomic dynamics ([10]). We also consider the action of two policy makers: the government and the central bank. The government hires a fraction of the population as public workers, so providing an additional component of the aggregate demand. Moreover, the public sector taxes private agents and issues public debt. The central bank sets the policy rate and manages the quantity of money in the system. Furthermore, in our framework the central bank is committed to buy outstanding government bonds.

We analyse the dynamics of the model by means of computer simulation; some macroeconomic properties endogenously emerge: business cycle fluctuations, nominal GDP growth, the Phillips curve, leverage cycles and credit constraints, bank defaults and financial instability, and the importance of government as an acyclical sector which stabilise the economy. In particular, banks' capitalisation plays a relevant role in determining credit conditions, so influencing firms' leverage and, in general, the macroeconomic evolution. The presence of an acyclical sector, that is the government, has a fundamental role in sustaining the aggregate demand and in mitigating output volatility. Another interesting feature of the model is that credit mismatch (that is the difference between banks' credit supply and firms' credit demand) tends to follow the cycle of banks' net worth: when banks are poorly capitalised this results in credit rationing for firms; in this case, the central bank intervenes providing credit to banks; on the contrary, when banks are well capitalised they are able to fulfil all credit demand. Accordingly, firms' mean leverage is influenced by credit availability.

Furthermore, we observe that in some cases large crises can appear in the macroeconomic system. Indeed, the macroeconomic system evolves towards an "extended crisis" scenario, where the private sector tends to disappear, that is almost only public workers remain employed. In this case, differently from the usual business cycle mechanism, the decrease of wages due to growing unemployment does not reverse the cycle, but rather amplifies the recession due to the lack of aggregate demand. In other words, the self-adjustment mechanism which spontaneously reverses the business cycle (e.g., the rise of the unemployment rate reduces the real wage and then the resulting increase of profits makes room for an expansionary production phase) does not work. Indeed, real wage lowers excessively boosting a vicious circle for which the fall of purchasing power prevents firms to sell commodities, then firms reduce production, unemployment continues to rise, and the system moves towards a large crisis.

In order to assess the role of inequality on financial conditions and macroeconomic dynamics, in this paper we consider heterogeneous consumption behaviours. As a matter of fact, rich people may accumulate higher wealth while poor people may suffer from low consumption, so creating negative consequences at the macroeconomic level, as a lack of aggregate demand, so increasing the likelihood of observing a crisis with large unemployment.

The paper is organised as follows. In Section 2 we provide a brief description of the agent based macroeconomic model. We discuss simulation results in Section 3. Finally, Section 4 concludes.

2 The Model

This paper is based on the model reported in [9]. Here we sketch some of the modelling properties which characterise our macroeconomic framework.

The system is composed of households ($h = 1, 2, ..., H$), firms ($f = 1, 2, ..., F$), banks ($b = 1, 2, ..., B$), a central bank, and the government, and it evolves over a time span $t = 1, 2, ..., T$. Agents are heterogeneous, live in an incomplete and asymmetric information context, follow simple behavioural rules, and use adaptive expectations. Four markets compose the economy: (i) credit market; (ii) labour market; (iii) goods market; (iv) deposit market. In what follows we describe the working of the goods market in more detail. For details about the working of markets, see [9].

The interaction between the demand (firms in the credit and labor markets, households in the goods market, and banks in the deposit market) and the supply (banks in the credit market, households in the labor and deposit markets, and firms in the goods market) sides of the four markets follows a common decentralised matching protocol: a random list of agents in the demand side is set, then the first agent in the list observes a random subset of potential partners and chooses the cheapest one. After that, the second agent on the list performs the same activity on a new random subset of the updated potential partner list. The process iterates till the end of the demand side list. Subsequently, a new random list of agents in the demand side is set and the whole matching mechanism goes on until either one side of the market (demand or supply) is empty or no further matchings are feasible.

2.1 Credit Market

Firms and banks interact in the credit market. Firms' credit demand depends on their net worth and the leverage target. The leverage target varies according to expected profits and inventories: if expected profits are above expected interest rate and there is a small amount of inventories, then the firm increases target leverage, and viceversa. Banks set their credit supply depending on their net worth, deposits, the quantity of money provided by the central bank, and on some regulatory constraints.

2.2 Labour Market

Government, firms and households interact in the labour market. On the demand side, first of all, the government hires a fraction of households. The remaining part is available as workers in the private sector. Firms' labour demand depend on available funds, that is net worth and bank credit.

On the supply side each worker posts a wage which increases if he/she was employed in the previous period, and viceversa. Moreover, the required wage has a minimum related to the price of a good.

As a result of the decentralised matching between labour supply and demand, a fraction of households may remain unemployed. The wage of unemployed people is set equal to zero.

2.3 Goods Market

Households and firms interact in the goods market. On the demand side, households set the desired consumption on the basis of their disposable income and wealth as follows:

$$c^d_{ht} = c_1 \cdot w_{ht} + c_2 \cdot A^{c_3}_{ht} \tag{1}$$

where w_{ht} is the wage gained by household h, $0 < c_1 \leq 1$ is the propensity to consume current income, $0 \leq c_2 \leq 1$ is the propensity to consume the wealth A_{ht}. In this paper, we add the parameter c_3, that was implicitly equal to 1 in [9]. Accordingly, for $0 < c_3 < 1$ consumption increases less than proportionally whith wealth, that is the saving rate is higher for wealthier agents. We will investigate the role of the parameter c_3 below, by means of computer simulations, trying to assess the effects of heterogeneous consumption behaviours on macroeconomic dynamics. In particular we consider two different scenarios, one with $c_3 = 1$, that is the *baseline* scenario, and one with $c_3 = 0.5$. Accordingly, when $c_3 = 1$ we have the following consumption function:

$$c^d_{ht} = c_1 \cdot w_{ht} + c_2 \cdot A_{ht}$$

Instead, if $c_3 = 0.5$ the consumption function is given by:

$$c^d_{ht} = c_1 \cdot w_{ht} + c_2 \sqrt{A_{ht}}$$

Moreover, if the amount c^d_{ht} is smaller than the average price of one good $\bar{p}$ then $c^d_{ht} = min(\bar{p}, w_{ht} + A_{ht})$.

Given the number of hired workers, n_{ft}, firms produce consumption goods:

$$y_{ft} = \phi \cdot n_{ft} \tag{2}$$

where $\phi \geq 1$ is a productivity parameter (equal for all firms and time-invariant).

Firms try to sell their current period production and previous period inventories. The selling price increases if in the previous period the firm managed to sell all the output, while it reduces if it had positive inventories. Moreover, the minimum price at which the firm want to sell its output is set such that it is at least equal to the average cost of production, that is *ex-ante* profits are at worst equal to zero.

2.4 Deposit Market

Banks and households interact in the deposit market. Banks represent the demand side and households are on the supply side. Banks offer an interest rate on deposits according to their funds requirement: if a bank exhausts the credit supply by lending to private firms or government then it decides to increase the interest rate paid on deposits, so to attract new depositors, and viceversa. However, the interest rate on deposits can increase till a maximum given by the policy rate, which is both the rate at which banks could refinance from the central bank and the rate paid by the government on public bonds.

Households decide their savings to be deposited in banks as the desired savings plus the residual cash at the end of the interaction in the consumption market. Moreover, they set the minimum interest rate they want to obtain on bank deposits as follows: a household that in the previous period found a bank paying an interest rate higher or equal to the desired one decides to ask for a higher remuneration. In the opposite case, she did not find a bank satisfying her requirements, thus she kept her money in cash and now she asks for a lower rate. We assume that a household deposits all the available money in a single bank that offers an adequate interest rate.

3 Simulations

We explore the dynamics of the model by means of computer simulations. We refer to [9] for the details about the parameter setting, the initial conditions, and the results of the baseline model that we now sum up.

3.1 Baseline Scenario

Simulation results show that endogenous business cycles emerge as a consequence of the interaction between real and financial factors. When firms' profits are improving, they try to expand the production and, if banks extend more credit, this results in more employment; the decrease of the unemployment rate leads to the rise of wages that, on the one hand, increases the aggregate demand, while on the other hand reduces firms' profits, and this may cause the inversion of the business cycle. Indeed, there is a significant cross-correlation between the unemployment rate and the firms' profit rate. First of all, there is a high positive correlation at lag 0: the profit rate is high when unemployment is high given that firms save on production costs (e.g., wage bill) but, at the same time, the aggregate demand does not decrease proportionally, because of public workers' expenditure and consumption due to wealth, thus firms can sell their commodities (including inventories) in the goods market. However, the presence of unemployed people, the tendency of wages to decrease due to the high unemployment rate, and the reduction of households' wealth, cause the fall of next period aggregate demand that, in turn, reduces firms' profits. Indeed, there is a negative correlation at lag +1. Instead, the negative correlation at lag -1 means that increasing profits boost the expansion of the economy and then a fall of the unemployment rate follows. Then, there is a dynamic relation between unemployment and the profit rate underlying the "real" economy, which gives rise to business cycle fluctuations that, in turn, are amplified by a financial accelerator mechanism. Business fluctuations are mitigated by the government, which acts as an acyclical sector, reducing output volatility through the stabilisation of the aggregate demand.

In some cases, differently from the usual business cycle mechanism, the fall of wages due to the increase of unemployment does not reverse the cycle, but generates a lack of aggregate demand that amplifies the recession in a vicious circle: indeed, the fall of purchasing power prevents firms to sell commodities, then firms decrease production, unemployment continues to rise, and the recession further deteriorates. In these cases, the system may remain trapped in a large unemployment crisis.

3.2 Heterogeneous Consumption Behaviour

In this subsection, we compare the result of the baseline model, obtained with a parameter $c_3 = 1$ in equation 1, with the results of the simulations in which $c_3 = 0.5$. In this way, we try to address the inequality topic in a symplified framework in which all households have the same skills and they all works in an economy with homogeneous goods. Thus, labour income does not vary much across households and capital income is distributed to households proportionally to their share (which in turn depends on households' wealth). Moreover, all households have a similar initial wealth. Indeed, in the baseline case ($c_3 = 1$), we obtain a wealth distribution that is negatively skewed, while in the real world it is highly positively skewed.

However, even in this symplified framework, a propensity to consume decreasing with wealth ($c_3 = 0.5$) modifies the household's wealth distribution. Indeed, in this case we can observe a larger wealth inequality: wealth distribution not only presents a left tail, but also a right tail, thus the richest agent has a higher wealth and the skewness becomes about zero (with a mean slightly higher than the median of households' wealth); the larger inequality is also signaled by an increase of the standard deviation; moreover, average wealth increases given that richest households save more. This analysis on wealth distribution is summarized in table 1.

Table 1. Statistics about wealth distribution at time T=500 in two simulation with different value of parameter c_3: $c_3 = 1$ and $c_3 = 0.5$

Statistic	$c_3 = 1$	$c_3 = 0.5$
Mean	1.38	1.61
Standard deviation	0.38	0.54
Skewness	-0.72	-0.08
Maximum	2.13	3.28

These results are robust both at different time steps (for instance we check the wealth distribution also at time t=150) and in different simulations. Indeed, we perform a Monte Carlo with 100 simulations on a time horizon $T = 500$ (again skipping the first 100 periods, then we analyse the last 400 time steps).

Table 2 reports some relevant macroeconomic features of the two Monte Carlo simulations with $c_3 = 1$ and $c_3 = 0.5$.

We can observe that in both cases the economy falls in a large crisis scenario, that is with a mean unemployment rate above 20%, 2 times over 100 simulations and the mean unemployment rate is the same. However, the unemployment volatility is much higher when $c_3 = 0.5$, that is whith larger wealth inequality, and the difference is statistically significant at 99% level. Therefore, the business cycle is "larger" with a lower minumum and a higher maximum for the unemployment rate. Indeed, while in the baseline case ($c_3 = 1$) we never find a time step with unemployment above 20%, but for the two large crisis scenario, in the inequality case ($c_3 = 0.5$) we detect 10 simulations in which the unemployment peaks above 20% (the two large crises plus other 8 simulations). If the policy maker considers the business cycle volatility as a problem to be stabilised, than the reduction of the wealth inequality seems to be an effective tool to reach this target. Especially, the policy maker could avoid large unemployment crises (e.g., with an unemployment rate above 20%) by reducing inequality. For instance, in an agent based macroeconomic setting, [3] find that more inquality leads to higher volatility, increasing the likelihood of unemployment crises; they also show that fiscal policy is an effective countercyclical tool especially when income distribution is skewed towards profits.

It is worth to note that the percentage of firm defaults increases in the case of $c_3 = 0.5$, probably due to higher macroeconomic volatility. Indeed, the mean

Table 2. 100 Monte Carlo replications for both $c_3 = 1$ and $c_3 = 0.5$ (data calculated on time span 101-500). The number of simulations with average unemployment rate and maximum unemployment rate above 20% are computed on all 100 simulations. Instead, the other statistics refers to simulations with average unemployment rate below 20% (that is 98 simulations for both cases); in brackets we add the standard deviation of the corresponding statistic among the simulations; the last column indicates the p-value of the test on the null hypothesis that the value of the statistic is equal between the two cases of $c_3 = 1$ and $c_3 = 0.5$.

Variable	$c_3 = 1$	$c_3 = 0.5$	p-value H_0=0
Sim. with mean(ur) < 20%	98	98	
Sim. with max(ur) < 20%	97	90	
Unemployment rate %	9.73 (0.87)	9.71 (0.61)	85.2%
Unemployment volatility %	1.84 (0.12)	2.17 (0.18)	0.0%
Firm default rate %	6.21 (0.70)	6.51 (0.60)	0.1%
Bank default rate %	0.40 (0.38)	0.32 (0.37)	13.5%
Firm mean leverage	1.23 (0.31)	1.10 (0.21)	0.1%
Firm leverage volatility	0.23 (0.07)	0.30 (0.07)	0.0%
Interest rate %	8.04 (0.81)	8.51 (0.77)	0.0%
Credit constraint %	4.15 (1.64)	8.79 (2.57)	0.0%
Wage share %	63.7 (0.3)	63.6 (0.2)	0.7%
Public deficit %	3.16 (0.05)	2.96 (0.06)	0.0%
Inflation rate %	1.99 (0.04)	1.99 (0.03)	100%

firm leverage is lower in this case, then the economy should be safer according to this indicator. Instead, this is not the case given that firm leverage is more volatile, giving rise to the already mentioned "larger" business cycle, with stronger leveraging and deleveraging processes. Moreover, the higher volatility, that highlights a riskier economic environment, goes along with higher interest rates charged by banks to firms, that in turn affects both the mean firm leverage (lower, because it is less favorable to ask money to banks) and the number of firm defaults (higher, because of the higher cost of the debt).

We calculate the credit constraint as the percentage of credit required by firms that firms do not obtain. Given that we observe a correlation above 50% between firm leverage and credit constraint, and given that in the case of $c_3 = 0.5$ firm leverage is more pro-cyclical, in this situation there are periods in which the credit constraint is stronger. We can confirm this analysis computing the average of the standard deviation and the average of the maximum credit constraint in the two Monte Carlo settings: when $c_3 = 0.5$ the average standard deviation is 8.16% and the average maximum is 35.08%, while if $c_3 = 1$ the average standard deviation is 5.75% and the average maximum is 26.97%. Given that the distribution is truncated at zero (and it often happens that firms receive all the required credit), the longer right tail of the distribution in the case of $c_3 = 0.5$ explains the higher mean credit constraint. Moreover, this is another further which can eplains why the mean leverage is lower when $c_3 = 0.5$. Finally, the relation between firm leverage and credit constraint has both causal direction: a higher firm leverage implies a higher probability of credit constraint, but also a higher credit constraint implies a lower leverage, because firms are not able to reach their desired leverage. The wage share is statistically different between the two Monte Carlo, but the difference is economically not significant. The similar wage share between the two Monte Carlo experiments implies a similar pressure to increase wages and thus similar effets on price dynamics. Indeed, the inflation rate is the same in both cases.

Instead, the public deficit is slightly lower in the case with $c_3 = 0.5$, given that in this case there are richer households and we assume the presence of a 5% tax rate on wealth (only above a certain wealth level). However, in both cases the public deficit remains on admissible values (compared to GDP).

To summarise, we observe a negative impact on the economy of a propensity to consume that decreases with wealth (that also creates an economic system with larger wealth inequality). Indeed, in this case the business cycle is "larger" and we count a higher number of simulations in which we detect large unemployment crises. This riskier economic environment with a stronger volatility implies a larger number of firm defaults, a higher mean interest rates, and a higher mean credit constraint.

4 Concluding Remarks

"While several authors have noticed that there might be a link between rising inequality and the crisis ([11], [14], [8]), there is as of yet little systematic analysis" ([12]). We proposed an analysis of the effects of wealth inequality on macroeconomic dynamics in an economy composed of heterogeneous households, firms, banks, and two policy makers, that is the government and the central bank. The main result is that growing inequality leads to more macroeconomic volatility, increasing the likelihood of observing large unemployment crises. However, this is just a first step towards a complex quest. Our aim is to further extend our analysis in order to understand the consequences of growing inequality on the evolution of the macroeconomy.

For instance, we may consider the introduction of credit consumption through which the saving of rich can finance the consumption of poor. But, in a context of growing inequality, debt accumulation may increase financial fragility, spreading in the system through credit interlinkages ([2]), eventually leading to a financial collapse. So finance may postpone the crisis due to the lack of aggregate demand, but it also creates the bases for a later financial crisis. However, we leave this complex topic for future works.

References

1. Aiyagari, S.R.: Uninsured idiosyncratic risk and aggregate saving. Quarterly Journal of Economics 109(3), 659–684 (1994)
2. Delli Gatti, D., Gallegati, M., Greenwald, B., Russo, A., Stiglitz, J.E.: The financial accelerator in an evolving credit network. Journal of Economic Dynamics and Control 34(9), 1627–1650 (2010)

3. Dosi, G., Fagiolo, G., Napoletano, M., Roventini, A.: Income distribution, credit and fiscal policies in an agent-based Keynesian model. Journal of Economic Dynamics and Control 37(8), 1598–1625 (2013)
4. Epstein, J.M., Axtell, R.L.: Growing Artificial Societies: Social Science from the Bottom Up. MIT Press/Brookings Institution (1996)
5. Flannery, M.J., Rangan, K.P.: Partial adjustment toward target capital structures. Journal of Financial Economics 79(3), 469–506 (2006)
6. Huggett, M.: The risk-free rate in heterogeneous-agent incomplete-insurance economies. Journal of Economic Dynamics and Control 17, 953–969 (1993)
7. International Monetary Fund: Spillovers and Cycles in the Global Economy. World Economic Outlook, ch. 5 (April 2007)
8. Rajan, R.: Fault Lines. How Hidden Fractures Still Threaten the World Economy. Princeton University Press (2010)
9. Riccetti, L., Russo, A., Gallegati, M.: An Agent Based Decentralized Matching Macroeconomic Model. MPRA paper No. 42211. University Library of Munich, Germany (2012)
10. Riccetti, L., Russo, A., Gallegati, M.: Leveraged Network-Based Financial Accelerator. Journal of Economic Dynamics and Control 37(8), 1626–1640 (2013)
11. Stiglitz, J.E.: Freefall: Free Markets, and the Sinking of the World Economy. Norton (2010)
12. Stockhammer, E.: Rising Inequality as a Root Cause of the Present Crisis. Working Paper No. 282, Political Economy Research Institute (PERI). University of Massachussets, Amherst
13. Tesfatsion, L.S., Judd, K.L.: Handbook of Computational Economics: Agent-Based Computational Economics, vol. 2. North-Holland (2006)
14. Wade, R.: The global slump. Deeper causes and harder lessons. Challenge 52(5), 5–24 (2009)

By the Numbers: Track Record, Flawed Reviews, Journal Space, and the Fate of Talented Authors

Warren Thorngate and Wahida Chowdhury

Psychology Department and Institute of Cognitive Science
Carleton University, Ottawa, Canada
warren_thorngate@carleton.ca,
WahidaChowdhury@cmail.carleton.ca

Abstract. We conducted a computer simulation of hundreds of competitions for limited journal space, varying (a) the correlation between the talent of authors and the quality of their manuscripts, (b) the correlation between manuscript quality and quality judged by peer reviewers, (c) the weights reviewers and editors gave judged quality versus number of previous publications (tract record), and (d) the proportion of manuscripts accepted for publication. The results show that even small decreases in the correlations, and small increases in the weight given to track record, quickly skew the outcomes of the peer review process, favouring authors who develop a track record of publications in the first cycles of journal publication while excluding many equally-talented or more-talented authors from publishing (the *Matthew Effect*; Merton, 1968). Implications for declines in the quality of published manuscripts and for wasting talent are discussed.

Keywords: peer review, reputation, track record, competition.

1 Introduction

Scientists frequently submit manuscripts for publication; grant proposals for funding, applications for jobs, tenure, promotion, sabbaticals, fellowships and other limited resources. Their submissions are usually assessed by other scientists as part of the venerated tradition called *peer review*. The assessments are important; many scientific careers are made or broken by the outcomes of competitions for these resources (Thorngate, Dawes, & Foddy, 2009; Thorngate, Liu & Chowdhury, 2011).

Peers can employ many kinds of information in making their assessments. Some information is intrinsic to the submission: the freshness of its ideas, the relevance of its literature, logical derivation of its predictions, soundness of its research methods, cogency, etc. Other information is extrinsic to the submission, including information about the reputation of its author. Reputation is often induced from a list of the author's previous scientific achievements (a curriculum vita), also known as the author's *track record*. The primary purpose of the present study was to determine how the outcomes of competitions for journal space are influenced by the relative importance peers attach to submission quality versus track record.

B. Kamiński and G. Koloch (eds.), *Advances in Social Simulation*,
Advances in Intelligent Systems and Computing 229,
DOI: 10.1007/978-3-642-39829-2_16, © Springer-Verlag Berlin Heidelberg 2014

There is likely a high correlation between submission quality and track record, so there is likely to be much redundancy in the information the two provide. The redundancy strengthens the argument of some peers that there is little to be gained from assessing track record in addition to submission quality. However, it also strengthens the argument of other peers that there is little to be gained by assessing submission quality in addition to track record. Variations of these arguments emerge in debates about whether research funding should support people or projects -- whether the proof of the pudding is in the eating or in the reputation of the chef. Some funding agencies resolve such debates by requiring peers to assess both submission quality and track record, giving weight to both (for example, Canada's Natural Sciences and Engineering Research Council, 2012).

Still, submission quality and track record are not the same. The judged quality of a submission depends largely on the submission's author and the peers who judge it. Judgments of track record depend on the author's submission history and the successes it brings. Peer reviews of a current submission are not supposed to be influenced by the author's previous submissions. In contrast, track records are defined by previous submissions, so the success or failure of an author's previous submissions can influence the fate of a current one.

The cumulative effect of track record on the fate of current submissions exemplifies a positive feedback loop amplifying the effects of chance as much as the effects of scientific talent (Thorngte & Hotta, 1995). The chance effects can be illustrated by considering two young scientists, Alice and Bob, who submit their first research grant proposals to a funding agency that weighs track record more heavily than proposal quality in choosing whom to fund. Peers judge Alice to be slightly more talented than Bob and her proposal to be somewhat better than Bob's. Bob, however, has nine publications, and Alice only seven, a deficit of two publications resulting from the accidental death of her supervisor. So if the agency can fund only one of their proposals, Bob is more likely to receive funding because Bob's has the better track record. As a result, Bob will have more resources for increasing his scientific output than will Alice, and thus will likely have an even more impressive track record than Alice's in any subsequent competition, further increasing his likelihood and decreasing her likelihood of new funding. As the future unfolds, Alice's career would likely flounder while Bob's career would soar, not because Bob was a better scientist than Alice, but because the funding agency could not fund both, because it weighed track record more than submission quality, and because Alice's supervisor had the temerity to die.

With funding, perhaps Alice would have made more substantial contributions to science than would Bob. Perhaps not. The relation between the talent of scientists and the quality of their submissions is almost certainly imperfect, so it is quite possible that scientists with lesser talent will sometimes produce better science than their betters. Talent, like any psychological characteristic, expresses itself statistically in a long run of independent opportunities. Track records break the rule of independence, so those who, by talent or by chance, establish track records before others do are likely to retain a career-long competitive advantage.

The tendency for early judgments of track record to snowball during a scientist's career was recognized by Merton (1968) in his analyses of interviews with Nobel Prize winners and diaries of other scientists. He reports, "[Nobel Prize laureates] repeatedly observe that eminent scientists get [disproportionate] credit for their contributions to science while [relatively] unknown scientists tend to get disproportionately little credit for comparable contributions" (p. 57). Merton concludes that, regardless of the true merit of a scientist's work, the reputation of a scientist, determined largely by track record, influences publication prospects and subsequent enhancement of reputation, leading to what Merton termed the *Matthew Effect*: The rich get richer and leave the poor behind.

Consistent with the Matthew Effect, a number of studies have found a positive correlation between the reputation of an institute and the productivity of its members, measured by the number of manuscripts they published and the number of citations their publications received (e.g., see Howard, Cole & Maxwell 1987; Morgan, Meier, Kearney, Hays & Birch, 1981). Other studies showed that an author's total number of publications, citations, and scientific achievements affect, and are affected by, the author's reputation or previous scientific achievements (e.g., see Bourne & Barbour 2011; Bornmann & Daniel, 2007; Dewett & Denisi 2004; Ferber 1986). Day & Peters (1994) found that academic journals with new and fresh ideas were referenced as long as the journals had an article by a renowned author. Ofori-Dankwa & Julian (2005) reported that adoption of a new theory depends, in part, on the reputation of the author of the theory and author's university affiliation.

Computer simulations of resource allocations have demonstrated a surprisingly large influence of resource scarcity, allocation rules, and chance on the distribution of limited resources, including grants and publications. Thorngate, Hotta and McClintock (1996), for example, showed that early winners in Bingo games were far more likely to survive future games than were early losers simply because their early winnings allowed them to remain longer in the game. The results suggest that lucky authors who establish a good reputation early in their career and are more likely to continue their career than unlucky ones. Squazzoni and Gandelli (2012) showed that the reliability of reviewer's manuscript assessments resource allocation rules strongly influence competitions for journal space. Thorngate, Dawes and Foddy (2009) further demonstrated that the outcomes of competitions for research grants are strongly influenced by resource scarcity and chance, especially when the competitions are repeated and the results of previous competitions are used in choosing current winners. The present study attempted to investigate the influences of resource scarcity, selection rules, and chance on the outcomes of competitions for limited journal space.

Most scientists know that many scientific manuscripts eligible for publication in scientific journals are nevertheless rejected. Resources are scarce and not everyone can be accommodated. The American Psychological Association, for example, reports publishing 28 journals in 2011 (APA, 2011). During that year the journal editors collectively received 11,635 manuscript and rejected 77% of them, in part because there was no more money to pay for additional journal pages.

The scarcity of pages forces editors to devise procedures for deciding which manuscripts will be printed and which won't. A few of these procedures are common (Thorngate, Dawes & Foddy, 2009). Almost all editors, for example, quickly scan manuscripts for the eligibility of their content, culling manuscripts with content far removed from the journal's themes, likely fictitious, or clearly lacking in scientific rigour. Manuscripts surviving the eligibility cull are then distributed to 2-3 peers for review, usually with prescriptions about what the reviewers should look for and how they should indicate their assessments. Prescribed indicators include category selections (for example, "publish without revisions, publish after minor revisions, ...reconsider after major revisions, reject"), numerical ratings, and written recommendations with justifications. Most editors now omit the name of the author before sending manuscripts for review, presumably to reduce the chances of personal bias. Yet it is frequently possible to guess the author from self-citations (for example, "This study extends my previous work on punctuation (Smith, 2010) ...") and use the guess to assess track record. Even when reviewers cannot identify the author, the editor can, and may use the knowledge to resolve disagreements among reviewers' assessments.

Reviewers frequently disagree in their assessments of manuscripts, sometimes a little, sometimes a lot. The correlation between reviewers' assessments of manuscripts is likely no greater than that achieved by reviewers of grant proposals: a modest but stubbornly consistent r = +0.50 (Thorngate, Dawes & Foddy, 2009; see also Petty, Fleming & Fabrigar, 1999; Whitehurst, 1984).

Editors have a wide variety of decision rules for resolving the disagreements -- options ranging from taking the advice of the most-credible reviewer to flipping a coin. Many of these options can be represented as a weighted average of the reviewers' assessments and the track record of the manuscript author. Suppose, for example, that Reviewer X gave Manuscript A her 2nd highest rating, and Reviewer Y gave it his 7th highest rating, out of 100 manuscripts. Suppose also that the editor counted the number of publications of A's author, compared the count to those of all other 99 manuscript authors, and calculated that A's author had the 5th best track record. The editor might then apply a simple formula such as the one below to resolve X and Y's disagreement about the rank of manuscript A.

Let:
wmq = the weight give to rank of manuscript quality;
wtr = the weight given to rank of track record; and
wmq + *wtr* = 1.00;

$$\textit{Weighted rank of Ms A} = wmq*(Xrank + Yrank)/2 + wtr*track\text{-}rank. \quad (1)$$

So if, for example, *wmq* = 0.7 and *wtr* = 0.3, then

Weighted rank of manuscript A = 0.7*(2+7)/2 + 0.3*5 = 3.15 + 1.50 = 4.65.

The editor could then compare this weighted rank of 4.65 with the weighted ranks of all other 99 manuscripts similarly calculated, and publish manuscripts from the highest-ranked downward until the journal ran out of space. If the editor set *wtr* = 0.0, then the resulting weighted rank would simply be the average of the two reviewers' assessments of manuscript quality. If the editor set *wtr* = 1.0, then the reviewer's assessments would be given no weight and only track record would matter. Our simulation addressed what happens at and between these extremes.

2 The Simulation

The simulation, written in the programming language R and available from the authors, examined how the above decision rule (Equation 1) for weighing two reviewers' judgments of manuscript quality and track record influenced three outcomes of competitions for journal space. These outcomes were:

1. the average percent of the most-talented authors having the greatest number of publications;
2. the average percent of the highest quality manuscripts that were published; and
3. the percent of winners of a first competition for journal space who accrued the highest track records in subsequent competitions.

We constructed our examination by creating 100 hypothetical journals, all of which we assumed were published four times a year for 25 years (100 issues). We also assumed that the editors of each journal received 100 new manuscripts per issue and sent each manuscript to two reviewers. To simplify programming, we additionally assumed that each manuscript was written by only one author, and that each author submitted one new manuscript per issue. Thus, the 100 authors submitting manuscripts to the 1st issue of Journal J were the same as the 100 authors submitting different manuscripts to the 2nd issue of Journal J, the 3rd issue, etc. One-hundred different authors submitted to Journal K, another 100 to Journal L, etc. Our selection of 100 for journals, issues, and manuscripts per issue was arbitrary. Preliminary runs of the simulation with variations of these numbers gave equivalent results.

In order to determine the effects of resource scarcity on the three simulation outcomes, we varied *npub*, the number of the 100 submitted manuscripts that could be published in each issue. Half the simulation runs allowed 40 out of the 100 manuscripts to be published; the other half allowed only 20 of the 100 manuscripts to be published.

For each journal, the simulation began by assigning a unique "talent" score to each of its 100 loyal authors. The talent scores were generated by random sampling from a normal distribution; in keeping with standard psychometric assumptions (e.g., Kaujman, 2009), each author's talent score, like an IQ score, remained constant across all her/his "career" of 100 manuscript submissions. All authors began their careers with no publications (so the ranks of all tracks records were tied), submitting their first manuscript for issue 1 of their assigned journal.

After their talent scores were assigned, the 100 authors each submitted a manuscript to the first issue of their assigned journal. The manuscripts varied in their *true quality*, which we defined as the average of what thousands of reviewers would judge the quality to be. A number representing the true quality of an author's current manuscript was calculated by adding or subtracting some normally-distributed, random error to each author's talent score. This error defined the first of two sources of chance that could influence the three simulation outcomes. We varied the amount of this random error by changing the correlation coefficient between talent and true quality. In half the simulation runs we set the talent/true-quality correlation, *rttq* = +0.75; it was rttq = +0.50 for the other half of the runs. The lower the correlation, of course, the greater the amount of random error was added.

Once the true qualities of the 100 manuscripts were assigned, each manuscript was assessed by two reviewers. The reviewers independently assigned a new number representing the *judged quality* of each manuscript they reviewed. The judged quality of a manuscript was based on the true quality plus or minus another dollop or normally-distributed random error. This was the second of the two sources of chance that could influence the three simulation outcomes. As above, the dollop of error was varied by changing the correlation coefficient between the true and judged quality of manuscripts. In half the simulation runs, *rtjq* = +0.75; for the other half it was *rtjq* = +0.50.

At this point in the simulation, each manuscript was given two judgments of quality, one from each of two reviewers. The judgments of the first reviewer were now ranked across all 100 manuscripts submitted for the upcoming journal issue. The judgments of the second reviewer were likewise ranked, then the two ranks of each manuscript were averaged. Finally, this average rank was combined with the rank of each author's track record according to the formula in Equation 1. The result was a single number for each manuscript+author pair representing the pair's overall *merit*.

In order to examine the effects of changing the weights given to manuscript quality versus track record on the three outcomes previously listed, we systematically changed the values of *wmq* and *wtr* (= 1 – *wmq*) in Equation 1. In different simulation runs, weight *wmq* was set to 1.0 (no weight given to track record), 0.9, 0.8, 0.7, 0.6, …, and 0.1 (almost all weight given to track record).

Once overall merit was calculated, the simulation proceeded to rank order all 100 merit scores. It then selected the manuscripts with the *npub* (20 versus 40) highest-ranked merit scores to be "published." The publication counters of the *npub* published authors were incremented by 1 to signify an increase of 1 in their track record. The publication counters of the 100-*npub* unpublished authors were, of course, not incremented.

To illustrate the simulation process, consider again Alice and Bob. Suppose Bob, assigned a talent score of 80, submits a manuscript with a true quality score of 73 to the first issue of Journal J. Similarly, suppose Alice, assigned a talent score of 83, submits a manuscript with true quality score of 76 for the first issue of Journal J. Reviewer X gives Bob's manuscript a judged quality score of 79 and Reviewer Y gives it a score of 71 for an average of 75. Reviewer X gives Alice's manuscript a judged quality score of 70 and Y gives it a score of 76 for an average of 73. Because Alice and Bob start with no publications, any weight given to track record does not effect their merit scores. If the editor of J had room to publish both manuscripts, then Alice and Bob would both add one publication to their track record. But if the editor could publish only one of their manuscripts, then Bob's higher-judged manuscript would prevail. Bob would then increase his/her track record from zero to one publication while Alice, despite superior talent and a better true-quality manuscript, would not.

The resulting difference of one publication in track record could compound itself in subsequent outcomes. Suppose Alice and Bob each submit a new manuscript for the second issue of Journal J. This time Reviewers X and Y both judge Alice's new manuscript to be better than Bob's, so the judged-quality ranking of her manuscript is higher than his. But, thanks to his publication in the first issue of J, Bob's track record is now higher than Alice's. Again, if the editor had room for both manuscripts, then Alice and Bob would both improve their track record. But if the editor had room for only one of their new manuscripts, a quality-versus-track-record ranking conflict would ensue. Giving more weight to manuscript quality would lead to Alice's first publication, equaling her track record with Bob's. Giving more weight to track record would put Bob's track record two publications ahead of Alice's, again preventing the second of Alice's superior true-quality manuscript from being published, and making him even more likely than Alice to be published in the remaining competitions.

In summary, the simulation attempted to determine the outcomes of the seemingly complex interactions among resource scarcity (journal space), chance (imperfect correlations between talent, true manuscript quality, and judged quality), and editorial decision rule (weights given to judged quality versus track record) as journals repeated their publication competitions. We did so to learn under what conditions the use of reputation information might improve the average quality of publications and the track record of those who author them.

3 Results

Figure 1 shows how the weight given to track record (X-axis) influences the average percentage of the N-best articles published when npub = 20 out of 100 (dashed lines) and when npub = 40 out of 100 (solid lines). The figure reveals three trends. First, increasing the weight given to track record results in a general decline of the proportion of best articles published. This can be seen in consistent drop of about 20% in each of the eight lines as track record weight increases from 0.0 to 0.9. Second, declines in percentage of best articles published reach a minimum asymptote when the weight given track record reaches about 0.7. Third, doubling the number of publications per journal issue from 20 to 40 increases the proportion of the 20 (40) best articles published only 10 to 20%, more as the weight for track record increases. This can be seen by comparing a solid line with the dashed line of the same colour.

Figure 1 also reveals two unexpected findings. First, increasing track-record weight up to about 0.4 slightly increases in the percentage of best articles published when the correlation between talent and article quality is higher than the correlation between true and judged article quality; this is revealed by the slight hump in the two green lines. Second, the pair of solid red and green lines, and the pair of dashed red and green lines, cross when the weight given track record reaches about 0.5. This indicates that track record has a somewhat different effect when the talent/true-quality correlations is varied than when true-quality/judged-quality correlation is varied. We found no immediate explanation for these results; their effects are small but worthy of further exploration. However, the general trend remains: When the correlations are moderate, increasing the weight given track record decreases the proportion of best articles published.

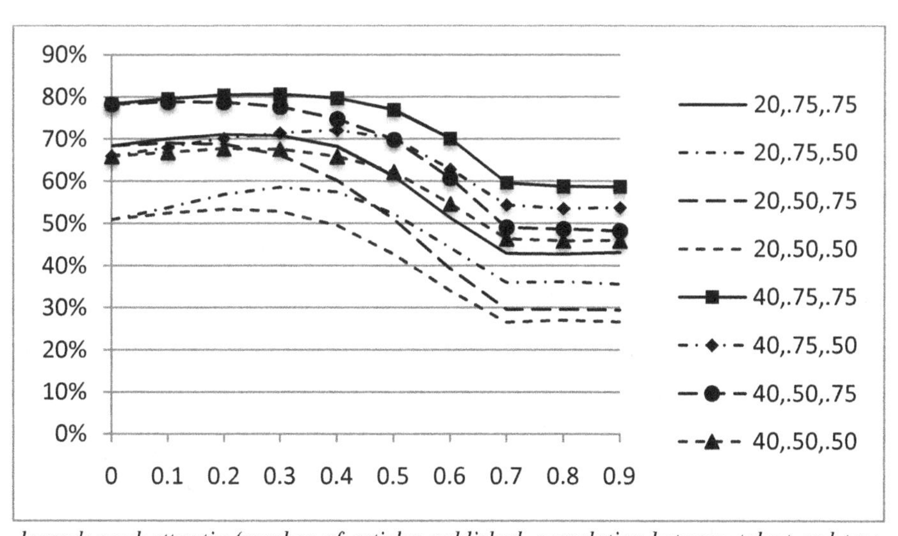

legend: npub,rttq,rtjq (number of articles published, correlation between talent and true manuscript quality, correlation between true and judged manuscript quality)

Fig. 1. Average percentage of the npub = 20 (40) best articles published for different weights of track record (0.0 – 0.9)

Figure 2 shows how the weight given to track record influences the percentage of the 20/40 best authors obtaining the 20/40 highest track records by the 100th journal issue. As the weight given to track record increases, the percentage of best authors who accumulate the best track records declines. This reduction accelerates as the track-record weight increase beyond about 0.2 and reaches an asymptotic minimum when the weight reaches about 0.7. Doubling the number of publications from 20 to 40 increases the proportion of the 20/40 best authors among those with the 20/40 best track by only 10-20%.

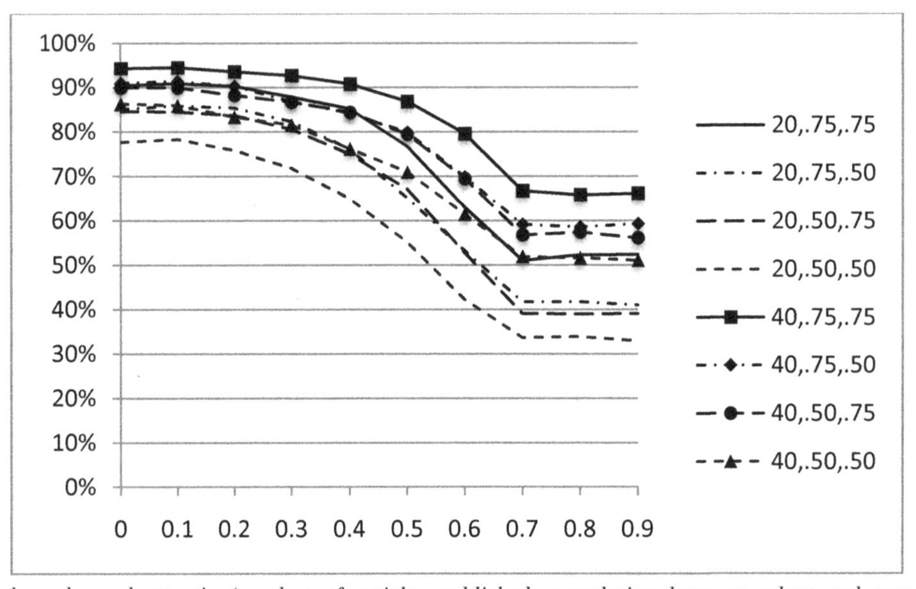

legend: npub,rttq,rtjq (number of articles published, correlation between talent and true manuscript quality, correlation between true and judged manuscript quality)

Fig. 2. Average percentage of the most talented authors accruing the best track records for different track-record weights (0.0 – 0.9)

Finally, Figure 3 shows how the weight given track record influences the percentage of the 20 (40) authors published in the first journal issue who accrue the best 20 (40) track records by the 100th journal issue. Not surprisingly, as the importance of track record in choosing manuscripts for publication increases, winners of the first round of the manuscript competition are more likely to win subsequent rounds. Indeed, once the weight given track record reaches about 0.7, almost 100% of published second and subsequent journal-issue authors are the authors published in the first round. Only imperfect correlations between talent and manuscript quality, and between true and judged quality, allow new authors to publish after the first issue – until the weight of reputation vitiates the effects of these imperfections as well.

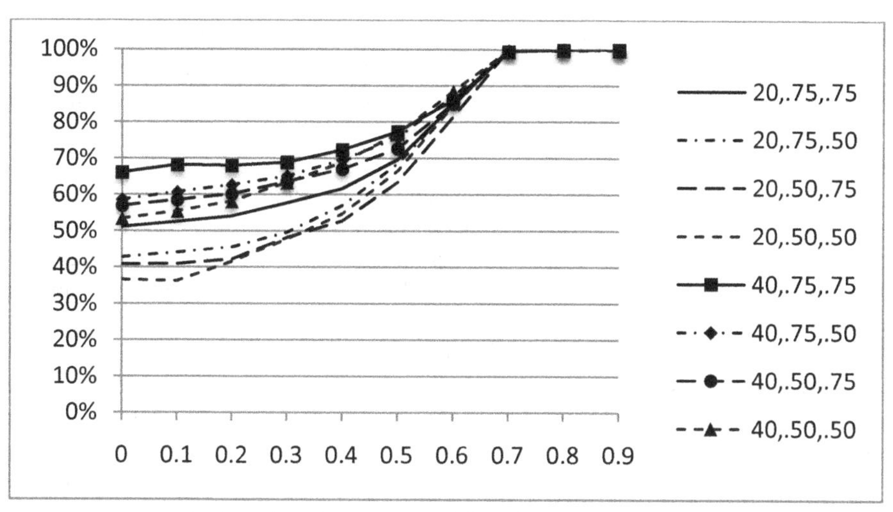

legend: npub,rttq,rtjq (number of articles published, correlation between talent and true manuscript quality, correlation between true and judged manuscript quality)

Fig. 3. Percentage of authors published in the first journal issue who accrued the highest 20 and 40 track records

4 Discussion

Our simulation examined how changing the relative weights given manuscript quality and track record would change the outcomes of competitions for a scarce resource: journal space. The results show that changing the weights has a strong effect on competition outcomes when there are imperfect relationships between the talent of authors and the quality of their manuscripts, and between the true and judged quality of the manuscripts. In particular, as more weight is given to track record, the proportion of highest-quality manuscripts published declines, and the proportion of most-talented authors with the best track records declines. Increasing the weight given to track record also increases the proportion of early-published authors who establish their reputation before others do, eventually creating a "closed shop" of first-published elite who leave no room for later arrivals.

Although our simulation models the editorial task differently than does the simulation of Squazzoni and Gandelli (2012), our general findings are consistent with theirs and with the empirical findings of others (e.g., Bornmann & Daniel, 2007; Bourne & Barbour 2011; Dewett & Denisi 2004; Ferber, 1986). The findings provide additional support for Merton's (1968) original observation of the Matthew Effect. And they prompt an important question: Why use track record at all?

The use of track record for making publication decisions can be justified when the correlations between talent and manuscript quality, and between true and judged quality, approach r = +1.0. At the extreme, if r = +1.0, the most talented authors would always produce the best manuscripts, the reviewers would always judge them as best, and the past would always predict the present. Under these conditions, manuscript quality and track record would be equivalent, but track record would be less time-consuming and more convenient and thus would be preferred. If the correlations were slightly lower, say, r = +0.98, an occasional decision error could occur, rejecting a superior manuscript while publishing an inferior one. Still, the speed and convenience of utilizing track record might outweigh the costs of a rare error.

However, the correlations between talent and manuscript quality, and between true quality and judged quality, have never been shown to approach r = +1.0. At more realistic levels, such as those used in our simulation, the deleterious effects of weighing track record cannot be ignored. With moderate correlations found in empirical studies (typically between r = +0.3 and r = +0.6; see Petty, Fleming & Farbigar, 1999; Thorngate, Dawes & Foddy, 2009; Whitehurst, 1984), for example, our simulation shows that increasing the weight given track record reduces the proportion of the best manuscripts published and of the most-talented authors with the best track records by 10-20%. The resulting errors of omission might not affect the eventual progress of science, but they can slow it down and can limit or end the careers of many talented scientists. Indeed, the errors can discourage more-talented but less-fortunate scientists from continuing their careers, while encouraging less-talented but more-fortunate scientists to continue publishing.

We note, however, the usual caveats. Our simulation selected a small set of factors we believed could influence the quality of articles published and the fate of prospective authors. Dozens of other factors are likely to have additional influences. Such factors include opportunities for rejected authors to resubmit their manuscripts, to find alternative journals, to submit manuscripts jointly authored by colleagues with established track records, or to improve their talent with practice. They also include factors indirectly related to peer review, for example, factors related to tenure, promotion, research funding priorities, and grant success. Interactive effects of these factors are likely to reduce the generality of our results. Conversely, some of our results are likely to moderate these interactive effects.

The results of our simulation suggest that, when less weight is given to track record, a higher proportion of talented authors are published, resulting in a higher proportion of good, published articles and greater number of talented authors with good track records. This leads to the prescription that track record should be given no weight at all. As many have noted, however (see Allesina, 2012), reliance on judgments of manuscript quality does not itself eliminate judgment errors, and the judgments themselves are much more demanding than judgments of track record, reducing the number of qualified reviewers eager to volunteer. The ends and means of peer review are thus in some conflict.

How might the conflict be resolved? Several proposals have been made and innovations tried. New journals, blind reviews, anonymous authorship, reviewer training, additional reviewers, shorter manuscripts, clearer editorial policies and explicit review criteria exemplify popular innovations. Alas, a vastly improved review system has yet to emerge from their use. Newer proposals, including many utilizing Internet technology and simulated with agent-based models, offer more possibilities (see, for example, Allesina, 2012; Grimaldo & Paolucci, 2013; Herron, 2011).

Future innovations, like previous ones, are likely to be limited by the Principle of Invidious Selection (Thorngate, 1988), which states that the merit of manuscripts will always become more homogeneous as authors copy current winning formulas, prompting the use of increasingly arbitrary review criteria. Even so, any innovations that increase reviewer consistency and reduce reviewer effort while improving the chances of publishing good articles and retaining talented authors are worth a try.

References

Allesina, S.: Modelling peer review: An agent-based approach. Ideas in Ecology and Evolution 5 (2012)

APA: Summary report of journal operations (2011), http://www.apa.org/pubs/journals/features/2011-statistics.pdf

Bornmann, L., Daniel, H.D.: Gatekeepers of science – Effect of external reviewers' attributes on the assessment of fellowship applications. Journal of Informetrics 1, 83–91 (2007)

Bourne, P.E., Barbour, V.: Ten simple rules for building and maintaining a scientific reputation. PLoS Computational Biology: a Peer Reviewed Open-Access Journal 7(6) (2011)

Day, A., Peters, J.: Quality Indicators in Academic Publishing. Library Review 43(7), 4–72 (1994)

Dewett, T., Denisi, A.S.: Exploring scholarly reputation: It's more than just productivity. Scientometrics 60(2), 249–272 (2004)

Ferber, M.A.: Citations: Are they an objective measure of scholarly merit? Signs 11(2), 381–389 (1986)

Grimaldo, F., Paolucci, M.: A simulation of disagreement for control of rational cheating in peer review. Advances in Complex Systems (2013), doi:10.1142/S0219525913500045

Herron, D.: Is expert peer review obsolete? Surgical Endoscopy 26, 2275–2280 (2012)

Howard, G.S., Cole, D.A., Maxwell, S.E.: Research productivity in psychology based on publication in the journals of the American Psychological Association. American Psychologist 42(11), 975–986 (1987)

Kaufman, A.: IQ Testing 101. Springer Publishing, New York (2009)

Merton, R.K.: The Matthew Effect in Science. Science 159(3810), 56–63 (1968)

Morgan, D.R., Meier, K.J., Kearney, R.C., Hays, S.W., Birch, H.B.: Reputation and productivity among U. S. public administration and public affairs programs. Public Administration Review 41(6), 666–673 (1981)

Natural Sciences and Engineering Research Council: Guidelines for the preparation and review of applications in engineering and the applied sciences (2012), http://www.nserc-crsng.gc.ca/NSERC-CRSNG/Policies-Politiques/prepEngAS-prepGenSA_eng.asp

Ofori-Dankwa, J., Julian, S.: From thought to theory to school: The role of contextual factors in the evolution of schools of management thought. Organization Studies 26(9), 1307–1329 (2005)

Petty, R., Fleming, M., Fabrigar, L.: The review process at PSPB: Correlates of interreviewer agreement and manuscript acceptance. Personality and Social Psychology Bulletin 25(2), 188–203 (1999)

R Core Team: R: A language and environment for statistical computing. R Foundation for Statistical Computing, Vienna, Austria (2012) ISBN 3-900051-07-0, http://www.R-project.org/

Squazzoni, F., Gandelli, C.: Saint Matthew strikes again: An agent-based model of peer review and he scientific community structure. Journal of Informetrics 6, 265–275 (2012)

Thorngate, W.: On the evolution of adjudicated contests and the principle of invidious selection. Journal of Behavioral Decision Making 1, 5–16 (1988)

Thorngate, W., Dawes, R.M., Foddy, M.: Judging Merit. Psychology Press, Taylor & Francis Group, New York (2009)

Thorngate, W., Hotta, M.: Life and luck: Survival of the fattest. Simulation & Gaming 26, 5–16 (1995)

Thorngate, W., Hotta, M., McClintock, C.: Bingo! The case for cooperation revisited. In: Tolman, C.W., Cherry, F., Van Hezewijk, R., Lubek, I. (eds.) Problems of Theoretical Psychology, Captus Press Inc., Canada (1996)

Thorngate, W., Liu, J., Chowdhury, W.: The Competition for Attention and the Evolution of Science. Journal of Artificial Societies and Social Simulation 14(4) (2011)

Whitehurst, G.J.: Interrater agreement for journal manuscript reviews. American Psychologist 39(1), 22–28 (1984)

Simulating Innovation: Comparing Models of Collective Knowledge, Technological Evolution and Emergent Innovation Networks

Christopher Watts[1] and Nigel Gilbert[2]

[1] Ludwig-Maximilians University, Munich, Germany
c.watts@lmu.de
[2] University of Surrey, Guildford, UK
n.gilbert@surrey.ac.uk

Abstract. Computer simulation models have been proposed as a tool for understanding innovation, including models of organisational learning, technological evolution, knowledge dynamics and the emergence of innovation networks. By representing micro-level interactions they provide insight into the mechanisms by which are generated various stylised facts about innovation phenomena. This paper summarises work carried out as part of the *SIMIAN* project and to be covered in more detail in a forthcoming book. A critical review of existing innovation-related models is performed. Models compared include a model of collective learning in networks [1], a model of technological evolution based around percolation on a grid [2, 3], a model of technological evolution that uses Boolean logic gate designs [4], the SKIN model [5], a model of emergent innovation networks [6], and the hypercycles model of economic production [7]. The models are compared for the ways they represent knowledge and/or technologies, how novelty enters the system, the degree to which they represent open-ended systems, their use of networks, landscapes and other pre-defined structures, and the patterns that emerge from their operations, including networks and scale-free frequency distributions. Suggestions are then made as to what features future innovation models might contain.

Keywords: Innovation; Novelty; Technological Evolution; Networks.

1 Introduction

Simulation models of innovation, including organisational learning, knowledge dynamics, technological evolution and the emergence of innovation networks may provide explanations for stylised facts found in the literatures in innovation, science and technology studies. As computer simulation models of social systems, they can provide one with something one cannot easily obtain from the system itself [8]. They offer a third approach to research, combining the ethnographer's interest in complex contexts and causal relations with the quantitative data analyst's interest in large-scale patterns. They can represent rigorously in computer code the micro-level interactions

B. Kamiński and G. Koloch (eds.), *Advances in Social Simulation*,
Advances in Intelligent Systems and Computing 229,
DOI: 10.1007/978-3-642-39829-2_17, © Springer-Verlag Berlin Heidelberg 2014

of multiple, heterogeneous parts, then demonstrate the consequences of these interactions and the circumstances in which they occur, including the emergence of macro-level patterns.

There are a number of stylised facts inviting explanation of which some of the most relevant to simulation models of innovation follow. Firstly, innovation can be progressive. New ideas and technologies solve problems, create new capabilities and render obsolete and replace old ideas and technologies. A second stylised fact may be found in the rate of quantitative innovation, that is, the rate at which a type of item becomes better, faster, cheaper, lighter etc. Perhaps the best known example of this is Moore's Law, which holds that the number of circuits that can be fitted on a chip increases exponentially over time, but examples of exponential growth rates exist for many other technologies, such as land and air transport, and the particular growth rates also appear to have grown exponentially over time since 1840 [9, 10]. Thirdly, there is the rate of qualitative innovation, that is, the rate at which qualitatively new types of good or service appear. One rough illustration of this is that humans 10000 years ago had a few hundred types of good available to them, while today in a US city there are barcodes for 10^10 types of goods [11]. Various stylised facts exist for the frequency distribution of innovation size, where measures of size include the economic returns from innovation [12, 13] and the number of citations received by a particular patent [14]. Given Schumpeter's famous description of the "perennial gale of creative destruction" [15], there is also interest in the size of the web of interdependent technologies and services blown away (rendered obsolete and uncompetitive, thereafter becoming extinct) by the emergence of a particular innovation. The innovation literature typically distinguishes between incremental and radical innovations [16], the former meaning a minor improvement in an existing technological approach, and the latter a switch to a new approach. In addition, it may be recognised that technologies' components are grouped into modules. This leads to the concepts of architectural and modular innovations [17], the former meaning a rearrangement of existing modules, while the latter means a change of a single module. Finally, emergent structures should be mentioned. Networks of firms, suppliers and customers emerge to create and make use of particular interlinked technologies. If empirical studies of such networks can identify regular structural features, then simulation models can investigate the circumstances under which these structures emerge.

This paper summarises some of the main points from a critical survey of several models of organisational learning, knowledge dynamics, technological evolution and innovation networks, undertaken as part of the ESRC-funded SIMIAN project (www.simian.ac.uk) and described in detail in a forthcoming book by the authors. Particular areas for model comparison include the ways these models represent knowledge and/or technologies, how novelty enters the system, the degree to which the models represent open-ended systems, the models' use of networks, landscapes and other pre-defined structures, and the patterns that emerge from the models' operations, primarily network structures and frequency distributions. In addition, based on our experiences with these models and some recent literature, suggestions are made about the form and features that future innovation models might contain.

2 The Innovation Models

Simulation models of innovation focus on the production, diffusion and impact of novel ideas, beliefs, technologies, practices, theories, and solutions to problems. Simulation models, especially agent-based models, are able to represent multiple producers, multiple users, multiple innovations and multiple types of interdependency between all of these, leading to some hard-to-predict dynamics. In the case of innovations, some innovations may form the components of further innovations, or they may by their emergence and diffusion alter the functionality and desirability of other innovations. All of the models surveyed below have these aspects.

Space permits only a few models to be surveyed here. Further models of technological evolution, with several points of similarity to the ones included here, may be found in Lane [18]. Also related are science models [19], models of organisational learning, for example March [20], models of strategic decision making, for example Rivkin and Siggelkow [21], and models of language evolution [22, 23]. Treatments of some of these areas can also be found in Watts and Gilbert [24, chapters 7, 5 and 4].

Space also does not permit more than brief indications of the functionality of the models in this survey. For more details the reader is directed to the original source papers and to Watts and Gilbert [24, chapter 7]. The brief descriptions of the models now follow.

Lazer and Friedman [1] (hereafter L&F) simulate an organisation as a network of agents attempting to solve a common complex problem. Each agent has a set of beliefs, represented by a bit string, that encode that agent's solution. Solutions are evaluated using Kauffman's NK fitness landscape definition [25], a moderately "rugged" landscape problem with N=20 and K=5. Agents attempt seek better solutions through the use of two heuristic search methods: learning from others (copying some of the best solution among the agent's neighbours), and trial-and-error experimentation (trying a solution different from your current solution by mutating one bit). The eventual outcome of searching is that the population converges on a common solution, usually a better solution than any present among the agents initially, and ideally one close to the global optimum for that fitness landscape.

Silverberg and Verspagen [3] (S&V) simulate technological evolution using nodes in a grid lattice to represent interlinked technologies, and percolation up the grid to represent technological progress. Technologies can be in one of four states: impossible, possible but yet-to-be-discovered, discovered but yet-to-be-made-viable, and viable. At initialisation, technologies are set with a fixed chance, to be possible or impossible. Technologies in the first row of the grid are then set to be viable. The best-practice frontier (BPF) is defined as the highest viable technologies in each column. Each time step, from each technology in the BPF, R&D search effort is made over technologies within a fixed radius. As a result of search some possible technologies within the radius may, with a chance dependent on the amount of effort divided by the number of technologies in the radius, become discovered. Any discovered technologies adjacent to viable technologies become themselves viable. Innovations are defined as any increases in the height of the BPF in one column. Innovation size is defined as the size of the increase. Since technologies may become viable because of horizontal links as well as vertical ones, it is possible for quite large jumps in the BPF, whenever progress in one column has obstructed by an impossible technology while progress continues in other columns from search radiuses can cover the column with the obstruction. The frequency distribution of these innovation sizes is recorded and plotted. For some values of the parameter search radius, this frequency distribution tends towards a scale-free distribution. In their basic [3] model, Silverberg and Verspagen represent the same amount of search as occurring from every column in the grid. Silverberg and Verspagen [2] extend this model with search agent firms who can change column in response to recent progress. The firms' adaptive behaviour has the effect of generating the scale-free distribution of innovation sizes without the need for the modeller to choose a particular value of the search radius parameter, and thus the system represents self-organised criticality [26].

Arthur and Polak [4] (A&P) also simulate technological evolution. Their technologies have a real-world meaning: they are designs for Boolean logic gates, made up of combinations of component technologies, beginning from a base technology, the NAND gate. Each time step a new combination of existing technologies is created and evaluated for how well it generates one of a fixed list of desired logic functions. If it replicates desired functions satisfied by a previously created technology, and is less expensive, where cost is defined as the number of component instances of the base technology, NAND, then the new technology replaces in memory the technology with the equivalent function and higher cost. The replaced technology may have been used as a component technology in the construction of other technologies, in which case it is replaced in them as well. The total number of replacements resulting from the newly created technology is its innovation size. As with the previous model, A&P find example parameter settings in which the frequency distribution of innovation sizes tends towards being scale-free.

The model for Simulating Knowledge dynamics in Innovation Networks (SKIN) [5, 27, 28] simulates a dynamic population of firms. Each firm possesses a set of units of knowledge, called kenes, and a strategy, called an innovation hypothesis (IH), for combining several kenes to make a product. Input kenes not possessed by the firm must be sourced from a market supplied by the other firms. Each kene is a triple of numbers, representing a capability, an ability and expertise. Products are created as a normalised sum-product of capabilities and abilities of the kenes in the IH, and given a level of quality based on a sum-product of the same kenes' abilities and expertise. Firms lacking a market for their products can perform incremental research to adjust their abilities, radical research to swap a kene in their IH for another kene, or enter an alliance or partnership with another firm to access that firm's kenes. Expertise scores increase in kenes when they are used, but decrease when not in use, and kenes with 0 expertise are forgotten by the firm. Partners are chosen based on past experience of partnership, customer and supplier relations, and the degree of similarity in kenes to the choosing firm. Regular partners may unite to form an innovation network, which then can create extra products in addition to those produced by its members. Products on the markets have dynamic prices reflecting recent supply and demand, research has costs, and firms' behaviour reflects their wealth.

The model of emergent innovation networks of Cowan, Jonard and Zimmermann [6] (CJZ) also simulates a population of firms with knowledge resources. Each firm's knowledge is a vector of continuous variables, representing several dimensions of knowledge. Pairs of firms can collaborate to create new amounts of knowledge, with the amount computed using a constant-elasticity-of-substitution (CES) production function. Each input to this function is a weighted sum of the minimum and maximum values in the corresponding dimension of the collaborating firms' knowledge vectors, the idea here being that if knowledge in each dimension is largely independent of the other dimensions, knowledge is decomposable into subtasks, and firms will be able to choose the best knowledge value for each subtask, but with interdependent knowledge dimensions, both firms may be held back by the weakest firm. If collaboration is by chance successful, the amount output from the production function will be added to one of the variables in a participant's knowledge vector. Evaluating potential collaboration partners is based on experience of recent success, including the evaluating firm's direct experience of the candidate partner (relational credit), and also the evaluator's indirect experience obtained from its other recent collaborators (structural credit). Once all firms have evaluated each other, a set of partnerships is formed using the algorithm for the roommate matching problem. Data on partnerships can be used to draw an innovation network of firms. The structural properties of this network can then be related to the main parameters, the weighting between collaborating firms' minimum and maximum knowledge inputs (representing the decomposability of knowledge) and the weighting between relational and structural credit.

Padgett's hypercycles model of economic production [7, 29, 30] draws upon the ideas from theoretical biology of hypercycles and auto-catalysis [25]. It simulates a population of firms engaged in the transformation and transfer of products. Each firm begins with randomly chosen production skills, called production rules. Inspired by Fontana's algorithmic chemistry these are of the form: given a product of type x, transform it into an output of type y. Each time step a new production run attempt is simulated. A product of a random type is drawn from a common environment by one randomly chosen firm and transformed using one of that firm's rules. The output product from transformation is then transferred to a randomly chosen neighbour in a grid network of firms. If a firm lacks a rule suitable for transforming the product it has received, then the product is dumped into the environment and the production run ends. Otherwise, the firm uses a compatible rule to transform it and transfers the output to one of its neighbours, and the production run continues. In addition to processes of transformation and transfer, firms learn by doing. Whenever two firms in succession in the production run have compatible rules to transform products, one of the firms increases its stock of instances of the rule it has just used. Meanwhile, under a process of rule decay, somewhere in the population of firms a randomly chosen rule is forgotten. Firms that forget all their rules exit the system, leaving gaps in the network. The effect of these four processes (product transformation and transferral, and rule learning by doing and rule decay) is that under various parameter settings a self-maintaining system of firms and rules can emerge over time through self-organisation. This system depends upon there being hypercycles of rules, in which constituent rules are all supplied by other constituent rules.

The models described are all capable of demonstrating that the emergence of some system-level pattern (e.g. convergence on a peak solution, scale-free distributions in change sizes, collaboration networks, self-maintaining systems) is sensitive to various input parameters and structures controlling micro-level behaviour (e.g. initial network structure, search and learning behaviour, knowledge structure).

3 Points of Comparison

The models are now compared for the ways they represent knowledge and/or technologies, how novelty enters the system, the degree to which they represent open-ended systems, their use of networks, landscapes and other pre-defined structures, and the patterns that emerge from their operations, including networks and scale-free frequency distributions. A summary is given in Table 1.

As may be clear from the above descriptions, the models differ widely in their representation of knowledge and technologies: there were bit strings (L&F), nodes in a grid (S&V), lists of components (A&P), kenes (SKIN model), vectors of continuous variables (CJZ) and algorithmic chemistry rules (hypercycles model). These were evaluated using NK fitness (L&F), connection to the base row and height of row (S&V), a list of desired logic functions and cost in terms of number of base components (A&P), and in terms of their ability to take input from and supply output to other model components (SKIN, hypercycles). It seems that later models tended to be better than earlier ones, and represent gains in knowledge.

How novelty enters the system varies in line with how knowledge or technologies are represented. L&F's use of heuristic search methods reflects the idea that there are two sources of novelty: new combination of existing parts, and mutation, during copying or experimentation. This view of novelty in innovation is found not only among Herbert Simon's disciples [31], but is also common among Schumpeterian evolutionary economists [32, 33]. The A&P model recombines existing parts when constructing new logic gate designs. The SKIN model's incremental and radical research processes have effects analogous to mutation of knowledge, while recombination of parts is made through alliance partnerships. The hypercycles model sees a self-maintaining system emerge via a process analogous to ant colony optimisation, as new production runs are constructed out of firms and rules. But novel choices of rule for transforming or route for transferring products can only be made between the rules and firms still present in the model. Once rules have been forgotten and firms have left the network, they do not return. So in the present version of the model there is no equivalent of a mutation process to reintroduce novel firms and rules. The CJZ model is based on the idea that innovation stems from collaborative use of existing knowledge resources, but it does not specify a mechanism for this. The constant-elasticity-of-substitution production function is a description of the pattern of innovation rather than its explanation. The S&V model also omits an explanation of how a new innovation has been made, beyond the simple concept of "search" from a position at the current best-practice frontier. The regular two-dimensional grid in which technologies lead on to other technologies is relatively simple. In contrast, the relationships between combinations of bits in the L&F model and between complex technologies in the A&P model form much more complex solution spaces / technology spaces.

Another view of novelty is as reinterpretation. In several of these models ideas and technologies can acquire new functions or values due to the appearance or disappearance of ideas and technologies. For example, a kene in the SKIN model can form an input in many different innovation hypotheses, some of which may only appear later in the simulation run. In the A&P model the discovery of a new node can connect other technology nodes and shorten paths to the best-practice frontier. In the hypercycles model the disappearance of a rule or firm alters the amount of production work for other rules and firms.

The hypercycles model simulates the emergence of novel, identifiable self-maintaining structures – a new type of thing with some significant degree of permanence or stability. Thus, whereas some models represent the application of heuristic search methods to combinations of pre-existing objects, models of emergent structures, such as hypercycles and auto-catalytic sets, have the potential to explain how there come to be objects in the first place.

The models vary in the degree to which they represent open-ended systems. L&F's model is relatively closed. Agents respond to their neighbours, while they differ in beliefs, and to their fitness evaluations. The landscape is static and fixed at the beginning, and does not respond to any events either inside or outside the model. Once a population has converged on a single peak in the fitness landscape, the model's dynamics come to an end. Change in A&P's model could also come to an end, since the list of desired functions is static, but in practice the list used by A&P is sufficiently long for the phenomenon of technology extinctions to continue throughout the length of their simulation runs. The models in both S&V and CJZ can permit indefinite extensions / additions of innovation, but S&V's percolation grid does not gain columns and CJZ's knowledge vector does not gain dimensions, so in both cases new qualities do not emerge. The SKIN model contains a limit on the number of products, but in practice this limit need not restrict the dynamics of the model. Firms and their kenes come and go throughout a simulation run. The hypercycles model reaches either a dead state in which the firm-rule network becomes too fragmented and no more learning by doing can occur, or it forms a self-maintaining system of firm-rule hypercycles. With no processes to introduce new firms or rules to the system, dead systems cannot be resurrected and live systems are unable to change much except by a particularly unlucky sequence of rule decay events. (Padgett and Powell [34] however contains proposals for future extensions to the basic hypercycles model that will make it more open-ended.)

Nearly all the models assume pre-defined structures. L&F assume a fitness landscape, for which they use Kauffman's NK fitness. S&V assume a network structure for their percolation grid technology space. A&P assume their list of desired functions. The hypercycles model assumes the initial structure for the network of firms, the initial allocation of rules to firms, and the various types of rule themselves (the "chemistry"). For this latter assumption Padgett's studies so far have restricted themselves to a simple cycle of rules, each rule having one input and one output product,

Table 1. Summary of model comparison

Model & key references	Unit of knowledge or technology	Unit representation	Value: pre-defined or endogenous	Sources of novelty	Emergent patterns
L&F [1]	Workers' beliefs / routines	Bit strings	NK fitness landscape	Recombination; Mutation	Convergence on peak / optimum
S&V [2, 3]	Technologies	Nodes in a grid network	Height & connection to baseline	Chance discovery from nearby nodes	Scale-free distribution of advances of connected nodes
A&P [4]	Technologies	Systems of NAND logic gates	List of desired logic functions	Recombination	Scale-free distribution of obsolescence sizes
SKIN [5, 27, 28]	Firms' "kenes": capabilities, abilities, expertise	Ordered triples of integers	Endogenous demand for inputs	Recombination of kenes; new abilities	Social networks; Scale-free distribution of production network sizes
CJZ [6]	Firms' knowledge	Vector of continuous variables	Endogenous demand for collaboration partners	Not specified	Social networks
Hypercycles [7, 29, 34]	Firms' production skills (rules)	Algorithmic chemistry	Role in hypercycles system	Emergence of self-maintaining system	Self-maintaining system of firms & rules

and each rule producing the input to the next rule in the cycle, with the final rule producing the input to the first rule. This is clearly a very abstract view of industrial production relations. The SKIN model uses modulo arithmetic (that is, its normalised sum-products) to control the relationships between kenes. This is simple to specify and compute, but remote from real-world applications. The CJZ model relies on its decomposability parameter and its CES production function to control innovation production, and no further description of the underlying knowledge or technological world is needed, but whether knowledge resembles constant elasticity of substitution remains to be demonstrated empirically.

Several models simulate the emergence of a scale-free frequency distribution representing the size of innovations (S&V, A&P and SKIN). In the first two cases innovation size is defined in terms of the number of technologies rendered obsolete or becoming extinct. In the case of SKIN the distribution applies not to the size of new kenes but to the size of new production networks [27]. L&F and A&P simulate systems that show progressive improvements, but with diminishing returns over time, as a peak or optimum is reached in the L&F model and all of A&P's list of desired functions become satisfied. SKIN, CJZ and the hypercycles model simulate the emergence of networks of firms. Only CJZ specifically analyse the structure of these in their paper, though the other models could certainly support analysis. But all three sets of authors are interested in discovering the sensitivity of this emergence to various parameter settings.

4 Towards Future Models of Innovation

Future models of innovation are likely to continue the concepts of innovation as heuristic search and recombination of parts, and to continue to describe the emergent networks and frequency distributions. As more data on industrial eco-systems, innovation networks and other networks of firms become available, the structures of these can be compared with the output from the models. Likewise, empirical frequency distributions can be analysed to find out to what extent they tend towards being scale-free, perhaps log-normal in shape, and with what parameters. Although some of the above models could generate scale-free distributions of change sizes, familiar from the theory of self-organised criticality [26], it is not yet known whether real-world creative destruction follows this distribution of sizes. Stylised facts concerning geometric growth in quality (better, faster, cheaper etc. – i.e. quantitative innovation) and the immense growth in number of types of goods and services (qualitative innovation) were not explained by the above models. Future models might address these and also the scaling laws relating innovation rates (number of patents per capita, number of entrepreneurs per capita) to city population sizes recently highlighted by Bettencourt, Lobo, Helbing, Kuhnert and West [35]. The chapters and models in Lane [18] represent a first step towards meeting this latter end.

Empirical studies should also inform the inputs and design of these models. For examples, fitness landscapes, the artificial chemistry of rules in the hypercycles model, and the initial firm networks in L&F and the hypercycles models play important roles in the models' behaviour, but abstract or arbitrary choices are currently used for these inputs. Some of the models made a distinction between incremental and radical innovation processes, as is common in innovation literature, but the further identification of modular and architectural innovations [17] was missing. This will require some thought as to how technology spaces and fitness landscapes should be structured. In addition, agent learning behaviour and market mechanisms could also be grounded in real-world cases.

None of the models translated their outputs into familiar economic terms beyond the SKIN model's use of market pricing. To provide insights into the role played by innovation in a whole economy or society, the models surveyed here need more connection between the results of innovation on the one hand, and the future ability of agents to innovate on the other. Knowledge and technologies serve purposes, while both human agents and firms have needs. Failure to meet their basic needs (food, capital) will make it harder to continue to engage in practices such as experimentation, learning from others and trading, thus slowing down the rate at which innovations are generated and diffuse. So far, innovation models have omitted the needs of their human components.

A key feature of technologies and practices is their ability to support more than one attribution of functionality. Technologies and routines developed for one purpose may be reinterpreted and developed to serve a new purpose, one not foreseen during earlier development and present seemingly only by accident. This is the process known in biology as exaptation. Villani, Bonacini, Ferrari, Serra and Lane [36] hold exaptive bootstrapping to be responsible for the continual growth in the number of types of thing, and present an early attempt at a model in which agents' cognitive attributions of functionality to artefacts enables the representation of this process. Several of the models surveyed above can simulate an innovation in the function or value of one thing, due to the appearance or disappearance of another, but they do not distinguish these changes in actual functionality from changes in agents' awareness of functionality. While the models are good at representing innovation as recombination, they are less clear at representing innovation as reinterpretation.

Introducing cognition raises the roles of imagination, analogy and metaphor in generating innovations. The range of logically possible combinations of all our current technologies is vast, and most of them seem nonsensical or unviable. We only have time to try out a tiny fraction of the combinations, so how do we manage to find any useful ones? Models that treat component technologies or beliefs as indivisible base units will lack a basis for their agents to anticipate how well two never-before-combined base technologies are likely to fare when combined for the first time. Gavetti, Levinthal and Rivkin [37] simulate the use of analogical reasoning in generating new strategies. Attention to reasoning processes and the circumstances in which they work well will be issues for modellers of other forms of innovation as well.

Acknowledgements. We acknowledge the support of the SIMIAN project (www.simian.ac.uk) funded by the UK Economic and Social Research Council's National Centre for Research Methods. A more detailed version will be included in a forthcoming book [24].

References

1. Lazer, D., Friedman, A.: The network structure of exploration and exploitation. Adm. Sci. Q 52, 667–694 (2007)
2. Silverberg, G., Verspagen, B.: Self-organization of R&D search in complex technology spaces. J. Econ. Interac. Coord. 2, 195–210 (2007)

3. Silverberg, G., Verspagen, B.: A percolation model of innovation in complex technology spaces. Journal of Economic Dynamics & Control 29, 225–244 (2005)
4. Arthur, W.B., Polak, W.: The evolution of technology within a simple computer model. Complexity 11, 23–31 (2006)
5. Ahrweiler, P., Pyka, A., Gilbert, N.: Simulating Knowledge Dynamics in Innovation Networks (SKIN). In: Leombruni, R., Richiardi, M. (eds.) Industry and Labor Dynamics: The Agent-Based Computational Economics Approach: Proceedings of the Wild@ace2003 Workshop, Torino, Italy, October 3-4, pp. 284–296. World Scientific, Singapore (2004)
6. Cowan, R., Jonard, N., Zimmermann, J.B.: Bilateral collaboration and the emergence of innovation networks. Management Science 53, 1051–1067 (2007)
7. Padgett, J.F., Lee, D., Collier, N.: Economic production as chemistry. Industrial and Corporate Change 12, 843–877 (2003)
8. Ahrweiler, P., Gilbert, N.: Caffe Nero: the evaluation of social simulation. Jasss-the Journal of Artificial Societies and Social Simulation 8 (2005)
9. Lienhard, J.H.: How invention begins: echoes of old voices in the rise of new machines. Oxford University Press, New York (2006)
10. Lienhard, J.H.: Some ideas about growth and quality in technology. Technological Forecasting and Social Change 27, 265–281 (1985)
11. Beinhocker, E.D.: The origin of wealth: evolution, complexity, and the radical remaking of economics. Random House Business, London (2007)
12. Scherer, F.M.: The size distribution of profits from innovation. Kluwer Academic Publishers, Norwell (2000)
13. Harhoff, D., Narin, F., Scherer, F.M., Vopel, K.: Citation frequency and the value of patented inventions. Review of Economics and Statistics 81, 511–515 (1999)
14. Trajtenberg, M.: A penny for your quotes: patent citations and the value of innovations. RAND Journal of Economics 21, 172–187 (1990)
15. Schumpeter, J.A.: Capitalism, socialism, and democracy. G. Allen & Unwin ltd., London (1943)
16. Abernathy, W.J.: The productivity dilemma: roadblock to innovation in the automobile industry. Johns Hopkins University Press, Baltimore (1978)
17. Henderson, R.M., Clark, K.B.: Architectural Innovation: The Reconfiguration of Existing Product Technologies and the Failure of Established Firms. Adm. Sci. Q 35, 9–30 (1990)
18. Lane, D.A.: Complexity perspectives in innovation and social change. Springer, Dordrecht (2009)
19. Edmonds, B., Gilbert, N., Ahrweiler, P., Scharnhorst, A.: Simulating the Social Processes of Science. Jasss-the Journal of Artificial Societies and Social Simulation 14 (2011)
20. March, J.G.: Exploration and exploitation in organizational learning. Organization Science 2, 71–87 (1991)
21. Rivkin, J.W., Siggelkow, N.: Balancing search and stability: Interdependencies among elements of organizational design. Management Science 49, 290–311 (2003)
22. Steels, L.: A self-organizing spatial vocabulary. Artificial Life 2, 319–332 (1996)
23. Steels, L.: Grounding symbols through evolutionary language games. Simulating the Evolution of Language, 211–226 (2002)
24. Watts, C., Gilbert, N.: Simulating innovation. Edward Elgar Publishing, Cheltenham (forthcoming)
25. Kauffman, S.A.: At home in the universe: the search for laws of self-organization and complexity. Penguin, London (1996)
26. Bak, P.: How nature works: the science of self-organized criticality. Oxford University Press, Oxford (1997)

27. Gilbert, N., Ahrweiler, P., Pyka, A.: Learning in innovation networks: Some simulation experiments. Physica A 378, 100–109 (2007)
28. Pyka, A., Gilbert, N., Ahrweiler, P.: Simulating knowledge-generation and distribution processes in innovation collaborations and networks. Cybernetics and Systems 38, 667–693 (2007)
29. Padgett, J.F.: The emergence of simple ecologies of skill: a hypercycle approach to economic organization. In: Arthur, W.B., Durlauf, S.N., Lane, D.A. (eds.) The Economy as an Evolving Complex System II, pp. xii, 583p. Advanced Book Program/Perseus Books, Reading (1997)
30. Padgett, J.F., McMahan, P., Zhong, X.: Economic Production as Chemistry II. In: Padgett, J.F., Powell, W.W. (eds.) The Emergence of Organizations and Markets, pp. 70–91. Princeton University Press, Princeton (2012)
31. Cyert, R.M., March, J.G., Clarkson, G.P.E.: A behavioral theory of the firm, Englewood Cliffs, N.J (1964)
32. Nelson, R.R., Winter, S.G.: An evolutionary theory of economic change. Belknap Press, Cambridge (1982)
33. Becker, M.C., Knudsen, T., March, J.G.: Schumpeter, Winter, and the sources of novelty. Industrial & Corporate Change 15, 353–371 (2006)
34. Padgett, J.F., Powell, W.W.: The emergence of organizations and markets. Princeton University Press, Princeton (2012)
35. Bettencourt, L.M.A., Lobo, J., Helbing, D., Kuhnert, C., West, G.B.: Growth, innovation, scaling, and the pace of life in cities. Proc. Natl. Acad. Sci. U. S. A. 104, 7301–7306 (2007)
36. Villani, M., Bonacini, S., Ferrari, D., Serra, R., Lane, D.: An agent-based model of exaptive processes. European Management Review 4, 141–151 (2007)
37. Gavetti, G., Levinthal, D.A., Rivkin, J.W.: Strategy making in novel and complex worlds: The power of analogy. Strategic Management Journal 26, 691–712 (2005)

Grand Canonical Minority Game as a Sign Predictor

Karol Wawrzyniak and Wojciech Wiślicki

National Centre for Nuclear Research
Hoża 69, 00-681 Warszawa, Poland
{kwawrzyn,wislicki}@fuw.edu.pl
http://agf.statsolutions.eu

Abstract. In this paper the extended model of Minority game (MG), incorporating variable number of agents and therefore called Grand Canonical, is used for prediction. We proved that the best MG-based predictor is constituted by a tremendously degenerated system, when only one agent is involved. The prediction is the most efficient if the agent is equipped with all strategies from the Full Strategy Space. Despite the casual simplicity of the method its usefulness is invaluable in many cases including real problems. The significant power of the method lies in its ability to fast adaptation if λ-GCMG modification is used. The success rate of prediction is sensitive to the properly set memory length. We considered the feasibility of prediction for the Minority and Majority games.

Keywords: Minority Game as a predictor, Financial Markets.

1 Introduction

The minority decision is defined as a function of a self-generated signal called aggregate attendance or aggregate demand [4] (Sec. 2). The MG can be potentially used as the predictor of any exogenous (fake) series, provided that dependencies in the signal reflect the patterns built in strategies[11,8,9]. The details of the current state of the art are presented further in Sec. 3. In Sec. 4 we presented the model and its configuration. Then, in section 5, we verified the quality of the predictor using the time series generated by the well understood, autoregressive stochastic process. After analyzing our numerical results, we provided a methodology for tuning the parameters. Intriguingly, the best results are achieved if the game is degenerated to only one single agent equipped with all strategies from the whole strategy space. This new discovery seems to stay in contradiction to commonly used optimization techniques [11,8,9]. Additionally, in Sec. 5 we presented some new insights which allow us to improve the model. For example, it was proved that if the exogenous signal is exploited, then there is no qualitative difference between minority and majority game. We also introduced a modification, the so called λ-GCMG, that is well suited for quasti-stationary signals. Finally, in the Sec. 5, the properly tuned MG model was applied as a forecaster of assets prices on financial markets.

B. Kamiński and G.Koloch (eds.), *Advances in Social Simulation*,
Advances in Intelligent Systems and Computing 229,
DOI: 10.1007/978-3-642-39829-2_18, © Springer-Verlag Berlin Heidelberg 2014

2 The Formal Definition of the Minority Game

At each time step t, the nth agent out of N $(n = 1, \dots, N)$ takes an action $a_{\alpha_n}(t)$ according to some strategy $\alpha_n(t)$. The action $a_{\alpha_n}(t)$ takes either of two values: -1 or $+1$. An aggregated demand is defined as $A(t) = \sum_{n=1}^{N} a_{\alpha'_n}(t)$, where α'_n refers to the action according to the best strategy, as defined in eq. (1) below. Agents do not know each other's actions but $A(t)$ is known to all agents. The minority action $a^*(t)$ is determined from $A(t)$: $a^*(t) = -\mathrm{sgn}A(t)$. Each agent's memory is limited to m most recent winning, i.e. minority, decisions. Each agent has the same number $S \geq 2$ of devices, called strategies, used to predict the next minority action $a^*(t+1)$. The sth strategy of the nth agent, α_n^s $(s = 1, \dots, S)$, is a function mapping the sequence μ of the last m winning decisions to this agent's action $a_{\alpha_n^s}$. Since there is $P = 2^m$ possible realizations of μ, there is 2^P possible strategies. At the beginning of the game each agent randomly draws S strategies, according to a given distribution function $\rho(n) : n \to \Delta_n$, where Δ_n is a set consisting of S strategies for the nth agent.

Each strategy α_n^s, belonging to any of sets Δ_n, is given a real-valued function $U_{\alpha_n^s}$ which quantifies the utility of the strategy: the more preferable strategy, the higher utility it has. Strategies with higher utilities are more likely chosen by agents. There are various choice policies. In the popular *greedy policy* each agent selects the strategy of the highest utility

$$\alpha'_n(t) = \arg \max_{s:\, \alpha_n^s \in \Delta_n} U_{\alpha_n^s}(t). \tag{1}$$

If there are two or more strategies with the highest utility then one of them is chosen randomly. Each strategy α_n^s is given the *payoff* depending on its action $a_{\alpha_n^s}$

$$\Phi_{\alpha_n^s}(t) = -a_{\alpha_n^s}(t)\, g[A(t)], \tag{2}$$

where g is an odd *payoff function*, e.g. the steplike $g(x) = \mathrm{sgn}(x)$ [5], proportional $g(x) = x$ or scaled proportional $g(x) = x/N$. The learning process corresponds to updating the utility for each strategy

$$U_{\alpha_n^s}(t+1) = U_{\alpha_n^s}(t) + \Phi_{\alpha_n^s}(t), \tag{3}$$

such that every agent knows how good its strategies are.

The presented definition is related to genuine MG [4]. If game is used as the predictor then the feedback effect is destroyed and μ is derived from using exogenous time series.

3 Relation to Other Models

As it is known from other works [4,2] the standard MG exhibits an intriguing phenomenological feature: a non-monotonic variation of the volatility when the control parameter is varied. There are two mechanisms potentially responsible for it: the feedback effect and the quenched disorder [4]. The incorporated feedback effect couples input and output signals in such a way that a minority decision at time t constitutes the basis for future agents' decisions. The quenched disorder is related to an initial, random realization of agents' strategies in their strategy space. In theoretical papers [3] it is discussed how the feedback mechanism affects observed behavior of MGs. For us it is important that the lack of the feedback does not influence the population's predictive power which is exclusively driven by the quenched disorder.

Some other authors applied the model to the exogenous, real data, assuming an existence of patterns in these data and using MG as a predictor of its future value [11,10,8,9,7,12]. Although the MG-based predictor is able to forecast any time series assuming that the length of patterns suits agents' strategies, the commonly used exogenous time series are those related to asset prices. In [11,10] the authors performed an experiment where the time series of hourly Dollar $/Yen ¥ exchange rate was examined. Although the results are interesting and inspiring, there is nearly no details about the conditions of the experiment. Such parameters like m, S, N are not revealed, making the results unreproducible.

The prediction method used in Ref. [10] was further developed by others [8,9,7,12]. Above methods of optimization are based on comparison between two distributions of signals, i.e. the exogenous and predicted one. If the distributions are mutually close to each other, the model is considered as a well fitted to the object. As we presented in section 5, this technique, although interesting, does not assure that the success rate of the one-step prediction is maximized.

4 The Model

Here, we present the details about our implemented predictor and its configuration. We used, the Grand Canonical extension of the MG in all our simulations.

4.1 Grand Canonical Extension

In the standard MG all agents have to play at each time step t, even if all of their strategies are unprofitable. Looking for analogies to financial markets we see that in real life investors behave differently. If for some of them trading is not profitable, they withdraw from the market. Formally, staying apart from the market is realized by *zero strategy*. This additional strategy, marked as α_i^0, maps all μ to the $a_i = 0$ and does not influence the aggregate demand A. We assume a constant risk-free interest rate as being equal to $U_i^{\alpha^0}(t) = U_i^{\alpha^0} = 0$.

4.2 Configuration

Technically, the predictor works according to the diagram presented in Fig. 1. The object that we suppose to model is treated as a black-box stimulated by (i) the vector of previously generated signs of samples $\mathrm{sgn}\big(\overrightarrow{y}(t-1)\big) = [\mathrm{sgn}\big(y(t-1)\big) \ldots \mathrm{sgn}\big(y(t-n)\big)]$, where $n > 1$, and (ii) the external information $\xi(t)$. We

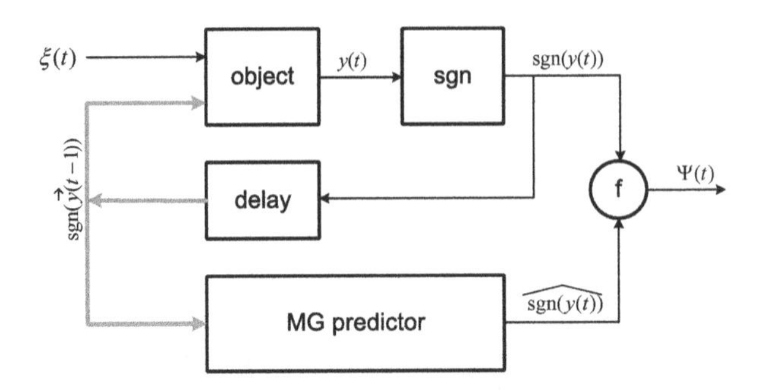

Fig. 1. Block scheme presents the MG in the configuration of prediction

assume that samples of ξ are Independent and Identically Distributed (IID). The model is supposed to retrieve dependencies between the past and future values of $\mathrm{sgn}(y(t))$. The delay block introduces a one-step delay to its every input sample $y(t)$ and forms the vector $\boldsymbol{y}(t-n)$ of $n > 1$ past samples. The MG model predicts the next sign of sample exploiting the information included in the signs of previous movements. The block f compares the predicted signal with a real output of the object and calculates the *correctness* $\Psi(t)$. The correctness $\Psi(t)$ represents the average success rate of prediction and is calculated as a percentage of properly predicted signs of y, provided that all samples up to time t are considered:

$$\Psi(t) = \frac{1}{t} \sum_{n=1}^{t} \delta(\mathrm{sgn}(y(n)), \widehat{\mathrm{sgn}(y(n))}) \tag{4}$$

where δ stands for Kronecker symbol.

Two types of objects were analyzed: the autoregressive stochastic processes and the time series of real prices of shares. The former is mainly used to demonstrate interesting properties of the model and to learn how to tune its parameters. The latter one is used as an example of practical application of the predictor. In the case of the autoregressive stochastic process, assuming that the definition of the object is known, the maximal theoretical level of Ψ can be calculated by placing $\mathbb{E}[y(n)]$ instead of $\widehat{\mathrm{sgn}(y(n))}$ in Eq. 4. The expected value $\mathbb{E}[y(n)]$ is calculated recursively for each n according to the process definition.

5 Optimization of the Parameters

Initially, we apply the predictor to the third order autoregressive time series, AR(3) defined as follows:

$$y(t) = 0.7y(t-1) - 0.5y(t-2) - 0.2y(t-3) + \xi(t), \tag{5}$$

where $\xi(t)$ is an instance of gaussian white noise. It was found numerically that the process is characterized by $\mathbb{E}[\Psi^{MAX}] = 0.77$, and $\mathbb{V}ar[\Psi^{MAX}] < 0.01$, where $t = 3000$ steps, and the average was taken over ten realizations.

5.1 Majority *vs.* Minority Game

The model extensively used in literature is based on the grand canonical minority game [11,10,8,9,7]. However, it was not obvious for us if the minority mechanism is better than the majority one. In fact we found that, in the case of prediction, both mechanisms are equivalent.

The algorithm of the majority game is very similar to that of the standard minority game. The only difference is a formula (2) which for majority game reads

$$\Phi_{\alpha_n^s}(t) = a_{\alpha_n^s}(t)\, g[A(t)]. \tag{6}$$

We consider two time series: the endogenous and exogenous. Considering first the game with endogenous time series we find a number of differences between the minority and majority game. In the minority game no one of strategies is permanently profitable provided that the game is large enough [16]. Hence, the number of winners and losers changes in time. On the contrary, in the majority game the number of winners and losers is stable and, on average, $N(1 - \frac{1}{2^S})$ agents are in majority. The reasoning behind is similar to that presented in Refs. [14,15,16] and utilizes our observation that the first large oscillation creates a comprehensive difference in utilities of strategies.

Intriguingly, both games are equivalent, regardless of the payoff, when series of decisions is exogenous. Originally, in the standard minority game, strategies predicting sign opposite to $y(t)$ are rewarded. Assuming that patterns in the exogenous signal exist, the individuals prefer more often strategies predicting $\operatorname{sgn}(y(t))$ incorrectly i.e. most of them fail with prediction. The predictor aggregates decisions of individuals and acts in opposition to the majority, and, predicts correctly. Contrary to the minority game, in the majority game, strategies that correctly forecast $\operatorname{sgn}(y(t))$ are rewarded and most of agents follow strategies more frequently recognizing patterns. Subsequently, the predictor acts according to action suggesting the majority and also correctly recognizes patterns. Given this, the majority and minority game should provide the same quality of prediction. This is confirmed by numerical simulations presented in Fig. 2 (right). As it is seen, there are no qualitative differences between them. Small distortions are due to random choice of strategies at the beginning of all simulations.

5.2 Tuning of m, N and S Parameters

Considering three parameters: m, N and S, at least m requires different optimization techniques than two others. The optimization of m is strictly related to analysis of time series properties, especially the analysis of the range of dependencies between values of Y. Therefore it depends on the researcher's knowledge about the object. There are different methods of finding this range.

If the process is explicitly known, m is equal to the order of this process, e.g. $m = 3$ for AR(3). The predictor with m lower than the order cannot work effectively because strategies would incorrectly recognize patterns. Larger values of m would introduce additional and unnecessary noise that would degrade the prediction. The latter effect is further illustrated and explained in the section 6.2.

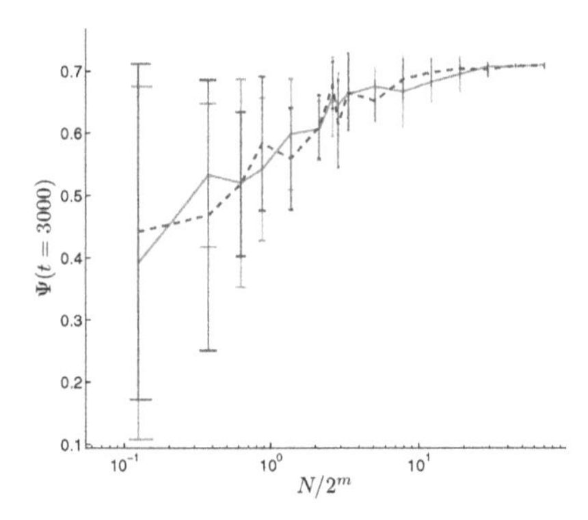

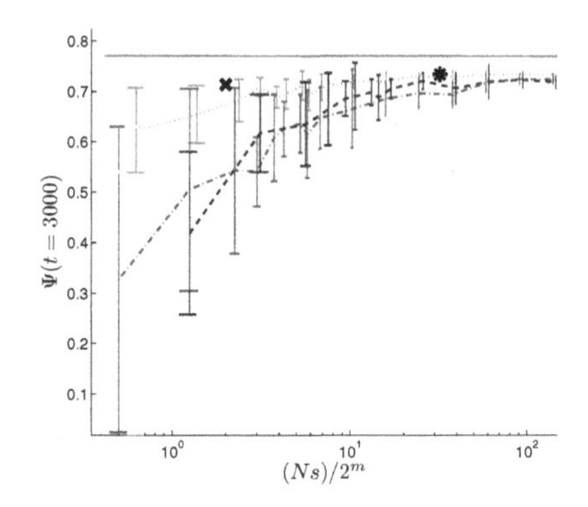

Fig. 2. Left: Comparison of correctness Ψ for the minority and majority games for $S = 2$ and variable N. Red dashed line corresponds to the minority game and green solid line to the majority game. Right: Solid grey line corresponds to the theoretical maximal value, green dotted line corresponds to MG for $N = 1$ and various S values, blue dashed line corresponds to MG for $S = 5$ and various N values, red dash-dotted line corresponds to MG for $S = 2$ and various N values. The 'x' mark corresponds to MG for $N = 1$ and $S = 16$ pairwise different strategies from the RSS.The star '*' corresponds to MG for $N = 1$ and $S = 256$ pairwise different strategies from the FSS.

If the order is unknown but the type of the process is known (e.g. autoregressive one) some techniques based on the autocorrelation analysis can be applied to detect the order, as presented in Ref. [1].

The problem is most difficult if the order and the system are unknown, as it is in many real cases (e.g. prices in financial markets). Still the correlation analysis can suggest the number of past samples be used. In this section we assume that the order of the examined AR process is explicitly given.

In order to find the proper technique for N and S optimization, let us assume that a certain pool of strategies of constant size has to be optimally assigned to agents. It means that, under the constraint $NS = const$, we are looking for the proportion between number of agents N and number of strategies per agent S that maximizes the correctness Ψ. We would also like to examine if the optimal proportion is sensitive to the constraint's change. Our numerical studies are presented in Fig. 2. The correctness is presented as a function of NS, as changing the size of the strategies' pool influences the results. Solid grey line corresponds to the maximal value that is reached by the best possible predictor (in this case linear filter). Blue dashed line corresponds to MG for $S = 5$ and various N values, red dash-dotted line corresponds to MG for $S = 2$ and various N values. These two curves show that the more strategies per agent the better the correctness Ψ, provided that the constraint $NS = const$ is preserved. Following further this reasoning, the most efficient is obtained for just one agent possessing given pool of strategies. Indeed, this is confirmed by the green dotted line corresponding to MG for $N = 1$ and various S values. As it is seen, the predictor with such configuration outperforms any other predictor.

In order to reason out of the presented results, assume there are many agents but only one of them has the best strategy. Even if this strategy suggests a correct prediction for itself the prediction of the whole system can be potentially incorrect, provided there is sufficiently many agents with bad strategies. Such situation is less probable when the number of strategies per agent is increased and concurrently the number of agents is decreased. This is impossible if there is only one agent because then the best strategy is always used. Similarly, if the constraint on NS is shifted towards larger values and number of agents is preserved, then the probability, that more and more agents have a correct strategy becomes large. Hence, in the limit $NS \to \infty$, all agents have the correct strategy and the efficiency of the group is the same as the efficiency of a single agent equipped with the aggregated pool of strategies.

Considering Ψ as a function of the S value, the larger value of S, the better results, as there is a larger probability that better strategy is drawn by an agent. However, if S is above some threshold, the pool of strategies is oversampled and many agent's strategies are the same. Since there is no particular gain if agent has more than one best strategy at his disposal, the success rate does not increase. Given this, one can wonder if a random draw is the most efficient way to generate strategies. Indeed it is not. Only the fully probed strategy space assures that, if the best strategy exists, then it is used in the game, provided $N = 1$. The fringe benefit is that there is also no redundancy between strategies, what reduces the computational costs. Hence, the best MG predictor is based on single agent equipped with all pairwise different strategies, i.e. all strategies covering Full Strategy Space (FSS). However, in cases when m is large, it can be hardly possible to generate so huge set of strategies. Therefore, for higher m it seems reasonable to use all strategies from Reduced Strategy Space (RSS). The RSS consists of only 2^{m+1} strategies which are pairwise uncorrelated or anticorrelated, i.e. the normalized Hamming distance between them is equal to 0.5 or 1 [6]. The RSS apparently reduces the complexity and still assures good quality of prediction as FSS is regularly probed. These statements are visualized in Fig. 2 where marks 'x' and '*' represent results for single agent and all strategies related to RSS (16 strategies) and FSS (256 strategies) respectively. Despite of a numerous differences between both pools the results are close, which indicates a superior power of RSS usage.

The presented above methodology differs significantly from approaches presented by other researchers [11,8,7]. In those papers the whole bunch of individuals is used and the strategy distribution is randomly initiated. Although those methods work either, the approach presented here is optimal.

5.3 Nonstationary Signals and λ-GCMG Predictor

The agents' behavior is determined by the strategies' utilities which they have at their disposal. Generally, the strategies that collect more frequently a positive payoff are characterized by a monotonously rising utilities. The utilities are thus potentially unbounded, what carries some considerable consequences if the exogenous process is a non-stationary one. Namely, if characteristics of a modelled object changes at time t then the group of best strategies would change either. But if, until t, some strategies collected many positive payoffs and after time t they are no longer profitable, then many steps with negative payoffs are required to lower their utilities. Hence, after time t the system still uses potentially ineffective strategies. The following example should provide some deeper understanding of this issue.

Let us assume that after 1500 steps the system described by Eq. 5 is replaced by the following one: $y(t) = -0.3y(t-1) - 0.2y(t-2) + 0.6y(t-3) + \xi(t)$. The new $AR(3)$ process is defined in such a way that the best MG strategies for this process constitute also the set of the worst strategies of the previous process. After time $t = 1500$ the predictor still uses strategy which is no longer efficient, but is characterized by the highest U. As a result, the Ψ as a function of time starts to decrease, as seen in Fig.3 (green dotted line).

In order to speed up the new appraisal of outdated U we introduced additional parameter λ to the rule (3)

$$U_{\alpha_n}(t+1) = \lambda U_{\alpha_n}(t) + \Phi_{\alpha_n}(t), \tag{7}$$

where $\lambda \in [0,1]$. If $\lambda = 1$ the strategies have an infinite memory of all previous rewards, what corresponds to standard MG. If $\lambda = 0$, then there is no memory effect. The intermediate values $0 < \lambda < 1$ preserve increasing of U to infinity, when the process is a stationary one, what considerably speeds up the time of adaptation. The cost of the introduction of the additional parameter is a need of its optimization. The heuristic analysis of correctness as a function of λ is presented in Fig. 4. The value $\lambda = 0.97$ assures the best results, although other values, that are close to it, work effectively either. Summing up, the λ-GCMG model effectively follows changes in the predicted signal and significantly outperforms the GCMG for a non-stationary process.

6 Sign Prediction of Assets' Returns

In this section we present the assumed model of movements of share prices. Given this we apply the λ-GCMG predictor to retrieve dependencies between past and future samples of stocks prices taken from various markets.

6.1 Price Movements Model

We assume that the dynamics of returns is driven by two factors where one is endogenous and another one exogenous. The model of signs of return rates is

$$\mathrm{sgn}(\mathrm{r(t)}) = f[\mathrm{sgn}\ r(t-1), \ldots, \mathrm{sgn}\ r(t-m)] + \xi(t). \tag{8}$$

where the first term reflects a dependence on previous returns and the second term represents the influence of external events. We are interesting only in signs

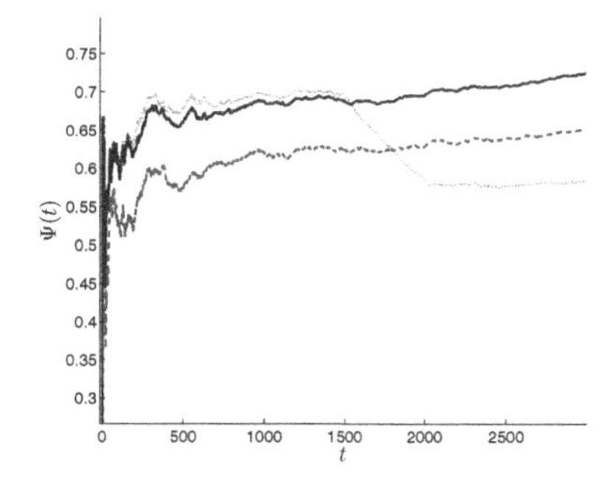

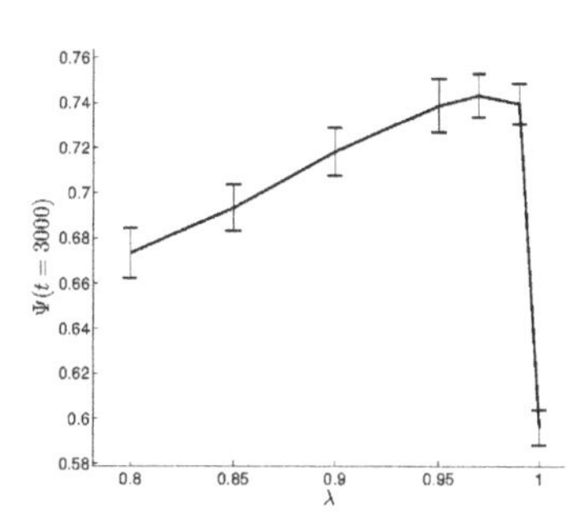

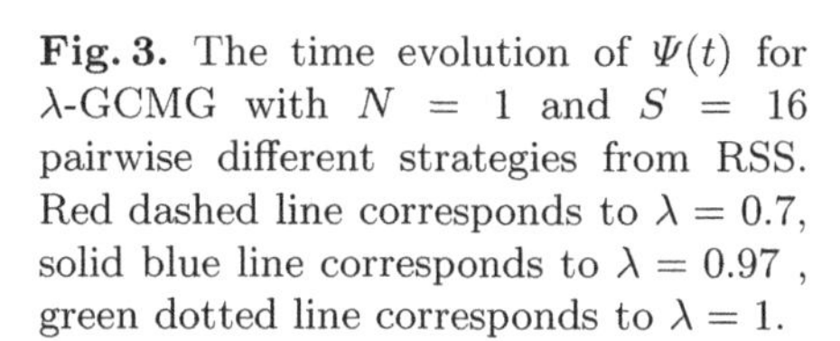

Fig. 3. The time evolution of $\Psi(t)$ for λ-GCMG with $N = 1$ and $S = 16$ pairwise different strategies from RSS. Red dashed line corresponds to $\lambda = 0.7$, solid blue line corresponds to $\lambda = 0.97$, green dotted line corresponds to $\lambda = 1$.

Fig. 4. The $\Psi(t)$ at time $t = 3000$ for λ-GCMG with $N = 1$, $S = 16$ pairwise different strategies from RSS and $\lambda = 0.97$. Error bars correspond to one standard deviation calculated over 10 realizations.

because MG predicts only the sign and not a value of sample. We assume that the exact form of function f is unknown and it can evolve over time. In case of the model in Fig. 1, the returns $r(t)$ correspond to signals $y(t)$. Considering the second term $\xi(t)$, there are many various exogenous factors, e.g. inflation, companies annual balances, etc. Investors react on them in various ways. Therefore, instead of considering reactions of individuals separately, we assume that signals of $\xi(t)$ are instances of IID process, with mean value equal to zero. The assumption seems more reasonable in the case of intraday data than for the end-of-day data [11]. This suggests that the majority in high-frequency data movements can be potentially self-generated, while the lower frequencies are dominated by exogenous factors.

6.2 Application of λ-GCMG

We look for a λ-GCMG model being able to estimate function f in Eq. (8). Using our previous results, we set the following parameter values: $N = 1$, $\lambda = 0.97$ and the number of strategies covering the RSS. The m value should be chosen from the range $1 - 10$ [13]. The precise choice of m requires some additional analysis. If we take a look at Fig. 5, where the λ-GCMG is applied as a predictor to FW20 time series i.e. futures contracts on WIG20 index that is the index of the twenty largest companies on the Warsaw Stock Exchange, then it is seen that better correctness is achieved for lower m i.e. $m = 1$. This result, at least at first sight, seems to be counter-intuitive. The RSS of higher order includes all strategies of RSS of lower order and additionally introduces the same number of extra strategies. The explanation is that for higher m the additional strategies, although do not capture properly dependencies, get from time to time higher

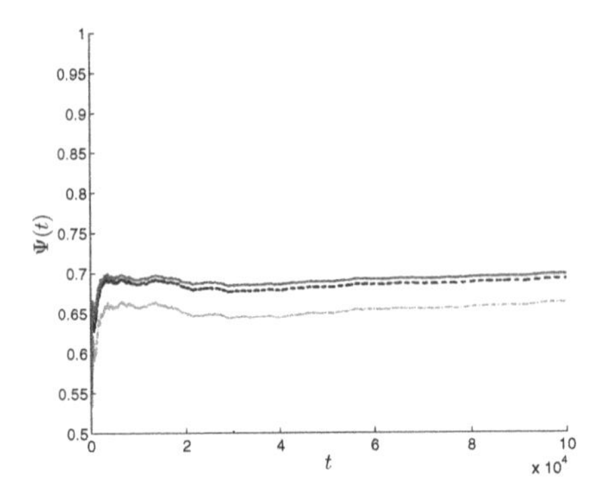

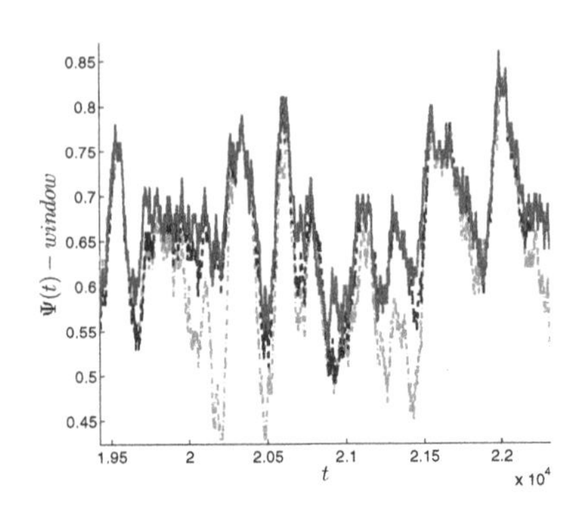

Fig. 5. The time evolution of $\Psi(t)$ (left) and $\Psi(t)$ calculated in sliding window of length equal to 100 (right) for λ-GCMG with $N = 1$, $\lambda = 0.97$ and strategies from RSS. Red solid line corresponds to $m = 1$, dashed blue line corresponds to $m = 2$, green dashed-dotted line corresponds to $m = 5$. The predicted signal is FW20.

utility than the basic ones correctly recognizing shorter patterns. This happens randomly, as sometimes samples are in order that is reflected by some of additional strategies. However, the pattern does not truly exist. In the next steps the forecast, based on one of additional strategies, is mostly wrong, what lowers the correctness. In other words, too long memory spoils the predictor introducing noise of unwanted strategies. The observed degradation of correctness for longer m indicates that there are no long-range and complicated patterns in the signal or that their nature is more subtle than MG strategies are able to capture. The first statement is supported by observations of the autocorrelation of returns, where the only distinct value is for $\tau = 1$. Similar premise is also included in Ref. [12].

The success rate for $m = 1$, it is mostly above 0.6, and from time to time even touches the level 0.8, if calculated in the sliding window. The mean value of correctness is 0.7 (Fig. 5 - left), what seems impressive, at the first sight. However, we checked that the correctness of the linear regression model of the first order is equal to 0.68. The use of higher-order filters does not improve the predictor.

The next question is, whether the results achieved for FW20 are specific only for this asset or they are more universal. We examined separately 5 stocks with the biggest impact on FW20 index (Fig. 6). The success rate decreases as a function of m, regardless of the analyzed year. So the results seem to be time and stock independent.

Another interesting issue is related to the analysis of the best strategy. The question is, whether there is only one strategy permanently outperforming other strategies or, maybe, different strategies lead at various moments? If there is only one, it would mean that patterns do not change over the game or that they do in a way the strategies are unable to capture. One permanently best strategy also would mean that the extended adaptive version of algorithm is, at least in this

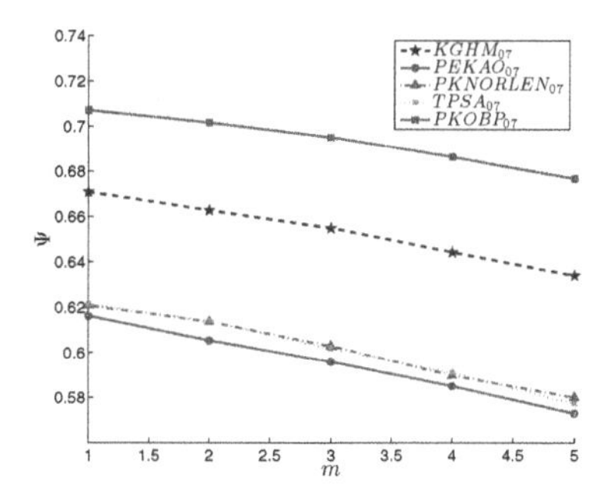

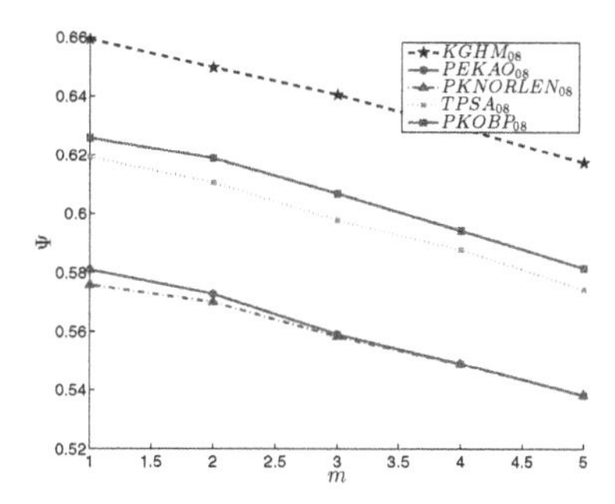

Fig. 6. The correctness Ψ as a function of the memory length m, for 5 companies with the biggest impact on index WIG20 for two different years 2007 (left) and 2008 (right).

case, as good as an ordinary GCMG or even as good as a linear filter. Indeed, we checked that one of the strategies permanently outperforms others. Interestingly, this strategy represents the mean-reverting approach, i.e. after history $\mu = 1$ it suggests $a = -1$ and after $\mu = -1$ it suggests $a = 1$. Accordingly, the opposite strategy representing a trend-follower approach, is the worst one. The results are consistent with autocorrelation analysis where for $\tau = 1$ the coefficient is negative. We checked that the results are general for all five examined firms.

The results for Polish market cannot be easily generalized to the London Stock Exchange market. For example, companies like Vodafone or Astrazeneca are not characterized by only one strategy being permanently better than others. The more so, for Astrazeneca, the best results with the correctness equal to 0.6 are achieved for $m = 2$. It is difficult to explain why such fundamental differences between stock markets exist. Their existence remains an open question.

The above analysis also shows that it is difficult to build a profitable investing system if one would capture patterns using only agents' strategies. If a sign of a next increment is known to the investor with encouraging probability, then the investor has to put the order. But placing the order introduces a perturbation to the forecast. If the order is executed, the system would move to the next time step and the transaction would be considered as entailing the positive or negative increment. The investor would predict only its own transactions what, of course, does not assure any profit. Given this, the prediction for at least two steps is required.

7 Conclusion

We applied the minority game as a predictor of an exogenous process. We found and explained that the degenerated game with only single agent and all strategies from FSS is the most efficient configuration. If using the FSS is computationally impossible, then the RSS is recommended. Considering the quality of prediction, minority and majority games are equivalent. Considering non-stationary signals, the parameter λ is introduced in order to speed up the model's convergence.

Applying the predictor to the intraday financial time series allows to effectively perform for one step where the correctness reaches 70% of properly recognized signs. The best strategy in the case of Polish stocks is a mean-reverting one. Only slightly worse predictions are attained by autoregressive systems. Unfortunately, these encouraging results are mostly useless, for building a profitable investing system. We explained that successful acting requires statistically significant forecasts for more than only one step forward.

References

1. Box, G., Jenkins, G.M., Reinsel, G.: Time Series Analysis: Forecasting & Control, 3rd edn. Prentice Hall (1994)
2. Challet, D., Marsili, M.: Phase transition and symmetry breaking in the minority game. Physical Review E 60, 6271–6274 (1999)
3. Challet, D., Marsili, M.: Relevance of memory in minority games. Physical Review E 62, 1862–1868 (2000)
4. Challet, D., Marsili, M., Zhang, Y.C.: Modeling market mechanism with minority game. Physica A 276, 284–315 (2000)
5. Challet, D., Zhang, Y.C.: Emergence of cooperation and organization in an evolutionary game. Physica A 246, 407–418 (1997)
6. Challet, D., Zhang, Y.C.: On the minority game: Analytical and numerical studies. Physica A 256, 514–532 (1998)
7. Chen, F., Gou, C., Guo, X., Gao, J.: Prediction of stock markets by the evolutionary mix-game model. Physica A 387, 3594–3604 (2008)
8. Gou, C.: Predictability of Shanghai Stock Market by agent-based mix-game model. In: Proceeding of IEEE ICNN&B 2005, pp. 1651–1655 (2005)
9. Gou, C.: Dynamic behaviors of mix-game models and its application. Chinese Physics 15, 1239 (2006)
10. Jefferies, P., Hart, M.L., Hui, P.M., Johnson, N.F.: From market games to real-world markets. The European Physical Journal B 20, 493–501 (2001)
11. Johnson, N.F., Lamper, D., Jefferies, P., Hart, M.L., Howison, S.: Application of multi-agent games to the prediction of financial time series. Physica A 299, 222–227 (2001)
12. Krause, A.: Evaluating the performance of adapting trading strategies with different memory lengths. In: Corchado, E., Yin, H. (eds.) IDEAL 2009. LNCS, vol. 5788, pp. 711–718. Springer, Heidelberg (2009)
13. Lillo, F., Farmer, J.D.: The long memory of the efficient market. Studies in Nonlinear Dynamics & Econometrics 8, 1–33 (2004)
14. Wawrzyniak, K., Wiślicki, W.: Multi-market minority game: breaking the symmetry of choice. Advances in Complex Systems 12, 423–437 (2009)
15. Wawrzyniak, K., Wiślicki, W.: Phenomenology of minority games in efficient regime. Advances in Complex Systems 12, 619–639 (2009)
16. Wawrzyniak, K., Wiślicki, W.: Mesoscopic approach to minority games in herd regime. Physica A 391, 2056–2082 (2012)

Agent-Based Models for Higher-Order Theory of Mind

Harmen de Weerd, Rineke Verbrugge, and Bart Verheij

Institute of Artificial Intelligence, University of Groningen, Groningen, Netherlands
h.a.de.weerd@rug.nl, {rineke,b.verheij}@ai.rug.nl

Abstract. Agent-based models are a powerful tool for explaining the emergence of social phenomena in a society. In such models, individual agents typically have little cognitive ability. In this paper, we model agents with the cognitive ability to make use of theory of mind. People use this ability to reason explicitly about the beliefs, desires, and goals of others. They also take this ability further, and expect other people to have access to theory of mind as well. To explain the emergence of this higher-order theory of mind, we place agents capable of theory of mind in a particular negotiation game known as Colored Trails, and determine to what extent theory of mind is beneficial to computational agents. Our results show that the use of first-order theory of mind helps agents to offer better trades. We also find that second-order theory of mind allows agents to perform better than first-order colleagues, by taking into account competing offers that other agents may make. Our results suggest that agents experience diminishing returns on orders of theory of mind higher than level two, similar to what is seen in people. These findings corroborate those in more abstract settings.

1 Introduction

In everyday life, we regularly interpret and predict the behaviour of other people by reasoning about what they know or believe. This *theory of mind* [1] allows us to understand why people behave a certain way, to predict future behaviour, and to distinguish between intentional or accidental behaviour. People also take this ability one step further, and consider that others have a theory of mind as well. This second-order theory of mind allows us to understand sentences such as "Alice doesn't know that Bob knows that she is throwing him a surprise party", by attributing to Alice the ability to have beliefs about Bob's knowledge. In this paper, we make use of agent-based computational models to explain why our ability to reason about mental content of others may have evolved.

The human ability to make use of higher-order (i.e. at least second-order) theory of mind is well-established, both through tasks that require explicit reasoning about second-order belief attributions [2, 3], as well as in strategic games [4, 5]. However, the use of any kind of theory of mind by non-human species is a controversial matter [6–8]. These differences in the ability to make use of theory of mind raise the issue of the reason for the evolution of a system that allows

B. Kamiński and G.Koloch (eds.), *Advances in Social Simulation*,
Advances in Intelligent Systems and Computing 229,
DOI: 10.1007/978-3-642-39829-2_19,

humans to use higher-order theory of mind to reason about what other people understand about mental content, while other animals, including chimpanzees and other primates, do not appear to have this ability.

A possible explanation for the emergence of higher-order theory of mind is that higher-order theory of mind is needed in situations that involve mixed-motive interactions such as negotiations or crisis management [9, 10]. In these situations, interactions are partially cooperative in the sense that an interaction can lead to a mutually beneficial outcome, but also partially competitive when there is no outcome that is optimal for everyone involved. For example, both the buyer and the seller of a house benefit from a successful sale. However, the buyer prefers a low sales price, while the seller prefers a high sales price.

In this paper, we consider agent-based computational models to investigate the advantages of making use of higher-order theory of mind in mixed-motive settings. We therefore model cognitively more sophisticated agents, in which there has been increasing interest in recent years [8, 11–13]. These agents perform actions based on their own desires and goals, but also take into account that the actions of other agents can influence their situation. By controlling the cognitive abilities of agents and monitoring their performance, we determine the extent to which higher-order theory of mind provides agents with an advantage over agents that are more restricted in their use of theory of mind. We have selected to study the interaction of cognitive agents in the Colored Trails setting, introduced by Grosz, Kraus and colleagues [14, 15], which provides a useful test-bed to study mixed-motive situations. Section 2 describes this setting in more detail.

We compare simulation results of agents of two different types. Agents of the first type base their beliefs on the iterated best-response. We also consider agents that use utility-proportional beliefs, which is more consistent with the behaviour of real life agents [16]. The results from the latter agents should provide insight in the effectiveness of higher-order theory of mind in mixed teams of agents and humans, which occur in an increasing number of domains [17–19]. Section 3 describes the two agent types and how these agents make use of theory of mind.

Section 5 presents the results of the simulations. These results are discussed in Section 6, in which we draw conclusions about whether or not mixed-motive situations may have contributed to the emergence of higher-order theory of mind in humans, as well as the extent to which higher-order theory of mind may be useful for computational agents that interact with people.

2 Colored Trails

To determine the effectiveness of higher-order theory of mind in mixed-motive settings, we have selected the Colored Trails (CT) setting. Colored Trails is a board game designed as a research test-bed for investigating decision-making in groups of people and computer agents [14]. The game is played by two or more players on a board of colored tiles. Each player starts the game at a given initial tile with a set of colored chips. The colors of the chips match those on the tiles of the board. A player can move to a tile adjacent to his current location by

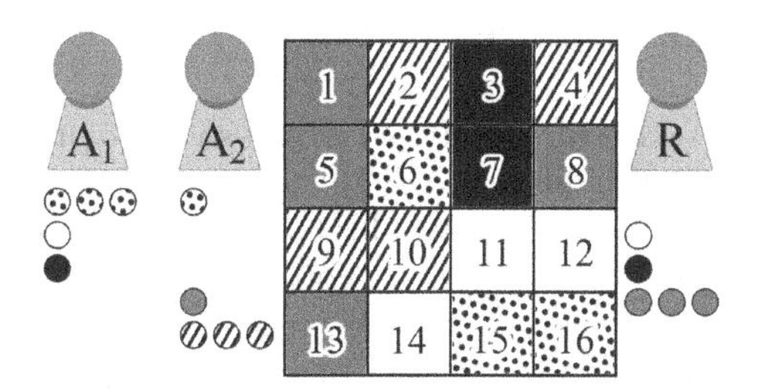

Fig. 1. An example of a Colored Trails game played by three agents. Agents A_1 and A_2 on the left are allocators. They both start at the tile marked 1, and aim to get as close to tile 16 as possible. Agent R on the right is a responder. She starts at tile 4 and tries to get as close as possible to tile 13.

handing in a chip of the same color as the destination tile. Each player is also assigned a goal location, which the player has to approach as closely as possible. To achieve this goal, players are allowed to trade chips among each other.

Figure 1 shows an example of a Colored Trails setting, in which there are three players, each with their own set of chips. Agent A_2 on the left, for example, has one dotted chip, one gray chip, and three striped chips. If agent A_2 is at the tile marked as 3, he can therefore move to tile 2 if he hands in one of his striped chips. However, if agent A_2 wishes to move to tile 7, he will have to make a trade with either agent A_1 or agent R to obtain a black chip.

Depending on the aspect of negotiation that is being investigated, scoring rules vary. Following [14], a player that reaches his goal tile is awarded 50 points. If a player is unable to reach his goal tile, he pays a penalty of 10 points for each tile in the shortest path from his current location to his goal location. To focus our research on the effectiveness of higher-order theory of mind in mixed-motive settings, players do not receive any points for unused chips. This way, players have to compete to obtain the chips they need to reach their goal location, and cooperate to find a mutually beneficial trade.

We consider a standard Colored Trails setup in which players are put into either the role of *allocator* or the role of *responder* [14]. An allocator can offer to trade some of his chips against some of the responder's chips. The responder does not make trades of her own. Instead, she chooses whether to accept an offer made to her by an allocator. We focus our attention on the scenario that includes two allocators and one responder. Here, allocators may benefit from considering the goal of the responder, as well as possible offers of the competing allocator.

The Colored Trails game is an example of a mixed-motive situation, in which players can generally improve their score by trading chips with another player. Since mutually beneficial trades may exist, an allocator may benefit from using theory of mind, and explicitly consider the goals of his trading partner. We expect that allocator agents capable of using theory of mind will outperform agents that are unable to consider the goals of other agents. Furthermore, we also expect that in cases where there are multiple allocators, allocator agents perform better when they are of a higher order of theory of mind.

3 Agents Playing Colored Trails

In our simulations, we consider repeated single-shot Colored Trails games, in which the set of players is divided into distinct sets of allocators and responders. Each allocator can offer to trade any subset of his own chips for any subset of chips belonging to one of the responders. For example, an allocator can give all his chips to the responder, or ask that the responder give all her chips to the allocator. The responder chooses whether or not to accept any of these offers.

3.1 Agent Types

We consider two types of theory of mind agents. Both types of theory of mind agents play the best-response given their beliefs about the behaviour of others, but they differ in the way they form these beliefs. Agents with iterated best-response beliefs (IBR) maximize their own expected payoff under the assumption that other agents do the same. This behaviour is similar to the iterated best-response models such as cognitive hierarchy models [20] and level-n theory [21]. IBR agents believe that other players will only choose an action that maximizes their expected score, and assign probability zero to the event that a co-player will perform any other action. This approach guarantees the best outcome when the agent's beliefs are correct. However, this approach ignores that other players may have different beliefs or a different understanding of the situation.

The assumption of iterated best-response models can be weakened by assuming that players choose better actions with higher probabilities, such as in t-solutions [22], quantal response equilibria [23], or utility proportional beliefs [16]. In addition to the iterated best-response agents described above, we also consider utility-proportional beliefs (UPB) agents in the setting of Colored Trails. The UPB agent believes that other allocators may choose any offer that would increase the allocator's score, but that the probability that he will make a certain offer is proportional to the expected utility of that offer. As a result, a UPB agent may perform better than an IBR agent when his beliefs are incorrect.

The following subsections illustrate the different orders of theory of mind reasoning involved in the game of Colored Trails. To avoid confusion, we will refer to allocators as if they were male, and responders as if they were female.

3.2 Responders

In the Colored Trails game, a responder is a player that does not offer to trade chips herself. Instead, she receives offers from other players, and decides whether to accept any of these offers. We assume that a responder refuses any offer that strictly decreases her score. If a responder is offered more than one acceptable trade, we assume that she chooses in a utility-maximizing way. That is, the responder selects the offer that allows her to reach her goal location as closely as possible without considering the score of the allocator. If multiple offers satisfy this condition, she selects one of these offers at random.

Once the responder has made a choice, a Colored Trails game ends. We do not consider learning across games, which may allow a responder to influence the behaviour of an allocator in future games. This means that for a responder, there is no additional benefit of predicting the offers an allocator is likely to make. The responders described here therefore do not consider the beliefs, desires, and goals of other agents, and as a result, do not make use of theory of mind.

3.3 Zero-Order Theory of Mind Allocator

A zero-order theory of mind (ToM_0) allocator understands the game, but is unable to attribute any mental content such as beliefs, desires, or goals to a responder. That is, although a ToM_0 allocator can determine what chips a responder would need to reach her goal location, he is unable to consider the possibility that she wants to reach her goal location. Instead, the zero-order theory of mind allocator considers the total set of chips that are owned by himself and the responder. He then determines the subset of chips C that will allow him to move to a tile as close as possible to his goal location. If there are multiple subsets of chips that satisfy this condition, the ToM_0 allocator selects one of these subsets at random. The allocator then offers to trade in such a way that he receives all the chips in the subset C, while leaving the remaining chips for the responder.

Since the ToM_0 allocator cannot attribute goals or beliefs to other agents, we assume that he does not make any predictions about the offers made by the competing allocator. Example 1 shows the behaviour of a ToM_0 allocator in a game with two allocators and one responder. In this example, agents form beliefs based on iterated best-response.

Example 1. Consider the setup illustrated by Figure 1. In this situation, the game is played on a 4 by 4 board with five different colors. There are two allocators, indicated by A_1 and A_2 to the left of the board, each with their own set of chips. Allocators are initially placed on the top left tile (tile 1) and aim to get as close as possible to the bottom right tile (tile 16). There is a single responder R, depicted to the right of the board. Unlike the allocators, she is initially placed at the top right tile (tile 4) and aims to reach the bottom left tile (tile 13).

Suppose that the agent A_1 is a ToM_0 allocator. Allocator A_1 cannot move with his own chips, but with the combined set of chips of agent A_1 and agent R, there are four possible paths for agent A_1 to reach his target, as depicted in Figure 2. The agent randomly selects one of these paths and makes corresponding offer that would yield him the chips that he needs to reach his goal.

3.4 First-Order Theory of Mind Allocator

The first-order theory of mind (ToM_1) allocator considers that responders and other allocators have beliefs and goals. While deciding to offer a trade to a responder, the ToM_1 allocator considers the viewpoint of the responder to determine whether he would accept if he were in her place. Concretely, the ToM_1 allocator does not make any offers that would decrease the score of the responder.

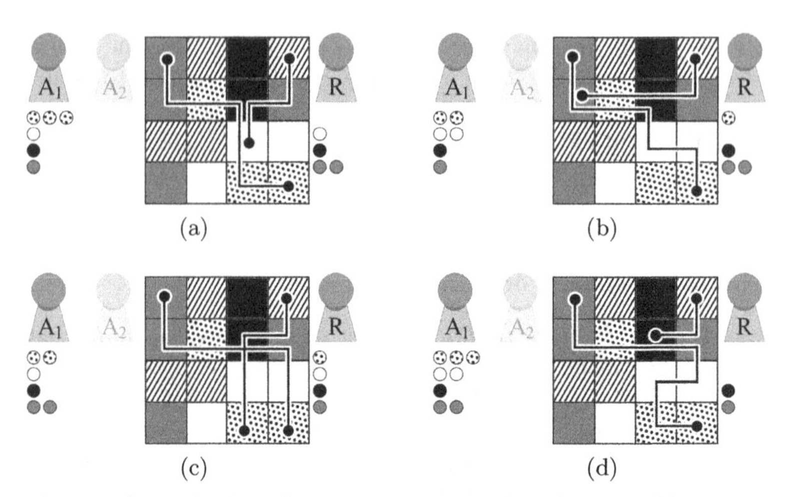

(a) (b)

(c) (d)

Fig. 2. If agent A_1 is a ToM_0 allocator as in Example 1, he is unable to consider the goals of responder R when making a trade offer. Instead, he offers to make a trade that would maximize his own score.

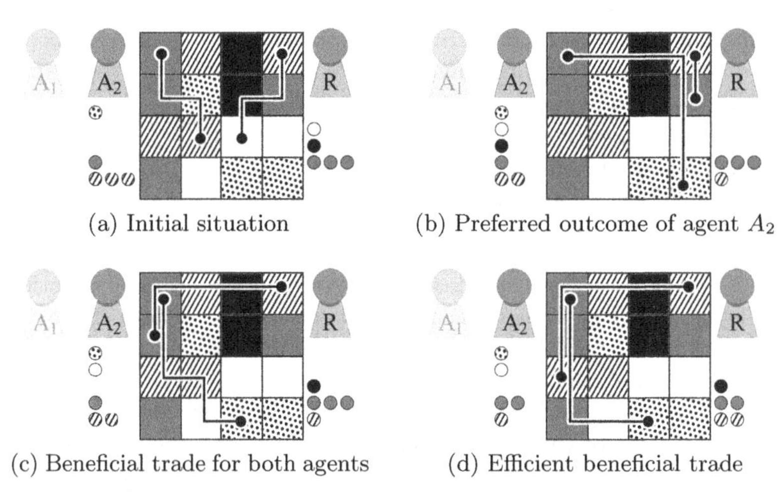

(a) Initial situation (b) Preferred outcome of agent A_2

(c) Beneficial trade for both agents (d) Efficient beneficial trade

Fig. 3. If agent A_2 is a ToM_1 allocator as in Example 2, he considers both his own goals, as well as the goals of the responder R and competing allocator A_1

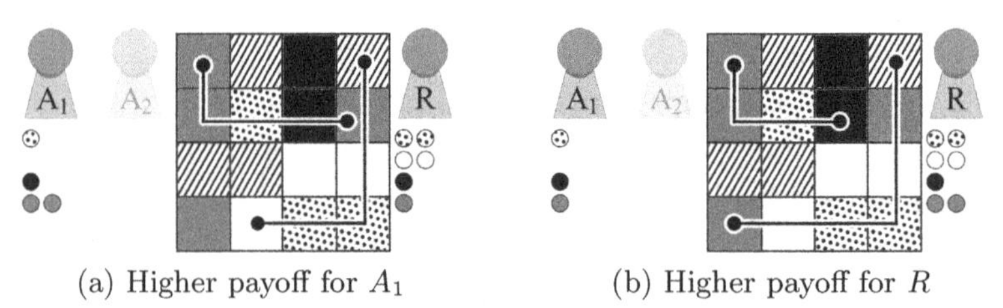

(a) Higher payoff for A_1 (b) Higher payoff for R

Fig. 4. If agent A_1 is a ToM_2 allocator as in Example 3, he believes that agent A_2 also considers the responder's goals when making an offer to her. In this case, the ToM_2 allocator A_1 chooses between two alternatives.

The ToM_1 allocator also considers the offers that he believes the competing allocator to make. However, the ToM_1 allocator does so without considering the possibility that the competing allocator is trying to predict the offer he is going to make himself. Instead, the ToM_1 allocator assumes that the competing allocator is a ToM_0 allocator. Using the procedure outlined in the previous subsection, the ToM_1 agent determines which offers the competing allocator is likely to make. The ToM_1 allocator then chooses to offer the trade that he expects will yield him the highest score. Example 2 illustrates the behaviour of a ToM_1 allocator.

Example 2. Consider the setup illustrated by Figure 1, and suppose that agent A_2 is a ToM_1 allocator. Initially, agents A_2 and R can each move three steps towards their own goal location (see Figure 3a). There is a trade that would allow agent A_2 to reach his goal (see Figure 3b), but responder R would only be able to move one step towards her goal in this case. Using his first-order theory of mind, agent A_2 concludes that responder R would not accept this trade.

It is not possible for agent A_2 to offer a trade that will allow him to reach his goal, and also increase the responder's score. However, the ToM_1 allocator can compromise by offering either the trade shown in Figure 3c or the one shown in Figure 3d. Although the ToM_1 agent is indifferent between these outcomes, he knows that the responder prefers the outcome of Figure 3d. Moreover, he knows that if agent A_1 makes an offer that allows the responder to move exactly four tiles closer to her goal, the offer shown in Figure 3c could be rejected by the responder, while the offer shown in Figure 3d would still be accepted. The ToM_1 allocator A_2 therefore chooses to make the offer as shown in Figure 3d.

3.5 Higher-Order Theory of Mind Allocator

Similar to the first-order theory of mind allocator discussed above, the second-order theory of mind (ToM_2) allocator forms beliefs about the trades that other allocators will offer, as well as the likelihood that a responder will accept a given offer. Note that since the responder does not make use of theory of mind, allocators do not benefit from considering the beliefs, goals, and intentions the responder may be attributing to others. As a result, both the ToM_1 allocator and the ToM_2 allocator believe that the responder will accept the trade that will yield her the highest score. The difference in performance of ToM_1 agents and ToM_2 agents is determined only by their ability to compete with other agents.

While the ToM_1 allocator believes that competing allocators offer a trade that maximizes their personal score, the ToM_2 allocator believes that competing allocators also take the point of view of the responder into account. That is, the ToM_2 allocator believes that competing allocators know that the goal of a responder is to approach her goal location as closely as possible. The ToM_2 allocator also believes that competing allocators try to predict the trade he is going to offer himself, and takes this into account when making his offer.

For increasingly higher orders of theory of mind, theory of mind allocators continue this pattern of forming increasingly deeper nested beliefs, and assuming that other agents are more sophisticated. In this paper, we restrict our investigation to ToM_i agents for $i = 0, 1, 2, 3, 4$.

Example 3. Consider the setup illustrated by Figure 1, and suppose that agent A_1 is a ToM_2 allocator. Following the process described in Example 2, agent A_1 concludes that agent A_2 is going to make the offer depicted in Figure 3d. This trade would allow the responder to move five tiles towards her goal location.

The ToM_2 allocator A_1 can choose to match this offer by making the offer shown in Figure 4a. If responder R were to accept this offer, allocator A_1 can move four tiles closer to his goal location, increasing his score by 40. However, allocator A_1 believes that allocator A_2 will make an offer that allows responder R to move five tiles towards her goal as well. In this case, responder R will randomly select which offer to accept, which means that there is a 50% probability that the responder will not accept the offer of allocator A_1. The ToM_2 allocator A_1 therefore assigns an expected gain of 20 to the offer shown in Figure 4a.

Alternatively, the ToM_2 allocator can make a better offer to responser R by allowing her to reach her goal location (Figure 4b). Allocator A_1 expects that responder R will accept this offer, allowing him to move three tiles to his goal location and increase his score by 30. Since this is the higher expected gain, ToM_2 allocator A_1 decides to make the offer shown in Figure 4b.

4 Simulation

We performed simulations of single-shot Colored Trails games, designed after the games in [14]. Games were played on a 4 by 4 board of square tiles. Each tile on the board was randomly colored with one of five possible colors. Players were allowed to move horizontally and vertically, but diagonal movements were not allowed. Each game involved two allocators and one responder. To make individual game settings more comparable, the responder was always initially located on the top right tile, while her goal was to reach the bottom left tile. As a result, the responder has 20 different possible paths to reach her goal, each using six chips. Both allocators were initially placed on the top left tile, while their goal location was the bottom right tile. In this setup, the goal of the responder overlaps partially with the goals of the allocators, but not completely.

At the start of the game, each player received an initial set of six randomly colored chips. Since each player needs at least six chips to reach his or her goal location, it is sometimes possible that after a trade, both the allocator and the responder can reach their respective goals. However, this is not always the case. To ensure that each allocator has an incentive to negotiate to increase his score, game settings in which some player can reach his or her goal with the initially assigned set of chips without trading were excluded from analysis.

To determine the effectiveness of theory of mind, we generated 10,000 random game settings. In each of these settings, we determined the score of a focal ToM_i allocator in the presence of a competing ToM_j allocator, for each combination of $i, j = 0, 1, 2, 3, 4$. The average score was measured for the same 10,000 game settings in each condition. This was both done for agents that base beliefs on iterated best-response, as well as for agents that hold utility-proportional beliefs.

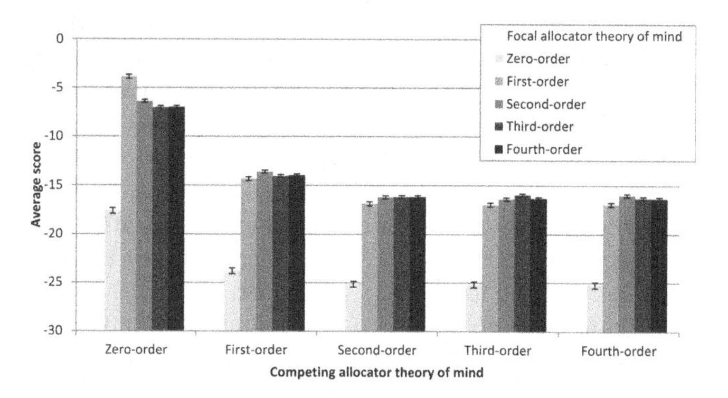

Fig. 5. Average scores of iterated best-response agents that differ in their order of theory of mind over 10,000 initial situations. Brackets indicate standard error.

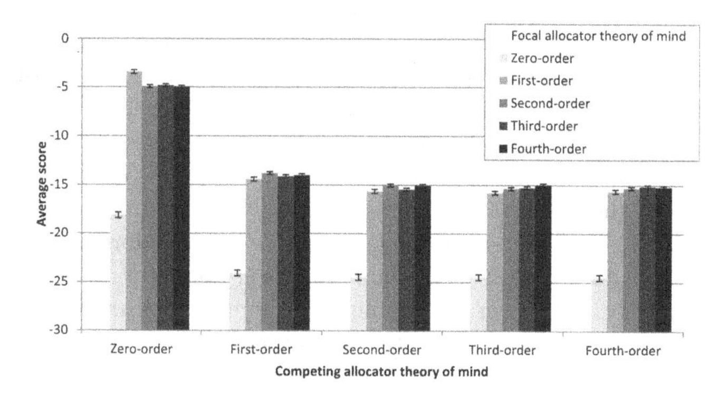

Fig. 6. Average scores of agents with utility-proportional beliefs that differ in their order of theory of mind over 10,000 initial situations. Brackets indicate standard error.

5 Results

We ran simulations of agents playing the Colored Trails for the two types of agents described in Section 3.1. The results for the agents who base their beliefs on iterative best-response are shown in Figure 5, while Figure 6 shows the results for agents that make use of utility proportional beliefs. Both figures summarize the average score of a focal allocator in the Colored Trails game as a function of his order of theory of mind and the order of theory of mind of the competing allocator. The figures show that, irrespective of the theory of mind ability of the competing allocator, focal ToM_1 allocators always score higher than focal ToM_0 allocators. When the competing allocator is a ToM_0 agent, the focal ToM_1 allocator also outperforms focal higher-order theory of mind allocators. Note that in this case, the focal ToM_1 allocator's assumption about the theory of mind abilities of the competing allocator are correct, while higher-order theory of mind allocators overestimate the competing allocator.

The focal allocator benefits from higher-order theory of mind when he competes with an allocator that can reason about the mental states of others. Paired t tests show that for any order of theory of mind of the competing allocator, performance of a focal ToM_2 allocator differs significantly from the performance of a focal ToM_1 allocator. However, this difference is low compared to the advantage the focal ToM_1 allocator has over the focal ToM_0 allocator. Orders of theory of mind higher than the second do not seem to benefit an allocator significantly. As a result, the average score of the focal ToM_3 allocator is not consistently higher than the average score of the focal ToM_2 allocator.

Figure 5 and Figure 6 summarize the performance of agents that differ in the way they form beliefs about other agents. The distributions differ significantly (K-S, $p < 0.01$), but one needs to look carefully to see the differences. For agents that form beliefs based on iterated best-response, a ToM_i allocator is correct in his beliefs when the competing allocator is a ToM_{i-1} agent. However, these agents do not consider the possibility that their beliefs may be wrong. In Figure 5, this results in a stronger advantage for having correct beliefs. In particular, when facing a competing ToM_0 allocator, the focal ToM_1 allocator performs best, while the average score of the focal ToM_2 allocator is the highest when the competing allocator is a ToM_1 agent.

Agents that form utility-proportional beliefs are never completely correct in their beliefs, since their beliefs reflect the possibility of mistakes. Figure 6 shows that as a result, the focal allocator has less of an advantage for being exactly one order of theory of mind higher than the competing allocator. Interestingly, this does not appear to cause lower performance of agents with utility-proportional beliefs compared to agents with iterated best-response beliefs.

6 Discussion and Conclusion

Many of the interactions that people engage in on a daily basis involve mixed motives, which are not fully competitive or fully cooperative. When the goals of interacting individuals overlap, there may be an advantage to considering the goals and beliefs of others explicitly, through a theory of mind. In this paper, we investigated whether agents benefit from the ability to reason about higher orders of theory of mind in the particular mixed-motive setting Colored Trails.

In the setting of Colored Trails we used, agents were put into the role of either allocator or responder [14]. The agents engaged in single-shot negotiations, where allocators made an offer which the responder could either reject or accept. Allocator agents were found to benefit greatly from first-order theory of mind, allowing them to consider the goals of other agents when making an offer.

Allocators could also benefit from higher-order theory of mind through competition with another allocator. Our results showed that second-order theory of mind benefits allocators whenever the competing allocator also has a theory of mind. By recognizing that competitors may also consider the point of view of the responder, second-order theory of mind allowed allocators to offer trades that the responder accepted more often than the offers of a first-order theory of mind allocator. These results are compatible with earlier research into the advantage of higher-order theory of mind in competitive settings [24, 25]. However, we did not find any benefit for the use of third-order theory of mind here. A possible explanation is that the settings in [24, 25] are zero-sum games, while allocators in Colored Trails compete for the opportunity to trade with a responder. In Colored Trails, an allocator of a higher order of theory of mind generally makes an offer that is more beneficial to the responder at the expense of his own score.

We compared the performance of allocators that made offers based on iterated best-response models with allocators that form utility proportional beliefs. Allocators that form beliefs based on iterated best-response models believe that every agent is a utility-maximizing agent, while an agent with utility-proportional beliefs takes into account that competing allocators may make mistakes. Interestingly, although our model did not include mistakes, iterated best-response agents did not outperform agents with utility-proportional beliefs.

Our results suggest that in mixed teams of humans and agents, agents that make use of theory of mind will perform better. Based on the experiments in Colored Trails, we expect that cognitive agents will suffer diminishing returns on higher orders of theory of mind. Interestingly, similar results are found for human participants [5], who do well on first-order theory of mind tasks, and have increasingly more difficulty with higher-order theory of mind tasks. In future work, we intend to compare the performance of human participants and theory of mind agents by letting them play directly against each other.

In future research, we aim to increase the emphasis on the mixed-motive nature of Colored Trails by allowing multiple rounds of negotiations. A responder that is allowed to make offers would benefit from considering the beliefs of others. This may give an allocator incentive to consider these beliefs of the responder in his initial offer. Higher orders of theory of mind may also become more effective when the game setting is not fully observable. In our setup, agents know the initial location, the goal location, and the chips in possession of every player. However, in everyday negotiation situations, the goals of the participants are usually not fully known [26]. Higher orders of theory of mind may be beneficial in determining the information available to each agent, as well as the information that agents may be revealing or trying to hide by making a specific offer.

Acknowledgments. This work was supported by the Netherlands Organisation for Scientific Research (NWO) Vici grant NWO 277-80-001 awarded to Rineke Verbrugge.

References

1. Premack, D., Woodruff, G.: Does the chimpanzee have a theory of mind? Behav. Brain Sci. 1(04), 515–526 (1978)
2. Perner, J., Wimmer, H.: "John thinks that Mary thinks that...". Attribution of second-order beliefs by 5 to 10 year old children. J. Exp. Child Psychol. 39(3), 437–471 (1985)

3. Apperly, I.: Mindreaders: The Cognitive Basis of "Theory of Mind". Psychology Press, Hove (2011)
4. Hedden, T., Zhang, J.: What do you think I think you think?: Strategic reasoning in matrix games. Cognition 85(1), 1–36 (2002)
5. Meijering, B., van Rijn, H., Taatgen, N., Verbrugge, R.: I do know what you think I think: Second-order theory of mind in strategic games is not that difficult. In: CogSci, Cognitive Science Society, pp. 2486–2491 (2011)
6. Tomasello, M.: Why we Cooperate. MIT Press, Cambridge (2009)
7. Penn, D., Povinelli, D.: On the lack of evidence that non-human animals possess anything remotely resembling a 'theory of mind'. Philos. T. R. Soc. B 362(1480), 731–744 (2007)
8. van der Vaart, E., Verbrugge, R., Hemelrijk, C.: Corvid re-caching without 'theory of mind': A model. PLoS ONE 7(3), e32904 (2012)
9. Verbrugge, R.: Logic and social cognition: The facts matter, and so do computational models. J. Philos. Logic 38, 649–680 (2009)
10. van Santen, W., Jonker, C., Wijngaards, N.: Crisis decision making through a shared integrative negotiation mental model. Int. J. Emerg. M. 6, 342–355 (2009)
11. Helmhout, J.: The Social Cognitive Actor. PhD thesis, University of Groningen (2006)
12. Wijermans, N., Jager, W., Jorna, R., van Vliet, T.: Modelling the dynamics of goal-driven and situated behavior. In: ESSA (2008)
13. Dykstra, P., Elsenbroich, C., Jager, W., de Lavalette, G.R., Verbrugge, R.: Put your money where your mouth is: The dialogical model DIAL for opinion dynamics. Journal of Artificial Societies and Social Simulation 16(3), 4 (2013)
14. Gal, Y., Grosz, B., Kraus, S., Pfeffer, A., Shieber, S.: Agent decision-making in open mixed networks. Artif. Intell. 174(18), 1460–1480 (2010)
15. van Wissen, A., Gal, Y., Kamphorst, B., Dignum, M.: Human–agent teamwork in dynamic environments. Computers Human Behav. 28, 23–33 (2012)
16. Bach, C., Perea, A.: Utility proportional beliefs (2011), `http://epicenter.name/Research.html` (accessed: September 27, 2012)
17. Kraus, S.: Strategic Negotiation in Multiagent Environments. MIT Press (2001)
18. Rosenschein, J., Zlotkin, G.: Rules of Encounter: Designing Conventions for Automated Negotiation Among Computers. MIT Press (1994)
19. Hiatt, L., Harrison, A., Trafton, J.: Accommodating human variability in human-robot teams through theory of mind. In: IJCAI, pp. 2066–2071. AAAI Press (2011)
20. Camerer, C., Ho, T., Chong, J.: A cognitive hierarchy model of games. Q. J. Econ. 119(3), 861–898 (2004)
21. Bacharach, M., Stahl, D.O.: Variable-frame level-n theory. Games and Econ. Behav. 32(2), 220–246 (2000)
22. Rosenthal, R.: A bounded-rationality approach to the study of noncooperative games. Int. J. Game Theory 18(3), 273–292 (1989)
23. McKelvey, R., Palfrey, T.: Quantal response equilibria for normal form games. Games and Econ. Behav. 10(1), 6–38 (1995)
24. de Weerd, H., Verbrugge, R., Verheij, B.: Higher-order social cognition in the game of rock-paper-scissors: A simulation study. In: Bonanno, G., van Ditmarsch, H., van der Hoek, W. (eds.) LOFT, pp. 218–232 (2012)
25. de Weerd, H., Verheij, B.: The advantage of higher-order theory of mind in the game of limited bidding. In: van Eijck, J., Verbrugge, R. (eds.) ROAM. CEUR Workshop Proceedings, pp. 149–164 (2011)
26. Raiffa, H., Richardson, J., Metcalfe, D.: Negotiation Analysis: The Science and Art of Collaborative Decision Making. Belknap Press (2002)

Adaptive Learning and Quasi Fictitious Play in "Do-It-Yourself Lottery" with Incomplete Information

Takashi Yamada and Takao Terano

Department of Computational Intelligence and Systems Science,
Interdisciplinary Graduate School of Science and Engineering,
Tokyo Institute of Technology
4259 Nagatsuta-cho, Midori-ku, Yokohama, Kanagawa, 226-8502 Japan
tyamada@trn.dis.titech.ac.jp, terano@dis.titech.ac.jp

Abstract. This study investigates a kind of guessing game, "do-it-yourself lottery" (DIY-L), with two types of players, adaptive learning and quasi fictitious play, by agent-based computational economics approach. DIY-L is a multi-player and multi-strategy game with a unique but skew-symmetric mixed strategy equilibrium. Here computational experiments are pursued to see what kind of game dynamics is observed and how each type of players behaves and learns in DIY-L by changing the game setup, learning parameters, and the number of each type of players. The main results are twofold: First a player who firstly and immediately learns to keep submitting the smallest integer becomes a winner in three-player games. Second, in four-player games, while the quasi fictitious play agent wisely wins when the other three players are all adaptive learners, one of the adaptive learners successfully makes advantage of the behaviors of quasi fictitious play agents when there are plural quasi fictitious play agents.

1 Introduction

J.D. Barrow introduces a guessing game called "do-it-yourself lottery" (DIY-L) where players choose and submit one positive integer and the one choosing the smallest number that is not chosen by anyone else is the winner [1]. When one rebuilds this lottery so that it has an upper limit and the number of players is fixed, (s)he can express it as a game. More precisely, DIY-L is a multi-player, multi-strategy and non-cooperative general-sum game with unique mixed strategy equilibrium. Besides, since this is a guessing game, it is usually iterated several times and players are expected to think how others will choose an integer not given foregone choices of others to win.

Multi-player and multi-strategy game situations are quite common in socio-economic systems. But, they have not been studied as deeply as 2×2 ones because multi-player games are considered more complex [3,11,13]. Or, one may have to

B. Kamiński and G.Koloch (eds.), *Advances in Social Simulation*,
Advances in Intelligent Systems and Computing 229,
DOI: 10.1007/978-3-642-39829-2_20,

take into account the possibility of coalition in such a situation [18]. Indeed, DIY-L has at least three players and usually three or more strategies.

The games with mixed strategy equilibrium are considered difficult to reach or, if luckily achieved, not stable [5,10,16]. In the corresponding game, MSE is almost impossible to calculate and depends on both the number of players and that of strategies. Besides, players in guessing games usually make their decisions with only the game results, not their past choices.

Östling et al. have implemented laboratory experiments of SL and found that there are mainly four kinds of behavioral rules employed (*random*, *stick*, *lucky*, and *strategic*) [12][1]. Their results may have the following three meanings: First, individuals are bounded rational and heterogeneous [9]. Second, some subjects try to make use of the information on past winning integers but others not, namely there is mixed evidence for whether individuals take care of foregone payoffs of others [8,17]. Third, some subjects use plural behavioral rules.

This study simplifies the findings of Östling et al. [12] and analyzes how players using simple adaptive learning model or quasi belief-based learning model behave and learn in a multi-player and multi-strategy game and consequently the whole game behavior by agent-based computational economics approach[2]. In particular, we are going to see whether the combination of two learning models affects the game results or which learning model is more adaptive in each setup.

The rest of this paper is organized as follows: The next section explains the basic framework of DIY-L. Section 3 presents experimental design and computational results. Finally, Section 4 gives concluding remarks.

2 Game Design

There are N players each of who chooses one positive integer from 1 to M (> 1). All of them know this setup. The player who submits the smallest integer that is not chosen by anyone else is a winner. The winner receives a positive payoff, usually normalized 1, and the losers do zero. If there is no uniquely chosen integer, all players become losers.

Here we consider DIY-L with $N \geq 3$ and $M \geq 3$. In case of bi-matrix game, there are three equilibria, (1) both players choose 1 and (2) one player chooses 1 and the other does 2. But, since one never makes one's opponent a winner so long as (s)he keeps on choosing 1 [12], this kind of game is not worth investigating.

[1] For Swedish Lottery (SL), Lowest Unique Bid Auction and Highest Unique Bid Auction, see the references therein.

[2] From the viewpoints of individual learning in agent-based computational economics literature, it seems that multi-player and multi-strategy games have not been intensively studied as Shoham et al. have pointed out [14]. Indeed, although Vu et al. have computationally analyzed such games [15], the number of studies is still small.

Table 1. Mixed strategy equilibrium in simplified DIY-L

N	M	1	2	3	4
3	3	0.464102	0.267949	0.267949	
3	4	0.457784	0.251643	0.145286	0.145286
4	3	0.448523	0.426330	0.125147	
4	4	0.447737	0.424873	0.125655	0.00173500

Then, each game form has a unique mixed strategy equilibrium. Table 1 gives the mixed strategy equilibria in cases of $(N, M) = (3, 3)$, $(3, 4)$, $(4, 3)$, and $(4, 4)$.[3]

3 Computational Experiments

3.1 Setup

There are two kinds of players, adaptive learning (AL) agent(s) and quasi fictitious play (QFP) agent(s), in DIY-L[4]. AL agents use only their attractions for their decision-makings, namely choose one pure strategy. On the other hand, QFP agents store the past possible, but not observable, plays of their opponents and then form their adaptive beliefs to make a decision:

- AL agents
  AL player i $(i = 1, \cdots, N)$ has a propensity $w_{i,k}(t)$ for k-th strategy (integer k) s_i^k $(k = 1, \cdots, M)$ at time t. Before the game, she is assumed to have non-negative propensities for all the strategies, namely $w_{i,j}(0) = w_{i,k}(0) \geq 0$ for $j \neq k$.
  At every turn, she chooses one pure strategy in accordance with the following exponential selection rule

$$p_{i,k}(t) = \frac{\exp(\lambda_a \cdot w_{i,k}(t))}{\sum_{k'=1}^{M} \exp(\lambda_a \cdot w_{i,k'}(t))}$$

  where $p_{i,k}(t)$ is the selection probability for strategy s_i^k and λ_a is a positive constant called sensitivity parameter [4,7].

[3] A perfectly rational player follows this table. For instance, in $(N, M) = (3, 3)$ DIY-L, she submits 1 *w.p.* 0.464102 and 2 and 3 *w.p.* 0.267949. Östling et al. have a succinct algorithm to calculate mixed strategy equilibrium in this setup [12]. The mixed strategy equilibrium in DIY-L with $N \geq 3$ and $M = 2$, namely binary choice game, is independent of the number of players; The mixed strategy equilibrium is 0.5 for each integer. Hence, we have omitted this kind of game setup.

[4] One of the anonymous referees questioned why we employed the learning models in economic literature, not in computer sciences such as LCS or XCS. It is true that LCS and XCS are quite powerful learning models for exploration, but such models do not explain real behaviors and learnings of individuals reported by Östling et al. [12]. Thus, in accordance with the discussions by Brenner [2], adaptive learning and quasi fictitious play are used.

After a turn, she updates her propensities as

$$w_{i,k}(t+1) = (1-\phi_a)w_{i,k}(t) + 1_{\{s_i^k, s_i(t)\}}(1-\phi_a)R$$

where R is a digital payoff, ϕ_a is positive constants called learning parameter, and $s_i(t)$ is player i's actually chosen strategy at t.

- QFP agents
  Cheung and Friedman propose a generalized belief-based learning model (usually called *weighted fictitious play*) in which players firstly expect what the others will do based on their own prior beliefs, which are usually ratios of the number of submitted plays to the whole moves [6].
  Let $L_i(t)$ be the total possible counts of past plays for player i ($i = 1, \cdots, N$) and $B^{\boldsymbol{s}}_{-i}(t)$ be her belief about her opponents will submit $\boldsymbol{s}_{-i} = (s_1^{k_1}, \cdots, s_{i-1}^{k_{i-1}}, s_{i+1}^{k_{i+1}}, \cdots, s_N^{k_N})$ where $\boldsymbol{s}_{-i}$ is a vector of the other player(s)' submission *s.t.* $\boldsymbol{s}_{-i} \in \boldsymbol{S} = \prod_{-i}\{1, \cdots, M\}$ with s_i^k being player i's k-th strategy and $\boldsymbol{s}_{-i}(t)$ is the strategy vector probably chosen at turn t *s.t.* $\boldsymbol{s}_{-i}(t) = (s_1(t), \cdots, s_{i-1}(t), s_{i+1}(t), \cdots, s_N(t))$. Then we write them as $L_i(t) = \sum_{\boldsymbol{s}\in\boldsymbol{S}} L^{\boldsymbol{s}}_{-i}(t)$ and $B^{\boldsymbol{s}}_{-i}(t) = \frac{L^{\boldsymbol{s}}_{-i}(t)}{L(t)}$ respectively, with $L^{\boldsymbol{s}}_{-i}(t) \geq 0$ and $L(t) \geq 0$. Note that the possible combination of other player(s)' submission is obtained by $s_i(t)$ and winning integer $v(t)$. For instance, when player i submits 1 and the winning integer is 0 (no winner) in $(N, M) = (3, \cdot)$ DIY-L, she can imagine that the others also choose 1. Or, when player i submits 1 and the winning integer is 1 (player i wins) in $(N, M) = (3, 3)$ DIY-L, she will expect that the possible submissions are $\{2, 2\}$ *w.p.* $1/4$, $\{2, 3\}$ *w.p.* $1/2$, and $\{3, 3\}$ *w.p.* $1/4$. Indeed, the numbers of combinations for each set ups are 6 in $(N, M) = (3, 3)$ DIY-L, 10 in $(N, M) = (3, 4)$, $(4, 3)$, and 20 in $(N, M) = (4, 4)$ respectively.
  After a turn, their beliefs are updated as

$$B^{\boldsymbol{s}}_{-i}(t) = \frac{(1-\phi_f)\cdot L^{\boldsymbol{s}}_{-i}(t-1) + \phi_f \cdot G(s_i(t), v(t), \boldsymbol{s}_{-i}, \boldsymbol{s}_{-i}(t))}{\sum_{\boldsymbol{s}'_{-i}\in\boldsymbol{S}}(1-\phi_f)\cdot L^{\boldsymbol{s}'}_{-i}(t-1) + \phi_f \cdot G(s_i(t), v(t), \boldsymbol{s}'_{-i}, \boldsymbol{s}'_{-i}(t))}$$

  where ϕ_f is a learning parameter and $G(s_i(t), v(t), \boldsymbol{s}_{-i}, \boldsymbol{s}_{-i}(t))$ is a function which determines the probability that the others probably choose a set of integers.
  Then the expected payoff for an integer k at turn t is calculated as

$$E_i^k(t) = \sum_{\boldsymbol{s}\in\boldsymbol{S}} B_i^{\boldsymbol{s}}(t)\pi_i(s_i^k, \boldsymbol{s}_{-i}(t))$$

  where $\pi_i(s_i^k, \boldsymbol{s}_{-i}(t))$ is player i's payoff for choosing integer k at turn t.
  Finally, strategy j at turn t is selected based on the exponential choice rule,

$$p_i^k(t) = \frac{\exp(\lambda_f \cdot E_i^k(t))}{\sum_{k'=1}^{M}\exp(\lambda_f \cdot E_i^{k'}(t))},$$

  where $p_i^k(t)$ is the probability that player i selects strategy k at turn t, and λ_f is the sensitivity of probabilities to expected payoffs.

Table 2. Game patterns in $(N, M) = (3, 3)$ lottery for the last 5,000 turns

a. One QFP vs. Two ALs					b. Two QFPs vs. one AL				
ϕ.	λ.	Pattern 1	Pattern 2	Pattern 3	ϕ.	λ.	Pattern 1	Pattern 2	Pattern 3
0.1	0.1	32	33	35	0.1	0.1	31	35	34
0.1	1.0	67	24	9	0.1	1.0	12	27	61
0.1	10.0	5	14	81	0.1	10.0	82	12	6
0.1	100.0	44	20	36	0.1	100.0	23	61	16

Table 3. Game patterns in $(N, M) = (3, 4)$ lottery for the last 5,000 turns

a. One QFP vs. Two ALs					b. Two QFPs vs. one AL				
ϕ.	λ.	Pattern 1	Pattern 2	Pattern 3	ϕ.	λ.	Pattern 1	Pattern 2	Pattern 3
0.1	0.1	52	27	21	0.1	0.1	22	35	43
0.1	1.0	94	6	0	0.1	1.0	0	5	95
0.1	10.0	1	12	87	0.1	10.0	76	14	10
0.1	100.0	60	0	40	0.1	100.0	15	85	0

With these learning algorithms, we run the computational experiments under the following conditions:

- Each game has at least one QFP agent and at least one AL agent. Hence, two kinds of three-person DIY-L and three kinds of four-person DIY-L are considered:
  - Three-person DIY-L
    - Two QFPs vs. one AL
    - One QFP vs. two ALs
  - Four-person DIY-L
    - Three QFPs vs. one AL
    - Two QFPs vs. two ALs
    - One QFP vs. three ALs
- It has 10,000 turns, iterated 100 times.
- Each player knows current turn, her previous submission, and the previous winning integer (if there is no winner, then this input is zero) for her decision-making. Which means, she does not directly know the previous submissions of others.
- Parameters are as follows: $\phi_a = \phi_f = 0.1$[5], $\lambda_a = \lambda_f = 0.1,\ 1.0,\ 10.0,\ 100.0$, $w_{i,k}(0) = 0.0$ for $k = 1, \cdots, M$, $L^{s}_{-i}(0) = 0$ and $L_i(0) = 0$ for $i = 1, \cdots, N$. If there are plural QFP or AL agents, they use the same initial conditions.

3.2 Result

The results presented in this section are from the simulation runs for the last 5,000 turns and we classified the simulation runs into several game patterns in accordance with the number of wins for each agent.

[5] We also pursued simulations with $\phi_a = \phi_f = 0.5$ and 0.9. But due to limited space, we report the results with $\phi_a = \phi_f = 0.1$ only.

Table 4. Game patterns in $(N, M) = (4, 3)$ lottery for the last 5,000 turns

a. One QFP vs. three ALs

ϕ.	λ.	Pattern 1	Pattern 2	Pattern 3	Pattern 4
0.1	0.1	29	21	21	29
0.1	1.0	64	23	10	3
0.1	10.0	92	7	0	1
0.1	100.0	96	4	0	0

b. Two QFPs vs. two ALs

ϕ.	λ.	Pattern 1	Pattern 2	Pattern 3	Pattern 4	Pattern 5	Pattern 6
0.1	0.1	23	15	12	23	10	17
0.1	1.0	50	19	15	10	1	5
0.1	10.0	0	0	0	23	24	53
0.1	100.0	0	0	0	100	0	0

c. Three QFPs vs. one AL

ϕ.	λ.	Pattern 1	Pattern 2	Pattern 3	Pattern 4
0.1	0.1	22	23	29	26
0.1	1.0	17	7	29	47
0.1	10.0	100	0	0	0
0.1	100.0	100	0	0	0

- Three-person DIY-L ($N = 3$)
  Each game setup has the following three game patterns in accordance with the ranking of QFP (or AL) player:
  - One QFP vs. two ALs
    Pattern x ($x = 1,\ 2,\ 3$) means that the QFP player is the x-th prize.
  - Two QFPs vs. one AL
    Pattern x ($x = 1,\ 2,\ 3$) means that the AL player is the x-th prize.
- Four-person DIY-L ($N = 4$)
  There are six game patterns in case that the players are equally split into two types while four patterns are considered otherwise:
  - One QFP vs. three ALs
    Pattern x ($x = 1,\ 2,\ 3,\ 4$) means that the QFP player is the x-th prize.
  - Two QFPs vs. two ALs
    Pattern x means that the ranks of QFP agents are 1st and 2nd ($x = 1$), 1st and 3rd ($x = 2$), 1st and 4th ($x = 3$), 2nd and 3rd ($x = 4$), 2nd and 4th ($x = 5$), and 3rd and 4th ($x = 6$).
  - Three QFPs vs. one AL
    Pattern x ($x = 1,\ 2,\ 3,\ 4$) means that the AL player is the x-th prize.

Tables from 2 to 5 summarize how many simulation runs there are observed in each game pattern for each game setup. The game patterns depend on both the game setup and the learning parameters: First, the games with a smaller λ. (= 0.1) shows a somewhat randomized behavior, but this is not surprising because this comes from the fact that the logit functions $\exp(\lambda_a \cdot w_{i,k}(t))$ and

Table 5. Game patterns in $(N, M) = (4, 4)$ lottery for the last 5,000 turns

a. One QFP vs. three ALs

$\phi.$	$\lambda.$	Pattern 1	Pattern 2	Pattern 3	Pattern 4
0.1	0.1	32	23	25	20
0.1	1.0	96	4	0	0
0.1	10.0	100	0	0	0
0.1	100.0	100	0	0	0

b. Two QFPs vs. two ALs

$\phi.$	$\lambda.$	Pattern 1	Pattern 2	Pattern 3	Pattern 4	Pattern 5	Pattern 6
0.1	0.1	21	23	17	16	13	10
0.1	1.0	83	16	0	1	0	0
0.1	10.0	0	8	10	41	26	15
0.1	100.0	0	0	0	100	0	0

c. Three QFPs vs. one AL

$\phi.$	$\lambda.$	Pattern 1	Pattern 2	Pattern 3	Pattern 4
0.1	0.1	16	20	26	38
0.1	1.0	0	1	10	89
0.1	10.0	100	0	0	0
0.1	100.0	100	0	0	0

$\exp(\lambda_f \cdot E_i^k(t))$ takes a value close to unity. Second, for $\lambda. = 1.0$, AL player(s) often became the loser(s). This implies that it it preferable for the players to have more information to win the game at this stage. Third, as sensitivity parameter $\lambda.$ goes larger, the game patterns become different between three-player DIY-L and four-player one. In four-player DIY-L, on the one hand, only one QFP (or AL) player could successfully dominated the game meanwhile one AL players is the 1st prize and the 4th and two QFP players share are the 2nd and the 3rd when DIY-L is something like a doubles game. This is independent of M and $\lambda.$. On the other hand, in three-player DIY-L, there is no apparent characteristics. It is true that AL player(s) have difficulties in winning the game for $\lambda. = 10.0$, but there is mixed evidence for $\lambda. = 100.0$, namely AL players won in some runs but not so in others.

Figures from 1 to 3 show what kind of integers each ranked player submitted and, if she is a winner, then won. Here, we provide the following characteristic results: $(N, M) = (3, 4)$ DIY-L with $\phi. = 0.1$, $\lambda. = 100.0$, and one QFP vs. two ALs (Figure 1), $(N, M) = (3, 4)$ DIY-L with $\phi. = 0.1$, $\lambda. = 100.0$, and two QFPs vs. one AL (Figure 2), and $(N, M) = (4, 3)$ DIY-L with $\phi. = 0.1$, $\lambda. = 100.0$ (Figure 1). In three-person DIY-L, a larger sensitivity parameter does not lead to only one game pattern (Tables 2 and 3). What Figures 1 and 2 indicate that a player persistently submitting the smallest integer becomes the 1st prize and giving up choosing such an integer has possibilities to be the silver

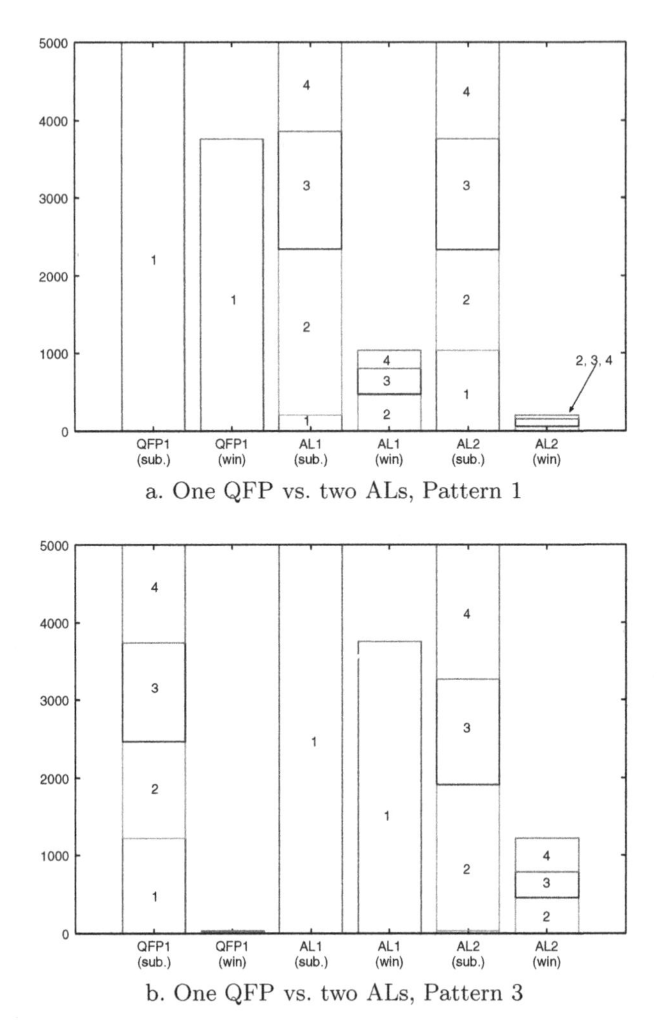

a. One QFP vs. two ALs, Pattern 1

b. One QFP vs. two ALs, Pattern 3

Fig. 1. Average submitted and winning integers for each ranked player in in $(N, M) = (3, 4)$ DIY-L with $\phi_{\cdot} = 0.1$ and $\lambda_{\cdot} = 100.0$

medalist[6]. Now that the integers 1 and one of the larger integers are occupied, the lowest ranked player has almost no chance to win, which makes her submit one integer randomly. A possible reason why no runs belong to pattern 2 in $(N, M) = (3, 4)$ with one QFP and two ALs is that the decision-making process of QFP players is more costly (Figure 1b and 2b)[7]. On the other hand, in

[6] Note that the silver medalist does not submit one of the integers except 1 with equal probabilities. In other words, the 2nd prize agent keeps submitting one integer, 2, 3, or 4.

[7] This is true for the games with $\lambda_{\cdot} = 10.0$, namely the reason why one of the adaptive learners is the 1st ranked is that she learns to submit 1 relatively more often than others.

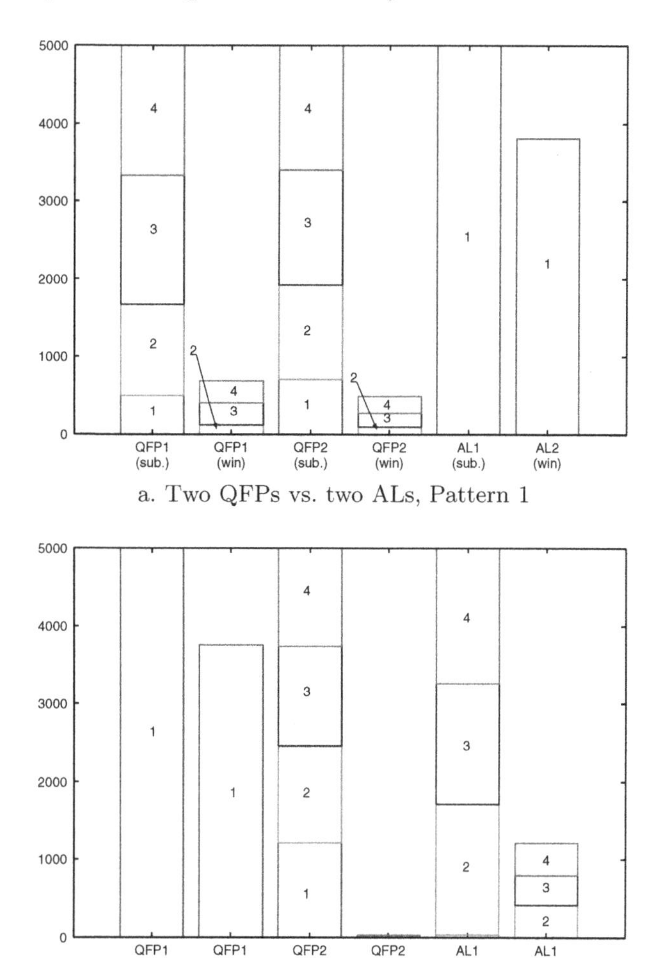

a. Two QFPs vs. two ALs, Pattern 1

b. Two QFPs vs. two ALs, Pattern 2

Fig. 2. Average submitted and winning integers for each ranked player in in $(N, M) = (3, 4)$ DIY-L with $\phi. = 0.1$ and $\lambda. = 100.0$

four-person DIY-L with $\lambda. = 100.0$, not only unique game pattern is reached for each setup but also different individual and aggregate dynamics are observed[8]. What is common to the panels in Figure 3 is that the QFP agent(s) tend to submit the smallest integer. When there is only one QFP agent in the lottery, she can be a winner because the AL players do not take into consideration the possible submissions of others and eventually avoid from submitting this integer. Then one of the AL agents adaptively and persistently chooses the second smallest integer, which makes her be the 2nd prize (Figure 3a). In this sense, this is similar to that in Figure 1a. But, when there are plural QFP players, their

[8] Similar results are obtained for $\lambda. = 10.0$.

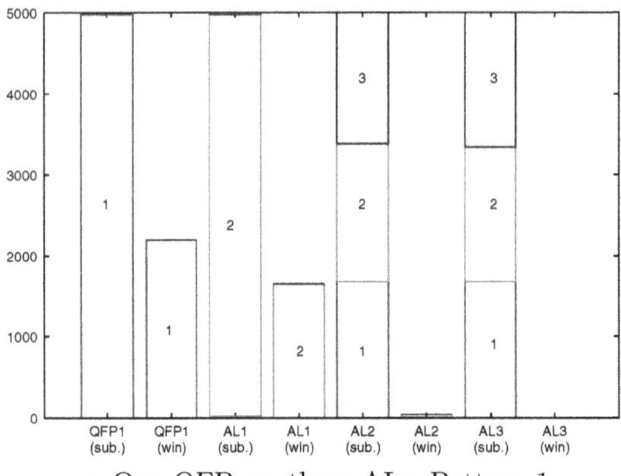

a. One QFP vs. three ALs, Pattern 1

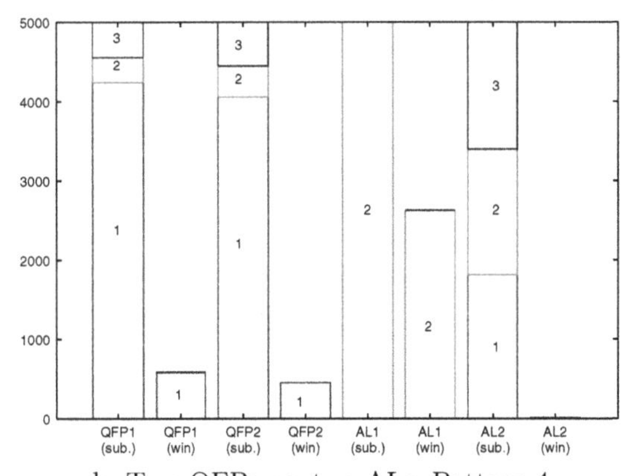

b. Two QFPs vs. two ALs, Pattern 4

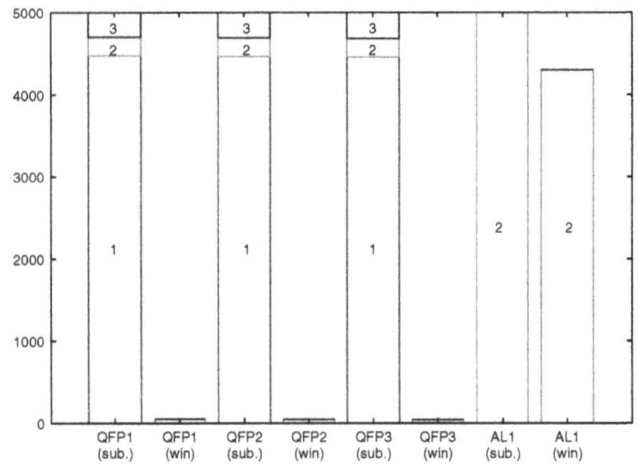

c. Three QFPs vs. one AL, Pattern 1

Fig. 3. Average submitted and winning integers for each ranked player in in $(N, M) = (4, 3)$ DIY-L with $\phi_{\cdot} = 0.1$ and $\lambda_{\cdot} = 100.0$

relatively sophisticated but not perfectly rational behaviors ironically prevent from winning the game; The QFP agents are likely to submit the smallest integer with the largest expected payoff. One AL player then learns that choosing 2, not 1, is more appropriate. As a result, unlike in three-player DIY-L with two (or more) QFP players, the way of thinking of AL players, but only for one of them, can make advantage of the cleverer thinking and behavior of QFP players in the end.

4 Concluding Remarks

This study investigates Barrow's "do-it-yourself lottery" with two types of players by agent-based computational approach. In a game with incomplete information, players cannot know the past plays of others but they can imagine how others behave from their own plays and the game results. Here we incorporate adaptive learning and quasi fictitious play into DYI-L and implement computational experiments by changing the game setup and the learning parameters. Our preliminary results show that sophistication works well only when other players are all adaptive in four-player games. In other words, adaptive player(s) can win the game as quasi fictitious play agents increase. On the other hand, three-player games are something like "on a first-come-first-served basis."

Acknowledgment. We thank two anonymous referees for useful comments and suggestions. Financial support from Japan Society for the Promotion of Science (JSPS) Grant-in-Aid for Young Scientists (B) (24710163), from Canon Europe Foundation under a 2013 Research Fellowship Program (Yamada), and from JSPS and ANR under the Joint Research Project, Japan – France CHORUS Program, "Behavioural and cognitive foundations for agent-based models (becoa)" (Terano) is gratefully acknowledged.

References

1. Barrow, J.D.: 100 essential things you didn't know you didn't know. Bodley Head (2008)
2. Brenner, T.: Agent learning representation: advice on modelling economic learning. In: Tesfatsion, L., Judd, K.L. (eds.) Handbook of Computational Economics: Agent-Based Computational Economics, vol. 2, pp. 895–947 (2006)
3. Broom, M., Cannings, C., Vickers, G.T.: Multi-player matrix games. Bulletin of Mathematical Biology 59, 931–952 (1997)
4. Camerer, C.F.: Behavioral game theory: experiments in strategic interaction. Princeton University Press (2003)
5. Crawford, V.P.: Learning behavior and mixed strategy Nash equilibria. Journal of Economic Behavior & Organization 6, 69–78 (1985)
6. Cheung, Y.W., Friedman, D.: Individual learning in normal form games: some laboratory results. Games and Economic Behavior 19, 46–76 (1997)

7. Erev, I., Roth, A.E.: Predicting how people play game: reinforcement learning in experimental games with unique, mixed strategy equilibria. American Economic Review 88, 848–881 (1998)
8. Goeree, J.K., Yariv, L.: An experimental study of collective deliberation. Econometrica 79, 893–921 (2010)
9. Haruvy, E., Stahl, D.O.: Equilibrium selection and bounded rationality in symmetric normal form games. Journal of Economic Behavior & Organization 62, 98–119 (2007)
10. Jordan, J.S.: Three problems in learning mixed-strategy Nash equilibria. Games and Economic Behavior 5, 368–386 (1993)
11. Matsumura, M., Ikegami, T.: Evolution of strategies in the three-person iterated prisoner's dilemma game. Journal of Theoretical Biology 195, 53–67 (1998)
12. Östling, R., Wang, J.T., Chou, E.Y., Camerer, C.F.: Testing game theory in the field: Swedish LUPI lottery games. American Economic Journal: Microeconomics 3, 1–33 (2011)
13. Platkowski, T.: Evolution of population playing mixed multiplayer games: Mathematical and Computer Modelling 39, 981–989 (2004)
14. Shoham, Y., Powers, R., Grenager, T.: If multi-agent learning is the answer, what is the question? Artificial Intelligence 171, 365–377 (2007)
15. Vu, T., Powers, R., Shoham, Y.: Learning in games with more than two players. In: Fifthe International Joint Conference on Autonomous Agents and Multi Agent Systems (AAMAS 2006). USB Memory (2006)
16. Walker, M., Wooders, J.: Mixed strategy equilibrium. In: Durlauf, S.N., Blume, L.E. (eds.) Game Theory, pp. 235–239. Palgrave Macmillan (2010)
17. Weizsäcker, G.: Do we follow others when we should? A simple test of rational expectations. American Economic Review 100, 2340–2360 (2010)
18. Wilkinson, N., Klaes, M.: An introduction to behavioral economics, 2nd edn. Palgrave Macmilan (2012)

Agent Based Simulation of Drought Management in Practice

Olivier Barreteau[1], Eric Sauquet[2], Jeanne Riaux[3],
Nicolas Gailliard[4], and Rémi Barbier[5]

[1] Cemagref, Montpellier, France
olivier.barreteau@cemagref.fr
[2] IRSTEA, Lyon, France
eric.sauquet@irstea.fr
[3] IRD, Tunisia
jeann.riaux@ird.fr
[4] IRSTEA, Montpellier, France
nicolas.gailliard@irstea.fr
[5] ENGEES, Strasbourg, France
rbarbier@engees.u-strasbg.fr

Abstract. Drought management in France is implemented locally. Due to discrepancies between assessment of drought situation by managing agency on one hand and water users on the other hand, as well as to uncertainty in measures and benchmarks, its efficiency is limited. We propose in this paper an agent based model designed to represent the suitable indicators of drought at the suitable spatial scale for any category of stakeholders. Initial test of the model show its suitability to explore sensitivity of efficiency of drought management setting according to its context: population of water users and their attitudes to water restriction rules as well as practical details of implementation.

Keywords: Drought management, rule enforcement, spatial indicators, exploratory simulation.

1 Introduction

The French water act institutionalizes a drought committee at county level. Such committee sets the rules characterizing a situation of drought and how to react when such situations occur. Characterization of a drought situation depends on two different activities: (i) defining benchmarks usually with thresholds and reference to past chronicles, and (ii) assessing current water levels to be compared to these thresholds. These both activities are in practice rather complex, due to several reasons including limited data sets across time and space and multiplicity of resources. Hence, actual characterization of a drought situation is controversial due to the salience of the issue for participants in such committees combined with the multiplicity of possible benchmarks, all incorporating uncertainty. The consequences of the practice of these activities as they are framed by the setting from the local decree are still unknown.

Through a consultancy for the French National Agency for Water and Aquatic Environments (ONEMA), we first made explicit these controversies [1], the origin of

B. Kaminski and G.Koloch (eds.), *Advances in Social Simulation*,
Advances in Intelligent Systems and Computing 229,
DOI: 10.1007/978-3-642-39829-2_21, © Springer-Verlag Berlin Heidelberg 2014

uncertainties making them possible [2] and the consequential mistrust in the implementation of local drought management acts [3]. Emergence of controversies is also fostered by the diversity of ways participants to drought committee meetings can assess drought situation by themselves: different places (e.g. where their well is located, or the bridge where they cross the river is) and different resources (groundwater or surface water). Each user comes also with his/her own indicator to characterize drought: water level but also length of riverbed without water.

All come then with their own view gathered in the single possible assessment: drought/No drought. Drought committee will rather ends up with a negotiated assessment. Implementation and respect of rules which is formally generated by this assessment will then depends on the adhesion of water users to the assessment, even more with the weakness of means for control. This raises an issue of effectiveness and fairness of these drought management acts according to the diversity of possible scenarios for implementing them. These initial empirical studies could make clear this concern of controversies in implementation while it is supposed to have all the rigor of science. To go further, we needed some simulation tool to explore contrasted scenarios of implementation of local drought management policy.

To create this tool, we took a pragmatic stance: explaining dynamics generated by this policy with "situated action" instead of planned action [4-5]. Agent Based Modelling has a suitable format for this [6]. More recently Guerrin [7] proposed a dedicated framework to represent action with a stance close to situated action paradigm. Hence we have decided to go for a virtual case study implemented in an ABM and empirically grounded through previous interviews and ethnographic analyses. The specific requirements for this virtual case include being spatially explicit enough to generate information about the water system according to the diversity of observations: place and type of data collected.

In this paper we present the simulation model that has been designed and implemented and the bottlenecks we had in building it in order to be able to represent knowledge coming from the ethnographic work and how we solved them. A first section comes back to the pragmatic situation of drought management facing controversies due to the diversity of indicators of water level and places to assess them among participants to drought committees (stakeholders and county administration). Second section defines the requirements it implies for modeling this process, with a focus on the perception interface between the natural system and stakeholders. Third section describes the model itself. In a last part we present first simulation outcomes in two situations: without uses in order to validate the environmental dynamics, and with irrigation uses and discrepancies in place of observation.

2 Drought Management Act in Practice

2.1 Main Features of Drought Management in France

Drought management at county level in France is a downscaling of a national frame set by the 1992 water act, leaving up to the county administration to enforce it in specific decrees adapted to local situation. These decrees set local protocols to anticipate for and handle water shortage situations. Major droughts in 2003, 2005 and

2006 made this issue crucial for close to all counties. Hence "drought action plans" have become generalized. These anticipate periods of water scarcity and propose rules to attenuate their consequences or to decrease their occurrence. These protocols involve three main institutions: the administrative authority legitimate to restrict water use (the Prefet and administrative services), a drought committee gathering various water users, and an infrastructure used for assessing the situation, including its rules of use and the knowledge about other users it encapsulates. The drought committee meets up in winter time, out of crisis period, to discuss and occasionally adapt the infrastructure. They also meet up during crisis to discuss the current situation, restrictions activated in consequences (according to what had been agreed upon in winter time), and possible derogations. Infrastructure entails deciding whether there is a crisis or not, and in case its severity. This implies formalizing:

- thresholds of water levels in specific places (exact place and choice between groundwater or surface water) characterizing situation of droughts,
- protocols to acquire the knowledge on water levels to be compared with thresholds,
- A partition of county into subsystems to cope with diversity in resource availability and uses within county, each of them having their own sets of thresholds of referential water levels.

2.2 Diversity of Indicators and Thresholds

In practice, several drawbacks occur, including information gathering and processing. According to farmers, the existing partition is not at a fine enough grain to cope with the diversity of perceptions. They push for a more subdivided partition, conflicting with another trend –and demand- for more solidarity and equity among water users at a larger scale. Farmers know water levels from the places where they pump water, and from what they see in their farm, where they pump water or along their way to the county main city (to go to the committee meeting for example). Stakeholders more concerned by ecological concerns, including representatives of fishermen, observe for example the length of river with no running water. Representatives of administration have a few automated monitoring places, but sometimes less than the number of areas in the partition, due to their cost. Thresholds are characterized from statistics on past chronicles. But due to lack of data, they are often engineered from proxy data and adapted. This situation leads to mistrust in the drought management institution and requests for postponing implementation of restriction rules.

2.3 Consequences for Enforcement of Rules

Hence, the implementation of these local "Drought action plans" consists permanently in crafting adjustments to reality of situations and needs, evolving with experience, needs and knowledge. Water users tend then to criticize or even disqualify the institution because of three reasons: observation of existing adaptation with the rules, uncovering of the uncertainty behind data used to implement the rules, disagreement with the suitability of drought observation compared to their own observations. This third source of disqualification maybe due to observation of rules applied to other water users for those located close to one border of the institutional zoning for drought management, but also to the existence of plurality of water resources (e.g. ground VS surface water) or misunderstanding of correlations between their use and availability of water elsewhere.

Paradigm of situated action [5] is then suitable to analyze this process. Attitudes of participants in the Drought Action Plan implementation depend mainly on the on-going context, based on their perception of water availability and needs as well as their perception of some fairness and ecological concerns. From a methodological point of view to organize observations in a way suitable with this paradigm, we considered Operational sequences [8] as a means to describe how stakeholders process for crucial activities such as assessing drought situations and drought references [2].

3 A Model to Explore Drought Management Patterns

We focus now on the issue of evaluation and agreement on evaluation among stakeholders, including administrative authority and water users. The challenge is the explicit representation of conflicting views on environment and its impact on the evolution on a socio-ecosystem.

3.1 Requirements for the Model

For the sake of simplification, we assume that water use can be limited to irrigation for it is the main quantitative use at low water times. We also assume the presence of a policy maker that helps to ensure suitability of drought management process.

With this model, we aim at analyzing sensitivity of the water system and water uses to various scenarios of institutional settings and individual behavioral patterns of water users in complying with restriction rules. The situation and focus described above generates the following requirements for model development:

- a spatially explicit representation of hydrological dynamics, with a granularity able to cope with the farm scale and distributed observation at each time step,
- representation of pumping at farm level according to perception of drought, and knowledge on rules in use at county level,
- dynamic representation of pumping consequences on surface discharge,
- representation of dynamics of implementation of rules at county level according to their efficiency on drought mitigation,

Finally we consider that stakeholders are first related to specific spatial objects. These can be specific places, such as a pump's location, a measurement station or any specific landmark. They can also be aggregated pieces of land, such as a river, with fishermen representatives for example assessing drought through the length of a river without flowing water.

Simulation of drought management policy in practice fosters then the need for specific modeling of resource dynamics: scale of observation of resource by stakeholders provides the scale of spatialisation of flow representation. Hydrological signal needs to be realistic at this spatial scale since it might be used for feedback in the water assessment component. Additionally, model needs to provide a dual representation of levels and flows. These both requirements are rather new for hydrological modeling. We now present how we adapted previous hydrological modeling frameworks to cope with them.

4 Existing Approaches for Hydrological Modeling

Hydrological modelling is currently divided into two main categories: conceptual models and physically based models. Conceptual models provide a simplified representation of the general behaviour of the catchment based on the continuity equation as well as additional mathematical relationships to simulate the links between rainfall and surface runoff. In this category GR models [9] for example are conceptual models based on relations between data series of inputs and outputs of water and calibrated on past series. Physically based models which try to represent the rainfall-runoff transformation based on the understanding of hydrological mechanisms which control the response through physically based equations. They aim at being explicit on water flows between surface, soil and ground water compartments which ends up in impacting on hill slope flows and river discharge [10]. Most of them are spatially distributed models accounting for the variability in the input variables as well as in the properties which influences the processes across the catchment. These distributed physically based models, such as in [11-12] among others, represent the dynamics of water flows on a landscape represented as a computing grid, from inputs due to rain to outputs including evaporation. Both categories of models are able to tackle connections between surface and groundwater. When dealing with whole river basins or territories equivalent to the size of a county, distributed models are still at a scale rather too large to cope with farm level, in order to represent the hydrology and the discharge at the outlet of the basin in an acute way.

More recently a few scholars have attempted to represent hydrological process in a fully distributed way, based on techniques such as cellular automata or agent based modeling. Delahaye and colleagues [13] have represented interactions between land use and flows with a cellular automaton featuring a topological graph with the only coded characteristics being the topography. Water then flows on this simulated landscape. A more extreme attempt has agentified “water bowls” in the RIVAGE model. In this model elementary particles of water move according to basic physical laws on a given landscape, they can meet up and aggregate in various form of water bodies [14]. These both innovative approaches have inspired our work. However they handle much more local scale than our need.

4.1 Integrated Hydrological Models and Agent Based Models

Agent based modeling is currently a common approach to represent the dynamic relations between a hydrological model and water uses. Le Page and colleagues review several of these in a recent chapter [15]. Berger uses them to represent impacts of technical innovation and policy changes in Chile, coupling economic impacts for farmers with new water policy setting in Chilean sub basins [16]. Van Oel and his colleagues use an ABM to represent dependence of land use decisions in arid northeast Brazil on practical water availability. Their model of water uses choice impact on a semi-distributed hydrological model with land cell and rivers represented as sequences of branches [17].

In several cases these models are considered useful for interaction with stakeholders, including because of their adaptability to explore various scenarios for example of water users' preferences or of their context of work such as climate [18]. This interactivity is best represented in participatory modeling cases such as the KatAWARE model designed in a south African basin [19]. These authors design an Agent based Model of a river basin taking in charge the suitable entities to cope with the various viewpoints of stakeholders according to their suggestions in the modeling workshops. Becu and colleagues [20] have proposed a whole method for eliciting conflicting views on what drives farmers in their practice up to including these heterogeneous representations within a single agent based model.

5 The GESPER Model

In this section we describe the GESPER model, based on a cellular automaton for the physical part and an ABM to include social and behavioral dimensions.

5.1 Physical Part of the Model

The physical layer is made of a grid of cells with a connectivity of four. We assume a cell representing a square of 500x500 m^2. Each square cell is made of three compartments: surface, sub-surface and groundwater. These cells are first described by their altitude, soil characteristics i.e. a soil water capacity, and ground water capacity. We apply to these elementary cells the algorithm of MERCEDES model [10] according to figure 1 below. This means adding further parameters to the cell characteristics to handle interfaces between compartments: infiltration rate, deep release rate and superficial release rate.

Surface transfer between cell is adapted from MODCOU algorithm [21-22], a physically distributed hydrological model with variable scale. This model needs to predefine two parameters of transfer of the cell, that depend on cell's hydrological status (part of a river or not). These two parameters, aDist and aVol, entail specifying the quantity of water is leaving a cell during one time step and the cell this "water pack" reaches. The distance of transfer (distanceTransfer) and volume of transfer (volTransfer) of this water pack during one time step is computed according to the equations below, where slope is the mean slope between initial and final cell and volStock is the current water level on the initial slope.

$$distanceTransfer = (timeStep * \sqrt{slope})\ /\ aDist$$
$$volTransfer = aVol * volStock$$

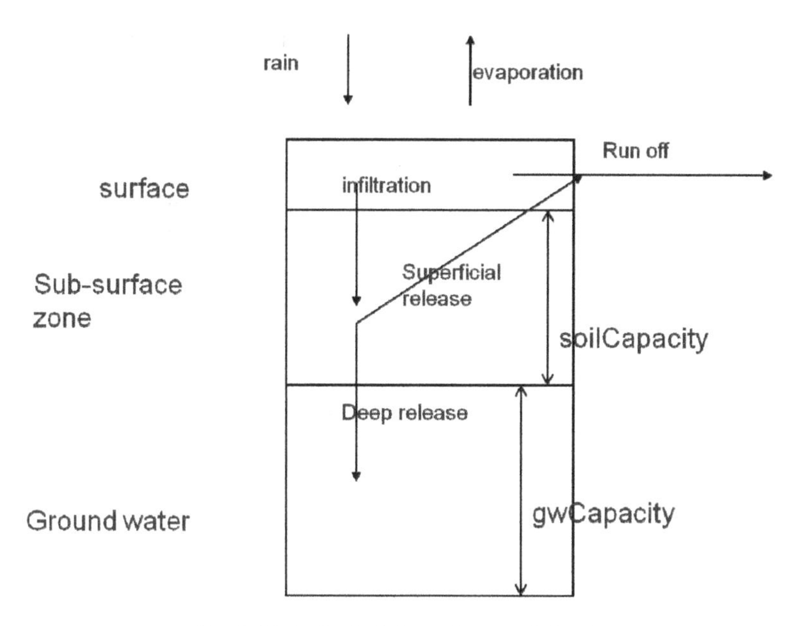

Fig. 1. Hydrological dynamics at cell level

When distanceTransfer is computed, the transfer path of the water pack is determined, according to the lowest altitudes. A key benefit of this modelling approach is the possibility to represent stocks and flows at the same time, since we can compute for each cell the quantity of water going through it during the time step and its time of residence in the cell. Any withdrawing along the flow can also be computed in any cell through decreasing the quantity of water flowing through it. Quantity withdrawn from each water pack is proportional to time of residence of this water pack on the cell, with constraint of the total quantity of water withdrawn during the time step and quantity of water available for each water pack.

Underground flow between cells is adapted from [23]. It features an additional attribute for Cell, deepTransfer, such that a cell c1 will transfer to a cell c2 deepTransfer * (c1 saturation – c2 saturation), if c1 saturation > c2 saturation and if c2 is the cell with the lowest saturation in c1 neighbourhood. A cell saturation is computed as the difference between its groundwater compartment capacity and its groundwater compartment content. Further each cell has a leakage from its groundwater compartment which is calibrated so that this compartment stays stable in average from one year to another without withdrawals. Calibration has determined this leakage parameter at 1.6mm for each cell. Adding this parameter is needed because there is no outflow from ground water.

Model handles specifically boundary cells to prevent from boundary effects. Each boundary cell goes through the same surface flow process if it can identified a target cell with the rules explained above for non boundary cells. If this identification is unsuccessful, the target cell is a virtual cell, assumed to be 1 m below emitting cell. Each boundary cell goes through the same underground transfer as explained above if it has a neighboring cell with a lesser saturation, otherwise a fix transfer (deepEmission) to a virtual neighboring cell is generated.

The model can take any map providing a topography as an entry. Only condition is a cell size of 500mX500m and the absence of endoreism. We used it for two counties in France: Drôme river in South East and Oise upstream subbasin in the North. However we rather consider them as contrasted virtual landscapes. Figure 2 below provides a view of this virtual landscape made with the Drôme river map.

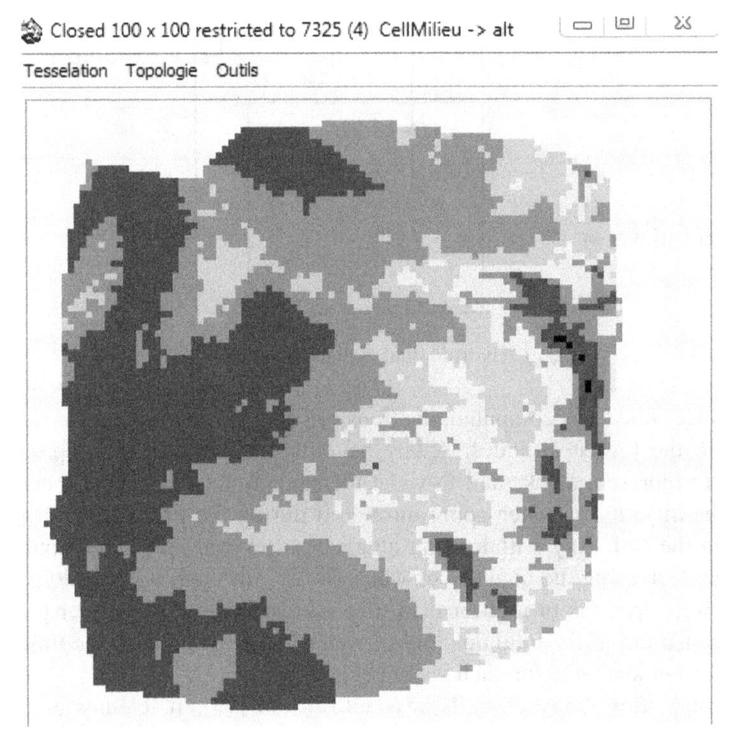

Fig. 2. View on altitudes in the virtual landscape. The darker green the lower, the darker brown the higher.

Climate is represented by potential evapotranspiration (PET) and rain. Even though we use in this example the same value for the whole area, the model is tailored to represent climate variability, with each cell having its own climate attribute.

5.2 Water Use Interface

The modeling work comes after a thorough ethnographic analysis of how stakeholders characterize a situation of drought, share this information with others and use all the gathered information in their actions [2] including patterns of negotiation in drought committee [3]. In the virtual landscape described above, each cell is either part of a farm or in public domain. All farms can be fed with water pumped either in surface water or in ground water through a pump entity (i.e. an instance of a class Pump) located on a cell belonging to a farmer and in a specific compartment in that cell (ground or surface). We assume that farmers have a full and perfect knowledge of water needs and are able to pump whenever they need if water is available. Farmers assess drought situation through water level at the location of their pump in the relevant compartment (surface or groundwater) with comparison to their own reference level.

Administration is represented by a single agent, an instance of PolicyMaker, with a limited set of reference points and threshold associated to them and uses. We start with only one reference point: the outlet of the largest basin in the area. When a threshold is crossed, the rule forbids pumping to farmers in the associated area for the stipulated water use. A major assumption for this agent is its incapacity of controlling respect of restrictions. He has to rely on the willingness to comply of water users. Other assumption regarding PolicyMaker is periodicity of its activity that is supposed to be weekly: drought situation is assessed only every 7 days.

Water users are endowed with three possible attitudes regarding restriction rule compliance: respect the rule (attitude := respect), respect the rule when agree on drought situation (attitude := ownAssessment), don't respect the rule (attitude := noRespect). At each time step (the day), they decide to pump water according to their needs, the rules, their attitude regarding the rule and their own assessment of drought situation.

We assume at this level there are no direct interactions between users and administration.

6 Model Implementation, Calibration and Verification

Model is implemented with Cormas platform (http://cormas.cirad.fr). It provides several indicators to check its realism to empirical data: outflow for the various sub basins, groundwater levels in each cell, but also water balance at cell and whole county level. Outflows can be compared to observed discharge levels in the river basin used to generate the maps for orders of magnitude as well as specific statistical description of these flows, such as average value, variance, or the annual monthly minimum flow with a return period of 5 years, QMNA5.

First level of model verification is checking that simulations comply with water conservation at any scale. Figure 3 below shows the aggregated balance for the whole basin along 3000 days. Absolute value of this balance is always below 0.2mm, which is less than 1% of the average discharge at the main sub-basin outlet.

We calibrated the leakage parameter in order to be realistic in order of magnitude of flow. Then we compared simulated mean flows and QMNA5, which is representative of low water period, with observed ones over a 40 years period for the Drome river valley and could calibrate the other hydrological parameters.

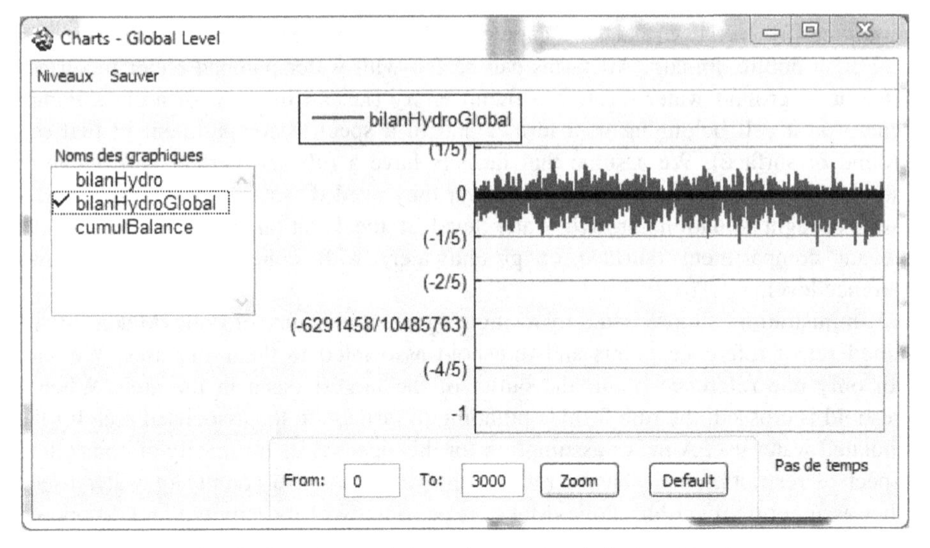

Fig. 3. Global water balance in mm. X-axis is time (days)

7 Simulation Outcomes

First simulation outcomes with this model show some sensitivity to farmers' attitudes with more acute crises when farmers respect the rules only when they agree on the assessment. The variation observed among the scenarios is not striking partly because we implemented a demand concentrated in a few points. Therefore the demand is first constrained by water availability in the place of pumping and not by the attitude of water user. Still we can see difference among scenarios with a higher amount of water pumped and less stress for corn when farmers use their own assessment than when they follow the restriction from the policy maker. This observation on simulation outcomes contributes to confirm model's validity. The impact on water level at the assessment place is relatively less important: it is more a delay in crisis when farmers respect the rule, but the crisis is not prevented. This is due to the existence of several sub basins. Several pumps actually have a very minor impact on the monitored sub basin, because they withdraw water from other sub basins while official assessment is in surface compartment. This might contribute to the present debate on subdivision of counties in a number of areas.

8 Conclusion and Perspectives

This experience shows the consequence on the representation of hydrology when we aim at exploring the enforcement of a water policy in practice: fine spatial distribution of representation and coupling of social and hydrological processes at this fine scale. We provided here the conceptual description of the model to meet this requirement, up to its implementation and initial tests. The verification of the model and the initial set of simulation show that the Gesper model is operational and proposes a sound basis to explore various scenarios of drought action plan and water uses. Capacity to reproduce realistic hydrologic patterns, including statistical description, makes it legitimate to be used to understand the dynamics induced by multiple uses and multiple assessments. In the example above farmers act upon their own qualification of water drought which can be different of qualification on the administration side. These qualifications are locally dependent but this location is encapsulated at each agent level. The outcome, on which the communication can occur, is whether there is a situation of drought or not. The model is fully spatially explicit and generates dynamically the hydrological state of each land cell. Granularity is still a little bit coarse (500m x 500m) for farm representation but it fits the data available for physical environment and allows keeping computing time low.

We expect this kind of tool to be useful at a meta-level. The water and aquatic environment agency considers that it will push water administration at county level to pay more attention to the assessment step in the enforcement of drought action plan. Discrepancies between assessments and the contestation which happens to occur in consequence are not only an issue of strategic game and acting in bad faith. It is also due to a true diversity of perceptions with potential consequences on effectiveness of the plan on drought situation.

Acknowledgement. This work has been done thanks to financial support of French environmental agency ONEMA.

References

1. Barbier, R., Barreteau, O., Breton, C.: Gestion de la rareté de l'eau: entre application négociée du « décret sécheresse » et émergence d'arrangements locaux. Ingénieres - EAT 50, 3–19 (2007)
2. Riaux, J., Barbier, R., Barreteau, O.: Construire et Argumenter des Enjeux de Vulnérabilité en Comité Sécheresse. In: Becerra, S., Peltier, A. (eds.) Risques et Environnement: Recherches Interdisciplinaires sur la Vulnérabilité des Sociétés. L'Harmattan, Paris (2009)
3. Barbier, R., Riaux, J., Barreteau, O.: Science réglementaire et démocratie technique. Réflexion à Partir de la Gestion des Pénuries D'eau 18, 14–23 (2010)
4. Conein, B., Jacopin, E.: Action située et cognition. Sociologie du Travail 94, 475–500 (1994)
5. Suchman, L.A.: Plans and situated actions: the problem of human-machine communication. Cambridge University Press (1987)
6. Ferber, J.: Multi-agent Systems. Addison-Wesley Longman, Reading (1999)
7. Guerrin, F.: Dynamic simulation of action at operations level. Autonomous Agent and Multi-Agent Systems 18, 156–185 (2009)
8. Lemonnier, P.: The study of material culture today: Toward an anthropology of technical systems. Journal of Anthropological Archaeology 5, 147–186 (1986)
9. Perrin, C., Michel, C., Andréassian, V.: Improvement of a parsimonious model for streamflow simulation. Journal of Hydrology 279, 275–289 (2003)

10. Bouvier, C., Delclaux, F.: ATHYS: a hydrological environment for spatial modelling and coupling with a GIS. In: HydroGIS 1996, pp. 19–28. IAHS Publication (1996)
11. Blöschl, G., Reszler, C., Komma, J.: A spatially distributed flash flood forecasting model. Environmental Modelling & Software 23, 464–478 (2008)
12. Beven, K., Freer, J.: A dynamic TOPMODEL. Hydrological Processes 15, 1993–2011 (2001)
13. Delahaye, D., Guermond, Y., Langlois, P.: Spatial interaction in the run-off process. Cybergeo: European Journal of Geography (2003)
14. Servat, D., Perrier, E., Treuil, J.-P., Drogoul, A.: When agents emerge from agents: Introducing multi-scale viewpoints in multi-agent simulations. In: Sichman, J.S., Conte, R., Gilbert, N. (eds.) MABS 1998. LNCS (LNAI), vol. 1534, pp. 183–198. Springer, Heidelberg (1998)
15. Le Page, C., Bazile, D., Becu, N., Bommel, P., Bousquet, F., Etienne, M., Mathevet, R., Souchère, V., Trebuil, G., Weber, J.: Agent-based modelling and simulation applied to environmental management: a review. In: Edmonds, B., Meyer, R. (eds.) A Handbook on: Simulating Social Complexity, pp. 499–540. Springer (2011)
16. Berger, T., Birner, R., Diaz, J., McCarthy, N., Wittmer, H.: Capturing the complexity of water uses and water users within a multi-agent framework. Water Resources Management 21, 129–148 (2007)
17. van Oel, P.R., Krol, M.S., Hoekstra, A.Y., Taddei, R.R.: Feedback mechanisms between water availability and water use in a semi-arid river basin: A spatially explicit multi-agent simulation approach. Environmental Modelling & Software 25, 433–443 (2010)
18. Valkering, P., Rotmans, J., Krywkow, J., Van der Veen, A.: Simulating Stakeholder Support in a Policy Process: An Application to River Management. Simulation 81, 701–718 (2005)
19. Farolfi, S., Müller, J.-P., Bonté, B.: An iterative construction of multi-agent models to represent water supply and demand dynamics at the catchment level. Environmental Modelling & Software 25, 1130–1148 (2010)
20. Becu, N., Bousquet, F., Barreteau, O., Perez, P., Walker, A.: A Methodology for Eliciting and Modelling Stakeholders' Representations with Agent Based Modelling. In: Hales, D., Edmonds, B., Norling, E., Rouchier, J. (eds.) MABS 2003. LNCS (LNAI), vol. 2927, pp. 131–148. Springer, Heidelberg (2003)
21. David, C.H., Habets, F., Maidment, D.R., Yang, Z.-L.: RAPID applied to the SIM-France model. Hydrological Processes 25, 3412–3425 (2011)
22. Ledoux, E., Girard, G., de Marsily, G., Villeneuve, J.-P., Deschenes, J.: Spatially distributed modeling: conceptual approach, coupling surface water and groundwater. In: Morel-Seytoux, H.J. (ed.) Unsaturated Flow in Hydrologic Modeling Theory and Practice, pp. 435–454. Kluwer, Dordrecht (1989)
23. Lanini, S., Courtois, N., Giraud, F., Petit, V., Rinaudo, J.-D.: Socio-hydrosystem modelling for integrated water-resources management—the Hérault catchment case study, southern France. Environmental Modelling & Software 19, 1011–1019 (2004)

Changing Climate, Changing Behavior: Adaptive Economic Behavior and Housing Markets Responses to Flood Risks

Tatiana Filatova[1,2] and Okmyung Bin[3]

[1] Centre for Studies in Technology and Sustainable Development, University of Twente, P.O. Box 217, 7500 AE Enschede, NL
t.filatova@utwente.nl
[2] Deltares, Postbus 85467, 3508 AL Utrecht, NL
[3] Department of Economics, Thomas Harriot College of Arts and Sciences, East Carolina University, Brewster A-435, Tenth Street, Greenville, North Carolina 27858-4353
bino@ecu.edu

Abstract. Spatial econometrics and analytical spatial economic modeling advanced significantly in the recent years. Yet, methodologically they are designed to tackle marginal changes in the underlying dynamics of spatial urban systems. In the world with climate change, however, abrupt sudden non-marginal changes in economic system are expected. This is especially relevant for urban development in coastal and delta areas where the probabilities of natural hazards such as catastrophic floods and hurricanes increase dramatically with climate change. New information about risks and micro-level interactions among economic agents alters individual location choices and impacts urban land markets dynamics potentially leading to the emergence of critical transitions from the bottom-up. We address this gap by incorporating adaptive expectations about land market dynamics into a spatial agent-based model of a coastal city. We build upon the previous research on agent-based modeling of urban land markets, and make a step forward towards empirical modeling by using actual hedonic study and spatial data for a coastal town in North Carolina, USA. Decentralized urban market with adaptive expectations about property prices in the areas with increasing hazard probabilities, may experience abrupt changes that shift the trends of spatial development and pricing.

1 Introduction

A major part of world population lives in coastal and delta areas. These areas are highly threatened by the adverse consequences of climate change as probabilities of severe disasters like Hurricanes Katrina and Sandy or European and Australian flooding of the last years increase Climate change might lead to a forced displacement of up to 187 million people in coastal zones (Nicholls et al. 2011). Yet, these risk are spatially correlated with rich amenities of those locations (Bin et al. 2008). Coastal and delta areas were historically developed due to their proximity to marine and river

B. Kaminski and G.Koloch (eds.), *Advances in Social Simulation*, Advances in Intelligent Systems and Computing 229,
DOI: 10.1007/978-3-642-39829-2_22, © Springer-Verlag Berlin Heidelberg 2014

transportation. Further developments are attracted to historic centers by agglomeration forces (Fujita and Thisse 2002) as well as by rich environmental amenities. As a result exposure and vulnerability in coastal areas rapidly increase due to the clustering of population and growth of property values in flood-prone areas (IPCC 2012). Land markets driven by individual preferences for locations play a crucial role in the formation of spatial patterns of activities and the economic value they receive (Randall and Castle 1985; Parker and Filatova 2008). Currently properties in coastal areas are expensive driven by coastal amenities and low subjective risk perceptions of traders in a housing market. As a matter of fact, many inhabitants of flood-prone areas worldwide have very low risk perceptions (Terpstra and Gutteling 2008; Ludy and Kondolf 2012). This is likely to reverse as climate change propagates causing potentially non-marginal changes in coastal land markets.

Modeling abrupt non-marginal changes in economic systems is challenging. Spatial econometrics and analytical spatial economic modeling advanced significantly in the recent years. Yet, methodologically they are designed to tackle marginal changes in the underlying dynamics of spatial urban systems (Varian 1992; Fujita and Thisse 2002; Hackett 2011). For example the result of spatial econometrics analysis is a hedonic function, which relates housing prices to a marginal change in the spatial and structural attributes of a property. For properties in hazard zone it also provides a marginal willingness to pay of a representative households for safety (i.e. or to avoid flood/erosion risk). Since a hedonic function is a snapshot of a market at a certain moment (Bockstael 1996), this willingness to pay for safety is not only constant across households but also over time despite the growth of climate-induced probabilities. Models tracing marginal changes are quite advanced and successful in projecting land-use trajectories and price dynamics along the existing trend when no abrupt irreversible changes in spatial environment or individual location preferences occur. However, in the world with climate change, abrupt sudden non-marginal changes in economic system are expected (Stern 2008). Personal experience of a disaster (Bin and Polasky 2004; Kousky 2010) as well as information about such shocking events elsewhere (Hallstrom and Smith 2005) affect individual expectations about future safety and their decisions where to locate. Hedonic price studies reveal that flood risk price discounts in hazard-prone areas are most evident immediately after a major hurricane and flood event due to updated risks perception. However, the information provided by the hazard event vanishes in 3-6 years, leading to changes in subjective risk perceptions and disappearing price differentials in risk zones (Lamond and Proverbs 2006; Bin and Landry Forthcoming). Thus, evolution of risk perception, which is triggered by hazard events (and possibly exacerbated through social interactions, i.e. opinion dynamics about flood risks in the area) plays a major role when buyers define their bid price on a property and when an eventual transaction takes place. As hazard probabilities grow with climate change we are to expect more frequent and more severe hurricane and flood events, leaving less chance of diminishing risks perceptions. As information about hazard event propagates and risk perception is updated, demand for properties which are at the greatest risk will fall leading to possible outmigration, price decrease and local housing bubbles in coastal cities.

In complex systems changes in individual expectations, especially driven by new information and emotions (Lux 2009; Anand et al. 2011) could lead to major abrupt shifts in the aggregated market dynamics. Specifically, certain currently very attractive coastal and delta areas might experience sudden out migration and housing market collapse. Pryce and Chen (2011) argue that conventional models of housing market dynamics in hazard-prone areas based on the historic data alone might not be able to shed light on how and under what conditions property prices and spatial patterns may change in a world with climate change. This also poses challenges for designing policies as decision support tools, which omit behavioral adaptation triggered by changing climate and emergence of potential regime shifts in economic systems, could be misleading. Thus, the strengths of conventional tools that are based on the decades of successful applications and validation could be reinforced by simulations, such as agent-based computational economics (ACE) (Tesfatsion and Judd 2006) to account for adaptive economic behaviour.

Current paper addresses this gap by integrating adaptive expectations about land market dynamics and hedonic analysis of housing market dynamics in flood-prone areas within a spatial agent-based land market model. As an ultimate goal we aim to incorporate evolution of individual risk perception into spatial ACE model and explore how the critical transitions in land markets emerge in spatial socio-economic systems from the bottom up as new information about growing coastal hazards diffuses.

2 Methods: Empirical Agent-Based Land Market Model with Adaptive Price Expectations

2.1 Model Assumptions

Our ACE combines the microeconomic demand, supply, and bidding foundations of spatial economics models with the spatial heterogeneity of spatial econometric models in a single methodological platform. We model a coastal city where both coastal amenities and flooding disamenities drive land market outcomes, facilitating separate analysis of the effects of each driver on land rents and land development patterns. We start with conventional urban economic model and gradually relax the assumptions of perfect rationality and homogeneity among households as well as the assumption of an instantly equilibrating land market. In particular, our ACE model is grounded in a monocentric urban model (Alonso 1964) enriched by coastal amenities following (Wu and Plantinga 2003; Wu 2006) and flood hazard probabilities following Frame (1998). Thus, spatial goods in this ACE market are quite heterogeneous differentiated by distance to CBD (D), coastal amenities (A), probability of hazard (P) and structural housing characteristics.

Heterogeneous household agents (buyers and sellers) exchange heterogeneous spatial goods (houses) via simulated bilateral market interactions with decentralized price determination. It is challenging to model price expectations in urban property markets characterized by high heterogeneity of goods, which are infrequently traded. While ACE has made a major progress on modeling markets of homogeneous goods

(Arthur et al. 1997; Kirman and Vriend 2001; Tesfatsion and Judd 2006) land is a good with very diverse attributes. The same house in a different location may have a disproportionally different price as do two houses with different structural charactecrizes in the same neighborhood. Modeling price expectations in housing markets needs an introduction of mediator who learns the efficient price of any unique house and who participates often in transactions of such infrequently-purchased good as a house (Parker and Filatova 2008; Gilbert et al. 2009; Ettema 2011; Magliocca et al. 2011). We build upon the previous research on agent-based modeling of urban land markets and introduce real estate agents who observe successful transactions and form price expectations. Adaptive expectations about property prices in the areas with increasing hazard probabilities, which real estate agents and households form, may experience abrupt changes that cardinally alter the trend of spatial development and the price trend.

The innovativeness of this paper is threefold: (i) in comparison to economic studies of land use our ABM explicitly simulates the emergence of property prices and spatial patterns under adaptive price expectations of heterogeneous agents, including the emergence of cardinally new trends in prices and spatial development, (ii) in comparison to other agent-based land markets, which are stylized abstract models (Parker and Filatova 2008; Gilbert et al. 2009; Ettema 2011; Magliocca et al. 2011), the current model makes step forward towards empirical modeling of ABM land markets by using actual hedonic studies and distribution of households preferences; (iii) in comparison to other empirical spatial ABMs modeling urban phenomena (Robinson et al. 2007; Brown et al. 2008) our ABM has a fully modeled land market with adaptive price expectations, which allows for the emergence prices and may lead to qualitatively different trends in spatial patterns (Parker et al. 2011).

2.2 Case-Study and Data

The model is applied to two coastal towns in Carteret county, North Carolina. The area is in general low lying and is prone to flooding with probability of 1:100 and 1:500 in certain zones. For ACE model initialization we employ spatially referenced data from multiple GIS data-sets on the locations of residential housing, coastal amenities (measured in terms of distance from coastal water and sound, and a boolean measure of waterfront), flood probabilities, distances to the CBD and national parks, and data on structural characteristics of properties (age, sq.ft, lot size, number of rooms and etc). The ACE land market model is programmed in Netlogo (Wilensky 1999) and vector data is uploaded using GIS extension. In addition, we use hedonic analysis (Bin et al. 2008) based on the real estate transactions from 2000 to 2004 after a period of active hurricane seasons from mid 1990s to 2003.

2.3 Buyers' Behavior

There are 3 main trading agents in the model (Figure 1): buyers, sellers and real-estate agents.

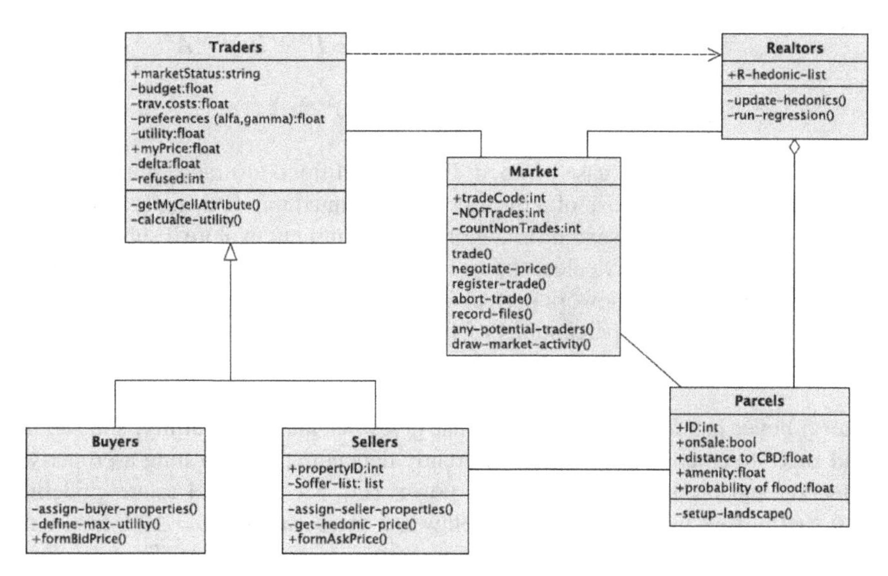

Fig. 1. UML class diagram of a coastal land market model

At the beginning of a trading period all active buyers start searching for a property that maximizes their utility. Household's utility depends on a combination of composite (z) and housing (s) goods which is affordable for her budget (Y) net of transport costs ($T(D)$):

$$U = s^{\alpha} z^{1-\alpha} A^{\gamma}$$

$$\text{or } U = s^{\alpha}(Y - T(D) - k_H H_{ask})^{1-\alpha} A^{\gamma} \tag{1}$$

Here kH is a coefficient to translate the asking price of a seller (H_{ask}) into an annual payment. Preferences for housing good (α) and amenities (γ) as well as exogenous incomes (Y) are heterogeneous across household agents.

When choosing a location in a costal town with designated flood zones, a household operates under the conditions of uncertainty. Thus, she in fact maximizes her expected **utility** (EU):

$$EU = P_i U_F + (1 - P_i) U_{NF} \tag{2}$$

where UF is households utility in case flood event occurs, U_{NF} is utility in the case of no flood, and P_i is a subjective risk perception of a buyer. In economic literature individual, possibly biased, risk perception is often formalized by means of altering objective probability of flooding (P)[1]. Thus, $P_i = P \pm \Delta$, where Δ is an individual bias that changes over time.

[1] We acknowledge that several alternative interpretations regarding the formalization of subjective risk perception may exist (i.e. misinformation about potential losses, misinformation about probabilities, level of risk or loss aversion, feeling of worry and dread, etc.). The implication of alternative formalizations of subjective risk perceptions is a subject for future work.

$$U_F = s^{\alpha}(Y - T(D) - k_H H_{ask} - L - IP + IC)^{1-\alpha} A^{\gamma} \quad (3)$$

$$U_{NF} = s^{\alpha}(Y - T(D) - k_H H_{ask} - IP)^{1-\alpha} A^{\gamma} \quad (4)$$

Here L is the damage in the case of flood, IP is annual flood insurance premium, IC is insurance coverage in the case of a disaster. It is assumed that housing search is costly, thus, households do not search for a global maximum but explore a subset of properties only, from which they select the one that delivers highest utility. Thus, buyers do not operate within a framework perfect information.

Buyers have subjective perceptions of flooding probability, which may be biased compared to the objective probability P. Risk perceptions could be dynamic over time.

After a buyer has found the property that gives her maximum utility, she submits her **bid price** to a seller. Buyers bid differently depending on how long a property is on a market and on their relative market power (Eq. 5 or 6). Real estate guidelines suggest that buyers bid between 3-5% below ask price, and up to 7-10% below ask price if they want to be aggressive and if a property is long on market. Thus:

$$H_{bid} \in [(H_{ask} - h); H_{ask}] \quad (5)$$

where h is a random number between 0-10% of the ask price of a seller.

If it is a sellers' market meaning that there is excess demand for certain areas then buyers need to be more strategic and bid high enough to assure they actually get the property that maximizes their utility.

$$H_{bid} \in [H_{ask}; (H_{ask} + h)] \quad (6)$$

However, in any case buyer's bid price should not exceed her **reservation price,** which is when translated into annual payment should not exceed 30% of her annual income.

2.4 Sellers' Behavior

At the model initialization stage some properties are for sale, i.e. each property has a seller (Figure 2). As simulation goes on, settled households may decide to move as their utility (Eq. 2) decreases compared to the original level.

At the beginning of a trading period active sellers announce their **ask prices**. They do so by requesting regression coefficients from the hedonic analysis of the current period and applying them on their property. At the initialization stage this hedonic function and coefficients come from (Bin et al. 2008). As model runs and new transactions occur real estate agents are rerunning the hedonic analysis. Regression coefficients may change as for example risk perceptions are evolving or new households with different preferences for locations are arriving to the city.

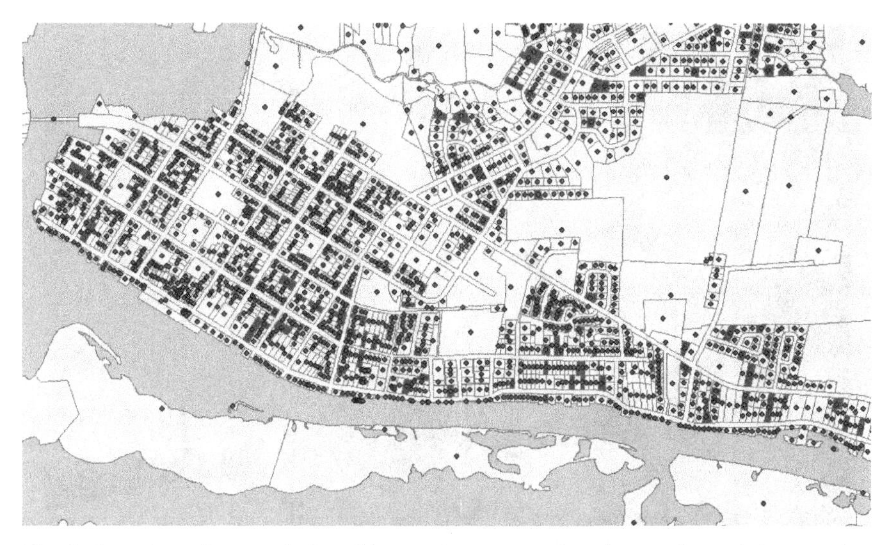

Fig. 2. A section of a coastal city with parcels contours. In red: recently-traded properties.

After buyers have submitted their bids, a seller checks how many bid offers he has received. He chooses the highest bid to engage in price negotiations. The **transaction price** is defined through a price negotiation procedure based on bid and ask prices and relative market power of traders.

2.5 Real Estate Agent

It is challenging to model price expectations in urban property markets characterized by high heterogeneity of goods, which are infrequently traded. We build upon the previous research on agent-based modeling of urban land markets and introduce real estate agents who observe successful transactions and form price expectations (Parker and Filatova 2008; Gilbert et al. 2009; Ettema 2011; Magliocca et al. 2011). At the end of each time step, which is equal to 1 month, all successful transactions got registered in a database together with all the attributes of the properties and traded agents. Each time step real-estate agents update expectations about prices based on these recent transactions. Specifically, real estate agent checks if there are enough transactions to run a comparable sales analysis. If yes, then he runs a hedonic analysis on the new transactions from the last 3 trading periods. If the number of realized transactions is not sufficient to capture the variation on housing prices in the regression analysis, then the horizon is extended yet for another month. Afterwards, these new coefficients got recorded into his memory. Then the real estate agent may decide to apply one of the price learning strategies to suggest final asking price for a seller. Following Magliocca et al (2011) we use some of the economic prediction models.

Regression analysis is realized by employing the R extension of Netlogo (Thiele and Grimm 2010), what makes it possible to have a direct coupling of R script and Netlogo ACE model.

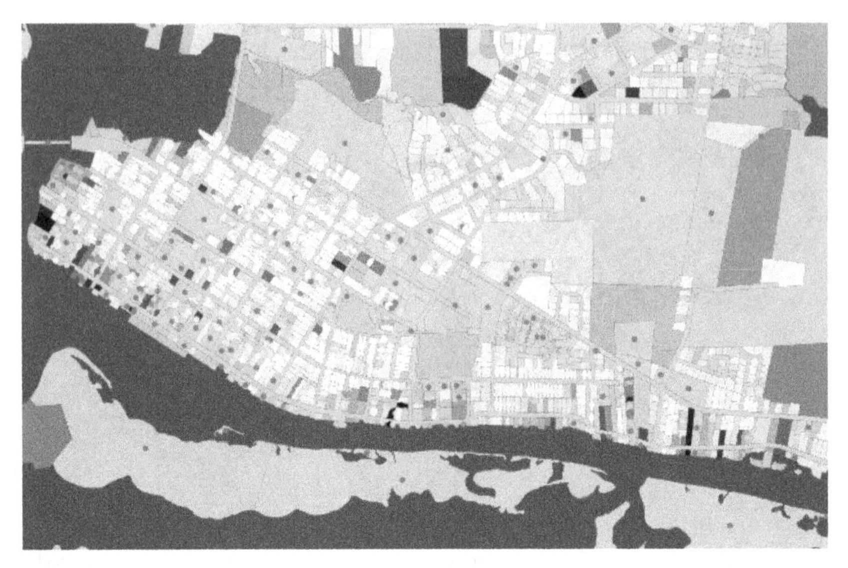

Fig. 3. Color-gradient of property prices in a section of a city after a sequence of trades

3 Discussions

The model produces an adaptive price dynamics in a costal town as new households arrive and search for locations and as some properties go for sale. Average price trend is monitored separately for flood-prone and safe areas. At the conference we will present and discuss the results in details.

The model presented above is the first stage of a four-year project. Thus, the presented model outline should be considered as a base model. Specifically, agents, which are designed to make decisions under uncertainty following expected utility approach, can be compared with agents operating according to the prospect theory logic (Kahneman and Tversky, 1979). This ongoing work will develop along the following two lines in the future: (a) an introduction of individual risk perception evolution based on the theories of opinion dynamics (Acemoglu and Ozdaglar, 2010) in addition to adaptive price learning dynamics; and (b) design and conduction of the parallel experiments with human subjects in the lab and ABM. Lab experiments will be used to acquire behavioural foundations about risk perception dynamics in a group when hazard probabilities change. ABM will help extending these behavioural patterns to larger (than in the lab) temporal and spatial scales.

References

Acemoglu, D., Ozdaglar, A.: Opinion Dynamics and Learning in Social Networks. M.I.T. Working Paper Series. MIT, Cambridge (2010)

Alonso, W.: Location and Land Use. Harvard University Press, Cambridge (1964)

Anand, K., et al.: Epidemics of rules, rational negligence and market crashes. The European Journal of Finance, 1–10 (2011)
Arthur, W.B., et al.: The economy as an evolving complex system II. Santa Fe Institute Studies in the Science of Complexity, vol. XXVII. Addison-Wesley (1997)
Bin, O., et al.: Flood hazards, insurance rates, and amenities: Evidence from the coastal housing market. The Journal of Risk and Insurance 75(1), 63–82 (2008)
Bin, O., Landry, C.E.: Changes in Implicit Flood Risk Premiums: Empirical Evidence from the Housing Market. Journal of Environmental Economics and Management (forthcoming)
Bin, O., Polasky, S.: Effects of flood hazards on property values: evidence before and after hurricane Floyd. Land Economics 80(4), 490–500 (2004)
Bockstael, N.E.: Modeling economics and ecology: The importance of a spatial perspective. American Journal of Agricultural Economics 78(5), 1168–1180 (1996)
Brown, D.G., et al.: Exurbia from the bottom-up: Modeling multiple actors and their landscape interactions. Geoforum 39(2), 805–818 (2008)
Ettema, D.: A multi-agent model of urban processes: Modelling relocation processes and price setting in housing markets. Environment and Urban Systems 35, 1–11 (2011)
Frame, D.E.: Housing, natural hazards, and insurance. Journal of Urban Economics 44(1), 93–109 (1998)
Fujita, M., Thisse, J.-F.: Economics of agglomeration. Cities, industrial location and regional growth. Cambridge University Press (2002)
Gilbert, N., et al.: An Agent-Based Model of the English Housing Market, Stanford, California, March 23-25. Association for the Advancement of Artificial Intelligence Spring Symposium Series. The AAAI Press, Menlo Park (2009)
Hackett, S.: Environmental and Natural Resource Economics: Theory, Policy, and the Sustainable Society (2011)
Hallstrom, D.G., Smith, V.K.: Market Responses to Hurricanes. Journal of Environmental Economics and Management 50, 541–561 (2005)
Field, C.B., Barros, V., Stocker, T.F., et al.: IPCC. Managing the Risks of Extreme Events and Disasters to Advance Climate Change Adaptation, Cambridge, England, 582 p. (2012)
Kahneman, D., Tversky, A.: Prospect Theory: An Analysis of Decisions under Risk. Econometrica 47(2), 263–292 (1979)
Kirman, A.P., Vriend, N.J.: Evolving Market Structure: An ACE Model of Price Dispersion and Loyalty. Journal of Economic Dynamics and Control 25, 459–502 (2001)
Kousky, C.: Learning from Extreme Events: Risk Perceptions after the Flood. Land Economics 86(3), 395–422 (2010)
Lamond, J., Proverbs, D.: Does the price impact of flooding fade away? Structural Survey 24(5), 363–377 (2006)
Ludy, J., Kondolf, G.M.: Flood risk perception in lands "protected" by 100-year levees. Natural Hazards 61(2), 829–842 (2012)
Lux, T.: Rational forecasts or social opinion dynamics? Identification of interaction effects in a business climate survey. Journal of Economic Behavior & Organization 72(2), 638–655 (2009)
Magliocca, N., et al.: An economic agent-based model of coupled housing and land markets (CHALMS). Computers Environment and Urban Systems 35(3), 183–191 (2011)
Nicholls, R.J., et al.: Sea-level rise and its possible impacts given a 'beyond 4 degrees C world' in the twenty-first century. Philosophical Transactions of the Royal Society a-Mathematical Physical and Engineering Sciences 369(1934), 161–181 (2011)

Parker, D.C., et al.: Do Land Markets Matter? A Modeling Ontology and Experimental Design to Test the Effects of Land Markets for an Agent-based Model of Ex-urban Residential Land-use Change. In: Heppenstall, A.J., Crooks, A.T., See, L.M., Batty, M. (eds.) Agent-based Models of Geographical Systems, pp. 525–542. Springer (2011)
Parker, D.C., Filatova, T.: A conceptual design for a bilateral agent-based land market with heterogeneous economic agents. Environment and Urban Systems 32(6), 454–463 (2008)
Pryce, G., Chen, Y.: Flood risk and the consequences for housing of a changing climate: An international perspective. Risk Management-an International Journal 13(4), 228–246 (2011)
Randall, A., Castle, E.N.: Land Resources and Land Markets. In: Kneese, A.V., Sweeney, J.L.S. (eds.) Handbook of Natural Resources and Energy Economics, vol. II, pp. 571–619. Elsevier Science Publishers B.V. (1985)
Robinson, D.T., et al.: Comparison of empirical methods for building agent-based models in land use science. Journal of Land Use Science 2(1), 31–55 (2007)
Stern, N.: The Economics of Climate Change: The Stern Review. Cambridge University Press, Cambridge (2008)
Terpstra, T., Gutteling, J.M.: "Households' Perceived Responsibilities in Flood Risk Management in The Netherlands. Water Resources Development 24(4), 551–561 (2008)
Tesfatsion, L., Judd, K.L.: Handbook of Computational Economics: Agent-Based Computational Economics, vol. II. Elsevier B.V. (2006)
Thiele, J.C., Grimm, V.: NetLogo meets R: Linking agent-based models with a toolbox for their analysis. Environmental Modelling & Software 25(8), 972–974 (2010)
Varian, H.R.: Microeconomic Analysis. W. W. Norton (1992)
Wilensky, U.: NetLogo. Center for Connected Learning and Computer-Based Modeling, Northwestern University, Evanston, IL (1999), http://ccl.northwestern.edu/netlogo/
Wu, J., Plantinga, A.J.: The influence of public open space on urban spatial structure. Journal of Environmental Economics and Management 46(2), 288–309 (2003)
Wu, J.J.: Environmental amenities, urban sprawl, and community characteristics. Journal of Environmental Economics and Management 52(2), 527–547 (2006)

Diffusion Dynamics of Electric Cars and Adaptive Policy: Towards an Empirical Based Simulation

Wander Jager[1], Marco Janssen[2], and Marija Bockarjova[3,4]

[1] University of Groningen, Groningen Center for Social Complexity Studies & Marketing
w.jager@rug.nl

[2] Arizona State University, School of Human Evolution and Social Change & Center for the Study of Institutional Diversity
Marco.Janssen@asu.edu

[3] Free University of Amsterdam, Faculty of Economics and Business Administration, Department of Spatial Economics
m.bockarjova@vu.nl

[4] University of Groningen, Faculty of Behavioral Sciences, Department of Social Psychology

Abstract. In this paper we apply the updated consumat approach to the case of diffusion of electric cars. We will discuss how data from a large sample can be used to parameterize a number of main behavioural drivers, and how these relate to behavioural processes. At this stage we explain how the data fit in the framework, and whereas a model is currently under development, first simulation results are to be available first during the ESSA conference.

Keywords: diffusion, electric cars, agent based modeling, human behavior, decision making, needs, consumat.

1 Introduction

Electric mobility is in its initial stage to penetrate the automobile markets as a step towards a more sustainable mobility. Electric cars may potentially address some pressing environmental and energy security problems. EVs offer zero tailpipe emissions of CO2, and, given sustainable energy production, much lower emissions over the whole lifecycle. In addition electric vehicles contribute to improved air quality (zero particulate matter emissions) and noise reductions in the urban areas. Finally, electric cars may provide storage capacity in a smart grid system.

In a large empirical study conducted in summer 2012 the motives and perceptions of 3000 Dutch drivers were investigated. This cross sectional dataset provides a detailed view of the respondents driving behaviour, preferences, values and uncertainties concerning fuel and electric cars. The data shows that some drivers are more favourable towards the adoption of an electric car than others. This is related to different reasons. For example, some respondents have a transportation pattern that can easily be met by an electric car, others like to show their innovativeness using new technology, and still others may prefer to drive electric because of environmental concern.

B. Kaminski and G.Koloch (eds.), *Advances in Social Simulation*,
Advances in Intelligent Systems and Computing 229,
DOI: 10.1007/978-3-642-39829-2_23,

Whereas this data contains information on how different segments of consumers perceive an electric car, and what their intentions are towards adoption of an electric car, these data do not show the social complexities associated with the diffusion of electric cars. With diffusion we refer to the adoption of a new product or behavior over time in a population. Diffusion remains a problematic theoretical concept because the precise definition of a successful or failing diffusion is not possible due to the open ending of the process. Whereas empirical analysis of diffusion processes require an ex-post analysis, only in the hypothetical case where 100% of a population adopts, one can speak of a completed diffusion process. Moreover, in many situations diffusions can be said to fail, e.g. when a new product is not being adopted and production is being discontinued. However, this does not exclude the possibility of a later success. In this paper we consider diffusion to refer to the process and factors surrounding the spreading of new products or behaviours in a population.

Social processes play a key role in the success or a failure of diffusion (e.g., Garcia & Jager, 2011), which may involve personal awareness of a new product that emerges from growing adoption of the product; as well as change in normative pressures that triggers acquisition of information about the new product, and so on.

Two complexities are of special interest in this context. First, due to the uncertainty and limited knowledge of many consumers concerning the electric car they are more likely to (be) discuss(ed), and opinions and attitudes may spread fast through the social networks. This diffusion of experiences and attitudes hence is very sensitive to both negative and positive experiences consumers have, and opinion leaders may have a strong impact on the success or failure of the diffusion process (see e.g. Van Eck, Jager & Leeflang, 2011). A recent example is the turmoil about the negative review of the Tesla car in the New York Times, which apparently resulted from a reviewer deliberately trying to exhaust the batteries to make the story dramatic (Broder, 2013). These stories may propagate in further discussions on the risk of ending up with a flat battery.

A second complexity arises from the fact that during a diffusion process (which is essentially a theoretical construct) the context of the decision making is changing. If a certain proportion of consumers adopts an electric car, the uncertainty on its performance will decrease, as well as the norms concerning its acceptability. Hence consumers that initially are very hesitant of switching to an electric car may at a later stage change their attitude.

Agent based models are known to offer a suitable tool for exploring these complexities. In this project our aims are to work towards what we call "adaptive policy" for complex systems. This implies that policy is developed for different segments, and assuming that the segments have a different switching moment and sensitivities, adaptive policy is tailor-cut and is aimed at identifying and targeting those consumer groups at the moment in time when they are at the brink of adoption. The theory of innovation diffusion of Rogers (1993) acknowledges that the adoption of novel products can be driven by different needs. So, for some consumers applies that they have a strong need to distinguish themselves from the others (anti conformism), which in turn stimulates them to be among the first adopting an electric car, even (or especially) when the performance of this car is inferior to that of a traditional fuel car. Other consumers put a high priority on the environmental issues and consider an electric car as fitting their personal needs, whereas being not sensitive to (anti)conformist drives at all.

Obviously, both types of adopters are likely to be present in the early stages, and also mixed types can be envisaged. The situation changes when the diffusion proceeds. The more consumers adopt, the more a conformity pressure may stimulate social sensitive consumers to adopt as well. Hence the motivational setting concerning the adoption of an electric car will change during the process, and different types of consumer (segments) are more likely to adopt at different moments in the diffusion process. We propose that it is important in the context of policy supporting diffusion that at different times during the diffusion process different policy measures target different segments of consumers. For example, in the beginning of the diffusion it is more efficient to stress the uniqueness of the electric car, whereas at later stages other consumers can be approached with a message that an electric car is about to become the norm. Also the effects of improving the range of electric cars, shortening the charging time and changing fuel and electricity process will have serious effects on the diffusion, and should be incorporated in diffusion scenarios.

The dataset collected by Bockarjova (2012) is containing almost 500 variables describing 3000 respondents. Questions address a.o. transportation behavior on a very detailed level, perception of electric and fuel cars on many different attributes (environment, status, safety, range), valuations of different aspects (e.g. charging times, infrastructure, willingness to change) and many personal characteristics (e.g., knowledge, involvement, demographics). This allows for parameterizing an agent based architecture that elaborates on the needs and personal characteristics driving the decision making of agents. The consumat approach offers a framework describing agents' needs and abilities as drivers of decision making. Since the introduction of the consumat in 2000 (Jager, 2000, Jager et al, 2000; Janssen & Jager, 2002), the consumat approach is being used as a generic model of human behavior on the decisions people make in satisfying their basic needs in various settings. The consumat approach offers a simulation framework that captures some of the main behavioural principles as discussed in the literature on consumer behaviour. Whereas the model is not capable of simulating elaborate cognitive processes, logical reasoning or morality in agents, it does allow for simulating a number of key processes that ensemble capture human decision making in a variety of situations, such as consumers purchasing products, farmers deciding on a crop, citizens deciding where to live, and other situations where people select a behavior from a set of possibilities. Computer simulations are being increasingly used in a policy context; proposed consumat framework may become a relevant tool to serve the purposes of policy and practice as a framework that captures the main drivers and processes of human behavior altogether.

In 2012 (Jager & Janssen, 2012) we updated the consumat approach to improve the decisional strategies and to address network issues more explicitly. In the following section we will discuss the different elements in the consumat architecture and show what empirical data will be used to parameterize these elements.

2 Connecting the Empirical Data with the Consumat Framework

In the following Figure 1 the basic architecture of the consumat framework is presented. The core of the model is the consumer decision-making. The type of decision-making an agent engages in depends on its satisfaction and uncertainty level. For example, an agent that is satisfied with a fuel car (an opportunity), and not feeling uncertain about this because everybody else also uses a fuel car will engage in repetition. In this situation this implies continue using the same car. No new information about the attributes of fuel or electric cars will be collected, hence the memory will not be updated. When an agent becomes dissatisfied with the current car, for example because the variable costs (fuel) rise rapidly, and also uncertain because they increasingly see agents with electric cars and are more often confronted with information about electric cars, the agent may switch towards inquiring, which implies the consultation of other agents concerning the attributes of different cars, and decide on the optimal car. Agents differ with respect to their uncertainty tolerance and ambition-level, which affects their tendency to use certain strategies. For example, a highly ambitious agent that has a high uncertainty tolerance is more likely to engage in optimizing. This causes this agent to be the first to find and adopt attractive alternatives. As such the consumat approach allows for "growing" phenomena that have been described in the innovation diffusion theory of Rogers (1993).

The aggregated choice of the agents has an impact on the environment, but also on culture. In particular it results in a social environment of the agent of other agents that adopted or not an electric car. Especially when an agent has a strong social need for conformity, this may result in a low social need satisfaction, stimulating the agent to consider an alternative.

In simulating the diffusion of electric cars we have to realize that we are entering a complicated market with many brands, models and fuel types. In the questionnaire respondents indicate what specific type of car (brand and model) they are interested in, and the questions address this specific type of car. More specifically, the respondents have to compare a fuel and electric version of a chosen car. This provides us with a direct comparison between electric and fuel versions of cars, which allows for the formalization of two opportunities in the simulation model: a fuel car and an identical electric car. Also respondents have indicated when they expect to purchase a new car, which also is relevant. The following Table 1 provides an overview of a number of key demographic variables, showing that the sample of 3000 respondents is representative for the Dutch population.

In the following sections we provide an overview of the architecture of the consumat, and describe how the data from the questionnaire can be used in parameterizing the different elements. Because of the many variables only in a few cases we will go into detail to give an example of the approach.

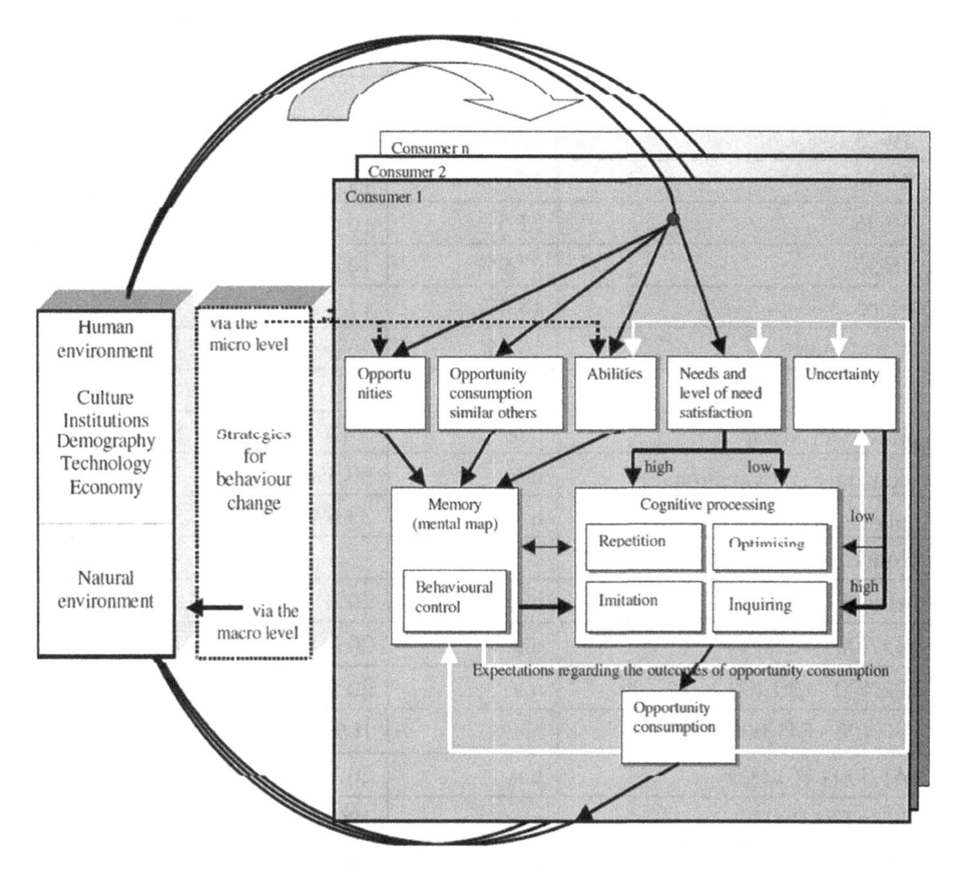

Fig. 1. The consumat architecture (revised from the 2000 version)

Table 1. Demographic data

	Car owners (%)[a]	**Dutch population**[b]	**Whole sample**
Gender (male)	n.a.	49.5%	49.6%
Age[c]			
19-25	6.7%	8.1%	6.4%
26-35	15.0%	15.7%	15.6%
36-45	21.5%	19.0%	25.9%
46-55	22.9%	19.6%	22.3%
56-65	18.4%	17.2%	15.9%
65 and older	15.5%	20.4%	14.0%
Education			
primary or lower	n.a.	5.1%	8.1%
secondary and vocational	n.a.	60.3%	56.6%
college and university	n.a.	33.6%	35.3%
Household income[d]			
below €15,900	n.a.	20%	9.9%
€15,900 - €22,400	n.a.	20%	14.9%
€22,400 - €30,400	n.a.	20%	17.4%
€30,400 - €41,000	n.a.	10%	11.0%
€41,000 - €51,000	n.a.	20%	14.6%
above €51,000	n.a.	10%	7.3%
unknown	n.a.	---	24.9%

[a] Source: BOVAG-RAI (2012).

[b] Source: Statistics Netherlands (CBS, 2011).

[c] Data for the Dutch population (Statistics Netherlands, 2011) is provided for the population of age 20 and older.

[d] Data on income is available from Statistics Netherlands (CBS, 2011) for disposable household income deciles (appear in the Table). Our data was gathered for household gross income using the following breakdown: below €15,000; €15,001 - €20,000; €20,001 - €25,000; €25,001-€30,000; €30,001 - €40,000; €40,001 - €50,000; €50,001 - €60,000; €60,001 - €70,000; €70,001 - €95,000; above €95,001.

2.1 Consumat Needs and Satisfaction with Opportunities

In the Consumat II we use 3 main need forces: existence, social and personality. These needs are all relevant concerning the need satisfying capacity of electric cars

Existence need relates to having means of existence, food, income, housing and the like, and thus basically refers to economical dimensions of existence. In terms of Goal Frame Theory here the gain motive is dominating.

To give an example of the data and formalization process we use this existence need as an illustration. The basic idea is that costs, range and charging time are crucial factors determining agent satisfaction. Whereas a limited range and longer charging time of an electric car will result in little disutility for an agent with short urban trips, an agent frequently making long trips will be more satisfied with a fuel car due to the longer range and short charging (fueling) time. Hence the transportation behavior of the agent influences his vehicle choices and has to be parameterized. For this we use a following question: "Indicate how often you make a trip of the following distance on a single day", and "how many kilometers a year do you expect to drive with your next car?" This resulted in the following tables 2 and 3:

Table 2. Frequency distribution of various daily trips (N=2533 for each row)

	Once, or few times a week	Once, or few times a month	Once, or few times a year	(almost) never
0 (not in use)	909	1187	336	101
1 -10 km	445	1531	445	112
10-50 km	109	1523	791	110
50-100 km	231	596	1150	556
100-200km	422	235	786	1090
200-500km	883	66	274	1310
> 500km	1642	13	39	839

Table 3. Frequency distribution of expected yearly mileage of a next car (N=2974)

	Frequency	%
below 5.000 km/year	164	5.51
5.000 to 10.000 km/year	729	24.51
10.000 to 15.000 km/year	882	29.66
15.000 to 20.000 km/year	539	18.12
20.000 to 25.000 km/year	301	10.12
25.000 to 30.000 km/year	120	4.03
30.000 to 35.000 km/year	80	2.69
35.000 to 40.000 km/year	47	1.58
40.000 to 45.000 km/year	30	1.01
45.000 to 50.000 km/year	43	1.45
50.000 to 100.000 km/year	23	0.77
Do not know	16	0.54
Total	**2974**	100.00

These data will be translated in an average number of kilometers driven on these trips because an agent has to drive a fixed number of km's , and an estimation of how often these trips are made (an agent cannot drive 1-10 km, but we parameterize this as 5 km). In figure 4 we show the individual score of a respondent translated in a unique empirical based "driving fingerprint" of the agent.

Table 4. The individual "driving fingerprint" of an agent

	100 times/year	24 times/year	2 times/year	0 times/year
0 km	*			
5 km			*	
25 km		*		
75 km		*		
150 km		*		
350 km			*	
800 km			*	

Given the range of a car it can be calculated how often it has to be fueled/charged. For example, given a range of 100 km of an electric car, a charging time (including detour) of 30 minutes & the transportation profile of the above respondent we conclude that this usage pattern with this electric car results in 28 trips a year over 100 km that require 30 minutes of charging during the trip: total charging time on-the-road = 14 hours charging a year. The same usage pattern with a fuel car with a range of 800 km results in 2 trips a year that require 5 minutes of tanking during a trip: total charging time on-the-road = 0.16 hours a year.

Obviously also financial costs play an important role in the existence need. Of particular interest are price differences in the purchase of electric versus fuel cars (for which data are available), prices of fuel and electricity (scenarios can be developed), and ownership type, as in the case of lease the bulk of costs will be the responsibility of the employer.

The importance of this existence need can be indicated by the answer on the question how acceptable different charging times at different locations are. For example, a respondent could indicate that charging at the highway should not take more than 15 minutes to be on the acceptable side. Also income and costs for the current mobility pattern will be used as indicators of the importance of the existence need.

This existence need combined with importance allows for calculating the impact of different types of vehicles on this existence satisfaction. We can compare how different types of cars, innovations concerning the range and charging speed of electric cars, and changes in the costs of mobility would impact the existence satisfaction of individual agents representing the respondents in the survey. In simulating the dynamics of innovation diffusion it thus is important to be capable of determining the impact of this innovation (even if it is incremental such as range) on individual agents. Whether an individual agent adopts a new product or not will also have an impact on other agents. This is related to the social needs of agents.

Social need relates to the interaction with others, belonging to a group and having a social status. The social need is composed of two drives: being similar, indicating a preference for cars matching those owned by one's social environment (conformity), and an upward drive, which in the car market usually was directed at bigger and more expensive cars, but nowadays also seems to be directed towards being "different" and being less dependent on a large car for deriving status (anti-conformity) (Festinger, 1954; Claidière & Whiten, 2012). Conformity implies liking the same type of car as its peers. This is formalized as the proportion of peers performing the same behavior weighed by the similarity of these peers. Concerning the anti-conformity drive an agent derives satisfaction from having a car that is deviating from the type of car that peers own. These two social drives imply that agents have to be equipped with a weighting function in their social need, balancing the similarity and superiority drive. Empirical data are available concerning the relative importance of both drives, thus allowing for a precise and detailed modeling of this social need. It is important to realize that e.g. the social satisfaction with an electric car depends heavily on both the (anti)-conformist orientation of the agent and the stage of the diffusion (how many adopted in the agent's social environment).

Personality relates to satisfying one's personal taste and engaging in activities and consumption behaviour one likes. A behavioral option may more or less fit with the personal preferences of an agent, which can be addressed as taste. The more a behavioural option matches the taste of an agent, the more satisfied it will be. Hence the taste of an agent can be defined as an (multiple) ideal point, and the agent prefers to minimize the difference between its ideal points and the corresponding scores of the behavioural option. In the case of cars there is a number of issues related to personal needs. First a number of questions address the personal joy of using innovative products. Next the personal perspective on environmental issues and energy security issues (dependency on oil producing countries) is relevant, and these scores are available.

Uncertainty of the agents, is important in the selection of a decision-making strategy. In our model uncertainty is coupled to the existence (Ne) and social need (Ns). For existence the uncertainty is expressed as the variability in the information obtained on the expected utility. Uncertainty concerning existence is measured with 4 questions addressing the (in)security of electric cars concerning life expectancy, endurance, safety, maintenance costs and depreciation. Social uncertainty addresses the proportion of friends using a different fuel than you.

2.2 Consumat Decision-Making and Personal Traits

The decision-making strategy the agents engage in depends on their level of need satisfaction and uncertainty. Four decision-making strategies are implemented:

Repetition (when satisfied and certain)**:** consider only the behaviour that one is performing now. Mind that if an agent that has a long time perspective gets new information on long term negative outcomes (e.g., environment), this will decrease its satisfaction and thus may stimulate the agent to search for alternatives. Repetition of satisfactory behaviour is the main mechanisms behind habitual behaviour. In the context of purchasing a car the repetition option refers to keep using the (same type of) car one owns.

Imitation (when satisfied and uncertain)**:** consider all behavioral options performed by peers (strong links). Considering that the social need is driving behaviour as well, this strategy implies that successful behaviour that is performed by a majority of others is most likely to be copied. However, also the behaviour of a successful peer that deviates from the group can be imitated, which complies to the principles of imitative processes (Bandura, 1977).

Inquiring (when dissatisfied and uncertain)**:** consider all behaviour performed by all other agents. This opens the possibility to explore behavioural opportunities that are being used by more distant agents (weak links).

Optimising (when dissatisfied and certain)**:** consider all possible behavioural options available. Here also opportunities are considered that are not yet used by other agents. Information can be found (e.g. on the web, news, advertisements) on e.g. an improvement in charging speed, which provides a new opportunity to be evaluated.

It is important to realize that after the purchase of a car a consumer can be satisfied for many years, depending on the relative importance of their needs and ambition level. The less satisfied an agent gets with the current car the more often an agent will engage in information collection. This information gathering stage (inquiring, deliberation) may continue until the satisfaction derived from the current car is outperformed by another model, and the agent purchases a new model.

The tendency of using certain decision strategies also depends on agents personality traits, which can be parameterized using data we have on ambition level and the tendency to take other consumers' behavior as an example. The following personal agent traits will be parameterised.

Ambition Level: Some agents are easier to satisfy than others. A question on the ambition level of the respondent is used to parameterize ambition level in the agents.
Uncertainty tolerance: This is being indicated by the question if someone describes him/herself as risk-taking with new cars versus being traditional and avoiding new technology.

Consumat Abilities: The abilities of the consumat first relate to its capacity to actually use particular behavioural options. Issues that are relevant are income and ability to charge at home and at work, for which we have data available.

Consumat Cognition: The consumat has a memory in which it stores information on the behavioural opportunities, and on the behaviour, attributes and success of other agents. Concerning the information on behavioural opportunities three types of information are available: own experience, experience of others and generic information (advertisement, media). Important here is that cognition is not updated when engaging in repetition or imitation. Hence an agent satisfied with a fuel car and having rejected an electric car earlier because of its limited range is not likely to find out that e.g. a new battery system is available significantly improving the range. We also add an expertise level to agents using questions on their knowledge on automotive technology. This offers an opportunity for more detailed modeling of expertise effects (outgoing influences). Further possibilities for model enhancement include making use of questions about outgoing influences to define the expertise and number of connections of an agent.

Social Interaction and Network: Agents in the consumat approach are more likely to interact the more similar they are. Similarity is based on data, such as age, income, critical opinions, and on physical proximity in the model (neighborhood effect). This will result in a network describing the frequency of interactions over time between different agents, however will not lead to an explicitly dynamic network as most of the variables determining the chance of interaction are fixed.

3 A Perspective on the Use of the Model: Towards Adaptive Policy

The previous section showed on a general level the model components and its parameterization. Due to the high number of model components and of questions in the dataset complete formalization would be too lengthy. Moreover, formalization is yet ongoing. The envisaged aim of the model is testing of adaptive policy. Theory proposes that in the process of diffusion (in this case, of electric cars), different consumers have different sensitivities for technical and social drivers in the system. Hence consumers that are most prone to adopt first are more likely to respond to different drivers than consumers that are expected to be the last adopters. This implies that the first group should be stimulated in a different manner than a potential second group of adopters. Failure or success in stimulating the first group will have second order effects on the next groups of consumers. For example, if a first group has negative experiences with range, this may be communicated to other consumers. Even if at a later stage the range is improved, the negative experiences of the first group will as yet be stored in the memory of other groups, and hamper the diffusion of electric cars as long as these consumers are still satisfied with fuel cars. Policy should thus be targeting a particular segment of consumers that is likely to respond positively on specified stimuli. Only after a critical part of this segment has adopted the product, a next group can be targeted. Approach to a next group may involve a different set of stimuli. For example the positive experiences of the first adopter group can be used as an argument, but the anti-conformist argument clearly becomes less relevant as diffusion progresses while the conformist argument grows in importance. Also it is of interest to whether the speed and the extent of diffusion are stronger influenced by incremental improvements of vehicle characteristics such as range and charging speed, or by less frequent "breakthroughs" which generate more attention to the product. Our aims with a fully parameterized model are thus the testing of the effects of both technological improvements and communication strategies directed at stimulating the diffusion of electric cars. First results are expected to be available for presentation at the conference.

References

1. Broder, J.M.: Stalled Out on Tesla's Electric Highway. New York Times (February 8, 2013), http://www.nytimes.com/2013/02/10/automobiles/stalled-on-the-ev-highway.html?ref=automobiles&_r=1& (retrieved on March 11, 2013)
2. Claidière, N., Whiten, A.: Integrating the Study of Conformity and Culture in Humans and Nonhuman Animals. Psychological Bulletin 138(1), 126–145 (2012)
3. Festinger, L.: A theory of social comparison processes. Human Relations 7(2), 117–140 (1954)
4. Garcia, R., Jager, W.: Introductory Special Issue on Agent-Based Modeling of Innovation Diffusion. Journal of Product Innovation and Management 28, 148–151 (2011)
5. Jager, W.: Modelling consumer behaviour. Doctoral thesis. Groningen: University of Groningen, Centre for Environmental and Traffic psychology (2000)
6. Jager, W., Janssen, M.A., De Vries, H.J.M., De Greef, J., Vlek, C.A.J.: Behaviour in commons dilemmas: Homo Economicus and Homo Psychologicus in an ecological-economic model. Ecological Economics 35, 357–380 (2000)
7. Janssen, M.A., Jager, W.: Stimulating diffusion of green products. Journal of Evolutionary Economics (12), 283–306 (2002)
8. Rogers, E.M.: Diffusion of innovations, 4th edn. The Free Press, New York (1993)
9. Van Eck, P.S., Jager, W., Leeflang, P.S.H.: Opinion leaders' role in innovation diffusion: A simulation study. Journal of Product Innovation and Management 28, 187–203 (2011)

Simulating Opinion Dynamics in Land Use Planning

Arend Ligtenberg and Arnold K. Bregt

Wageningen University, Laboratory for Geo-Information and Remote Sensing
arend.ligtenberg@wur.nl

Abstract. During a process of spatial planning multiple actors meet in a decision making process trying to satisfy as much as possible their individual desires and goals. Diffusion of opinions is considered an important factor in this process. A behavioural approach that represent the diffusion and distribution of opinions are absent in most spatial ABMs. This research presents a novel approach to simulating spatial opinions dynamics based on an agent based implementation of the Deffuant Weisbuch model. The model explicitly deals with spatial heterogeneity in opinions as well as heterogeneity in the willingness of agents to adapt their opinions. The model is demonstrated for a hypothetical spatial allocation problem in a study area in the south-east of the Netherlands. Three different scenarios representing different levels of willingness to cooperate within the population of agents are worked out. The results showed that explicit spatial opinion patterns can be simulated which are in line with what would be expected from the scenarios.

1 Introduction

Spatial planning aims to change the land use such that it fulfills the needs of the current and future demands of society. Traditionally spatial planning is considered a decision-making process in which a selected set of actors have to agree on a common goal to be achieved. Modern spatial planning slowly shifts from a linear process, based on coordinated steps, towards a continuous, multi-faceted process. Larger, more heterogeneous group of actors get involved in the decision-making. Decision-making in such heterogeneous, multi-faceted environment is highly complex and depends on continuous interactions between actors who are communicating and interchanging opinions, suggestions and ideas. Typically the result of such a process is not an optimal configuration of land use but one that satisfies the needs of the community of actors as a whole. This means that, considering spatial planning as a group-decision making process, the outcome can be considered an optimal one given a solution space determined by the continuous interactions of individual opinions of the actors, and the possibilities and constraint imposed by the physical and legislative characteristics of the environment. While most current land use models handle the pysical and legislative aspects of land use change well, representations of opinion dynamics are missing or only present at a aggregated level. One way of representing the

B. Kamiński and G.Koloch (eds.), *Advances in Social Simulation*,
Advances in Intelligent Systems and Computing 229,
DOI: 10.1007/978-3-642-39829-2_24,

development of opinions is to consider it a diffusion process. Such diffusion is characterized as a process where change is communicated through certain channels amongst members of a social system [1]. Diffusion embraces aspects such as learning, contagion, mimicry, trust, reputation, etc. [2]. The modeling of diffusion opinions has attracted a strong academic interest since the early 1960s especially to simulate market response to new products or services [3]. In spatial sciences research on the diffusion of opinion is very limited. Hagerstrand [4] was amongst the first to define clustered growth as spatial diffusion. More recent examples can be found in the work of Berger [5] who models the diffusion of new agricultural practices.

Traditionally, the diffusion process is modeled in an aggregated manner. For example, the acceptance of innovations is often modeled following the Bass model based on the theory of Rogers [1]. Important drawbacks of these aggregate models are that they cannot deal with heterogeneity within the population and with the spatial environment, are not behavioral based, and as such do not explain social change and social processes and are mainly good in explaining past behavior rather than forecasting future behavior [3].

To overcome these drawbacks Agent Based Modeling (ABM) is suggested. ABM differs fundamentally from aggregated approaches by not taking the system itself as the elementary modeling unit but the individual entities making up the system. ABM is a widely used technique to simulate land use change while taking into account the diversity in actors and various spatial and temporal scales. [6, 7]. Although most of current agent based land use simulation deal rather good with the spatial heterogeneity and scale issues on the behavior of agents, the social and behavioral processes itself are still underdeveloped. Most ABM apply simple reactive rules applied on the environment and other agents, they are however lacking theoretical grounded concepts of elementary processes like negotiation, formation of opinions, development of reputations and trust etc.

This paper describes a first attempt to implement a more elaborated model of opinion dynamics in the realm of spatial planning. Focus is on the dynamics of opinion as it can been seen as the most important driver for actors to accept a proposed change. The first part of this paper gives a brief overview of existing models of opinion dynamics. The second part describes a conceptual models of the development of opinions based on, an innovation diffusion approach, within a multi-actor, multi-goal spatial planning system. This model is implemented for a hypothetical case study of an area in the south-east of the Netherlands. The last sections presents an discussion on the results.

2 Background

Opinion dynamic models describe the change in opinion of an actor through time under influence of other actors. Most current models of opinion dynamics are somehow developed in analogy to physical models. Such a "socio-physics" approach can be beneficial for understanding the complexity of unknown processes. Simulating meaningful patters at this stage is more important than more exact predictions of systems behavior or accurate representations of real social processes. Most models of opinion dynamics take as a starting point the theory of social impact, first introduced by Latane [8]. Latane suggest that social impact results from social forces which operated in a social structure. These social forces are: strength, immediacy, and number of sources. Based on these assumptions social impact opinion models can be divided into two types: opinion models which are considered discrete, and continuous opinion models. Discrete models of opinion dynamics include for example Ising models, and the Sznajd models and define opinions in binary terms (like "yes", "no" or "accept", "reject"). Models of continuous dynamics consider a continuous opinion space. These models include for example the models of Deffuant-Weisburg (DW) [9, 10], and Hegselman-Krause (HK) [11–14] which are based on the concepts of bounded confidence, meaning that two agents only are willing to update their opinions at time $t+1$ if the difference between their opinions at time $t1$ are within a certain distance of each other. Based on these assumption various bounded opinion models are proposed such as the CODA model, based on Bayesian learning [15, 16], the innovation diffusion model of [17], the meta-contrast model of [18], or the already mentioned DW and HK type of models. In this research we focus on continous bounded opinion models.

The basic DW considers a population of N agents i who have a continuous opinion x_i on $[0..1]$. At each iteration two randomly chosen agents update their opinion according:

$$x(t+1) = x + \mu(x' - x) \tag{1}$$

$$x'(t+1) = x' + \mu(x - x') \tag{2}$$

An agent only adjust their opinions iff $\left| x - x' \right| < d$, where d is a threshold value. The idea behind d is that an agent only is willing to alter its opinion if the opinion of the other agent does not differ to much. The parameter μ controles the rate of convergence of the opinion.

The Hegselman-Krause model follows a similar structure. Given a population of n agents having an opinion $x_i(t)$ op agent i at time t the HK-model is defined as:

$$x_i(t+1) = \frac{1}{|I|} \sum_{j \in I} x_j(t) \tag{3}$$

Where I is $I(i, x, t)$ being the confidence set of agent i at time t. The confidence $I(i, x)$ consists of all agents j such that $x_i - x_j$ is in the confidence interval $[-\epsilon_i, \epsilon_i]$. $|I|$ is the number of agents in I.

The differences between the two approaches lies mainly in the way the communication is arranged. DW model is based on the exchange of information between two agents while the HK model assumes a global knowledge of each agent about the opinion of the other agents. The above models have an aggregate view on opinions which only depends on $t-1$.

Based on this, Martins [16, 19] proposes the CODA model which includes a simple model of Bayesian learning. Opinions are updated by a social interaction based learning process. The results of the CODA models are comparable to the

DW and HK models but allow for a better understanding of the approximations that need to be made in order to better understand the d and ϵ parameters of the DW and HK models.

The drawback of the above discussed models is that, while implemented at agent level they are limited in their explanation of the effect of space on the diffusion of opinion. Factors like location, shape and size of land use affect the opinions of actors. For example certain areas might be considered "non-negotiable" by actors because the have a strong emotional or economic meaning to an actor. This means that, to be able to explore opinion dynamics in a spatial environment, current opinion dynamics models need to be adapted to represent the effects of space on the rate of change and distribution of opinions both in space as well as between actors.

In this paper an agent based opinion dynamics model, based on the DW approach is proposed that explicitly represents the additional complexity of a spatial decision making systems. For most spatial problems the opinion space can be considered multi-dimensional as the opinion dynamics not only depend on time but also its location within a 2 dimensional geographic space. Moreover opinions are not only influenced by previous time steps but in addition also by the opinions agents have about other locations.

3 A Model for Spatial Opinion Dynamics

The proposed ABM represents a category of spatial decision-making problems where multiple actors having multiple desires and goals regarding the future development of an area have to generate a joint vision for future development i.e. a spatial plan. Each agent tries to realize its individual goals as good as possible within the context of goals of the other agents. The proposed model currently takes only one issue into account. This can be for example an allocation problem dealing with the question: where to realize a land use function (e.g. new urban areas). The model focuses on the dynamics of the opinions assuming that there is an intitial opinion for each agent, resulting from decision-making at individual level. As such it builds on previous work which described an agent based approach where land use change is purely driven by rational processes such as voting [20, 21]. In the proposed model, the dynamics of opinions are characterized by the following aspects:

- the initial opinion of an agent about a location;
- the social influence and;
- the spatial influence.

Including these three aspects the dynamics of opinion $O_{k,(t+1)}$ of agent k at time $t+1$ in a social-spatial system can be defined as:

$$O_{k,(t+1)} = f(Oi_{(k,g,t)}, \quad Qs_{(k,g,t)}, \quad Qg_{(k,g,t)}) \tag{4}$$

Where $Oi_{k,g}$ is the the individual opinion of agent k for land use g, $Qs_{k,g}$ the effect of the social influence of agent k for land use g, and $Qg_{k,g}$ is effect of the spatial influence on the opinion. O_k lies in the domain $[0, 1]$ where 0 is a maximum negative opinion, i.e. a full rejection of a location and 1 a maximum positive opinion.

The initial opinions of the agents purely results from individual decision-making based on the desires of an agent [21]. It is the result of a process where an agent evaluates the spatial and non-spatial aspects which it considers important for realizing its desires. It takes into account solely the agents current knowledge about the environment. In the ABM this internal decision-making is done following a classical rule based approach. Each agent has a number of desires, which are then translated to a number of rules. Executing these rules lead to a set of location specific opinions.

The social influence describes the effect of other agents in the social network. The social influence changes the opinion of an agent about, in this case, the suitability of a location for a specific land use function. This process of social influence (sometimes referred to as contagion) is well described in sociology. Social influence occurs through processes of communication or comparison [22]. The strength of the influence depends on the structure of the social network and especially on the number, length and strength of the links between actors [17, 22]. In spatial planning systems, social networks often show a high clustering of individual actors strongly connected based on similar interests of living in the same location. Examples of such clusters are environmental interest groups, neighborhood or farmers communities. As a key measure of social influence the concept of social-distance as found in the DW and HK models is adopted here. Social distance is defined as the difference between two agent for their opinions for a specific location:

$$S_{(k,l,i,j,t)} = o_{(k,i,j,t)} - o_{(l,i,j,t)} \tag{5}$$

Where: $S_{(k,l,i,j)}$ is the social distance between agents k and l for location i, j at time t and $o_{(k,i,j,t)}$ is the opinion of agent k and $o_{l,i,j,t}$ the opion of agent l about location i, j.

Social influence is defined as the ratio between the number of agents having an opinion within a certain treshold and the total number of agents having an opinion about that location:

$$Qsoc_{(k,g,i,j,t)} = \frac{\sum_{l=1}^{N} w_{(l,g,i,j,t)}}{N} \tag{6}$$

Where $Qsoc_{(k,g,i,j,t)}$ is the social influence factor for agent k for land use g at location i, j for time t, N is the total number of agents and $w_{(l,g,i,j,t)}$ is the dirac delta function for agent l:

$$w_{(l,g,i,j,t)} \begin{cases} 1 \text{ if } S_{(k,l,i,j,t)} < d_{(k,g,i,j,t)} \\ 0 \text{ otherwise} \end{cases} \tag{7}$$

Where $d_{(k,i,j,g)}$ is a threshold value for agent k for entering a negotiation.

The spatial influence, which is specific for land use systems, deals with the distribution of opinions in a geographic space. The assumption made here is that opinion of an individual agent not only depends on its location and type of land use but also on spatial factors like the size of an area, and the distance to areas having strongly deviating opinions. For example, small isolated spots in the spatial environment which have clearly a different opinion only supported by one or few actors might less likely survive in the opinion dynamics process than large areas of actors having similar opinions.

To represent the spatial influence on the opinion, a neighborhood representation is adopted. Based on the opinions of all agents in the simulation the spatial influence is alculated as:

$$Qspa_{k,g,i,j,(t+1)} = \frac{\sum_{(x=1)}^{X} \sum_{(k=1)}^{K} O_{k,x,t}}{N} \tag{8}$$

Where: $Qspa_{k,g,i,j,(t+1)}$ is the spatial influence factor of the opinion of agent k for land use g at location i, j, $O_{a,k,l,t}$ are the opinions of agent a stored in cell x part of the set of cells X of the neighborhood and $N = |X| * |K|$.

Based on the social and spatial influence factors a weighting factor is calculated which accounts for adaptation of the social distance threshold d. The weighting factor is calculated according:

$$Q_{k,g,i,j,t} = \pi * Qsoc_{(k,g,i,j,t)} + (1 - \pi) * Qspa_{k,g,i,j,t} \tag{9}$$

Where π is a parameter indicating the priority of the social influence versus the spatial influence. Finally, the threshold d is adapted according:

$$d'_{k,g,i,j,(t+1)} = d_{k,g,i,j,t)*(1-(|O_{k,g,i,j,t}-Q_{k,g,i,j,t}|)^{\tau})} \tag{10}$$

Where $d'_{k,g,i,j,(t+1)}$ is the adapted d for agent k at time $t + 1$. τ a parameter which defines the resilience of an agent against adaptation of its opinion. Figure 1 shows the effect of various values of τ (1, 3 , and 9) on the adaptation factor. An agent is only willing to update its opinion iff:

$$S_{(k,l)_{(i,j)}} < d_{k_{i,j}} \tag{11}$$

To update the agent opinions the Deffuant-Weisburg approach is adapted for a spatial context according:

$$O_{(t+1)_{(a,i,j)}} = O_{(t)_{(a,i,j)}} + \mu_a(O_{(t)_{(a,i,j)}} - O_{(t)_{(b,i,j)}}) \tag{12}$$

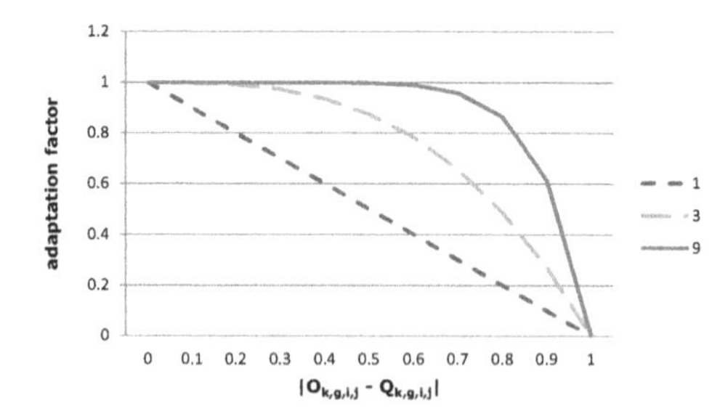

Fig. 1. Effect of different values for τ on the adaptation factor

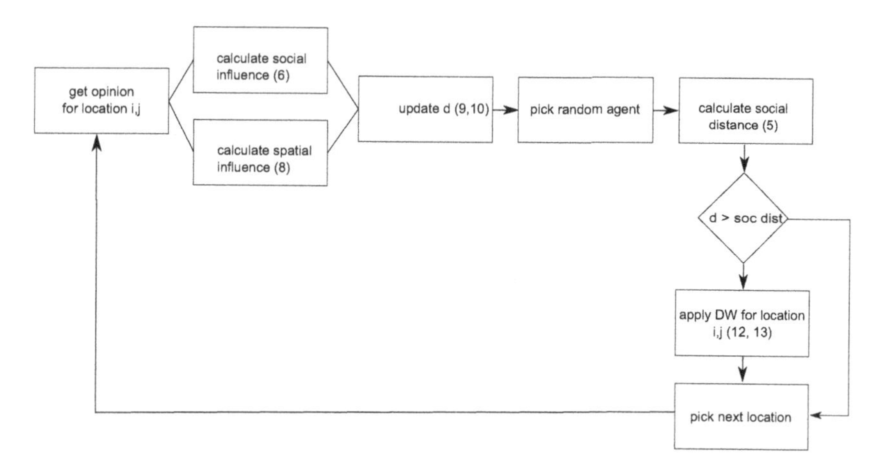

Fig. 2. Procedure carried out by each agent at each model iteration (numbers between brackets refer to the formulas in the text)

and

$$O_{(t+1)_{(b,i,j)}} = O_{(t)_{(b,i,j)}} + \mu_b(O_{(t)_{(b,i,j)}} - O_{(t)_{(a,i,j)}}) \tag{13}$$

The μ parameter is agent specific but currently independent of the location. The communication protocol is based on random encounters between two agents. Figure 2 shows an overview of the steps an agent carries out at every iteration. Each agent maintains a representation of the spatial environment defined by an ordered collection of cells $C = (c_{i,j}, c_{i+1,j}, c_{i+1,j+1}, \ldots, c_{i+n,j+m})$ where i, j are indexes that determine the location of the cell in a lattice. Each cell represents a discrete part of the area and contains information about the state of the environment, in this case the agents preference $p_{(i,g)}$ and the opinion of the agent at time t.

4 Case Study

An ABM implementation was realized in Repast Symphony for a case in the Netherlands. An area in the south-east of the Netherlands, "het Land van Maas en Waal" was picked to serve for the simulation of a hypothetical planning process. The case-study deals with the question where to allocate urbanization. The "Land van Maas en Waal" is a rural area under relative high pressure of urbanization because of the expanding cities Nijmegen and Arnhem. The model deals the problem related to multiple actors jointly developing a vision of where to allocate new urbanization. Each actor represents an organization or interest group that has own goals and desires. The remaining part of this section describes in detail the concept and design of the MAS for the above mentioned case. Although we illustrate the concept with a case study, the concept is generally applicable for spatial problems involving the allocation of a single land use. To illustrate the model three different types of actors were simulated by agents:

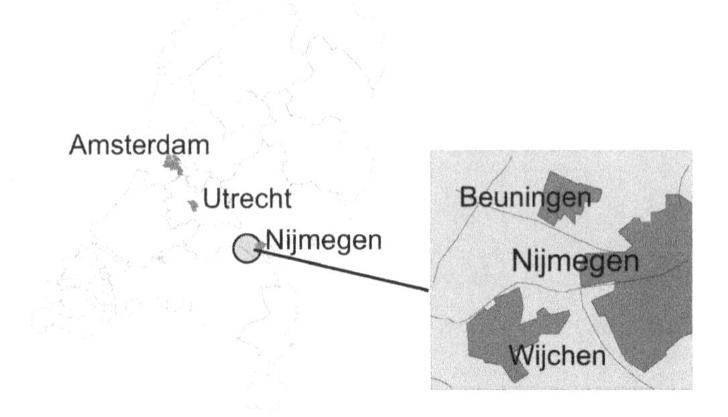

Fig. 3. Study area, the "Land van Maas and Waal"

"nature conservationist", "farmers", and "citizens representatives". Each agent has a number of desires which were implemented as spatial rules in the ABM (see for a detail description of how these rules are implemented [20, 21]). The rules for each agent are formulated hypothetically based on archetypical desires for each type of actor. Based on the rules each agent generated an explicit spatial representation of its opinion for realizing new urban areas in the study area. Of each actor type 15 agents were instantiated each having slightly different opinions accounting for the within group differences. These different opinions where accomplished by multiplying the initial opinions of each cell with a stochastic disturbance term :

$$v = (-1 * log\ r)^{\alpha} \tag{14}$$

Where v = the random perturbation and r [0, 1] is a random number, and α is a parameter to scale the pertubation and is given for this study a value of 0.1. Figure 4 shows the initial mean opinions for each actor group. These initial

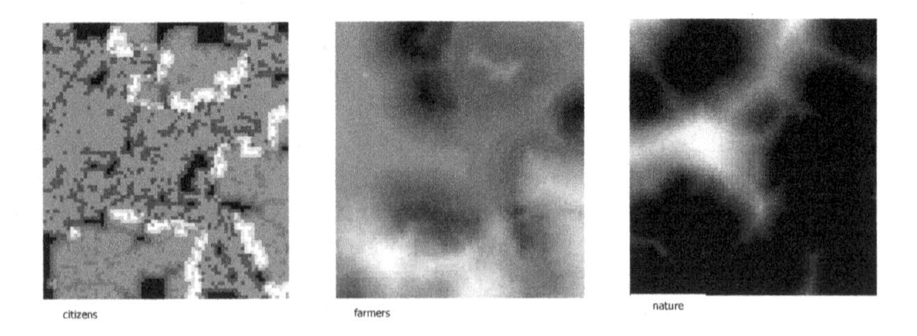

Fig. 4. Initial mean opinions of the citizens, farmers, and nature conservationists. The lighter the colors the more positive the opinions.

opinion serve as the initial input for the opinion dynamics process as depicted in Fig. 2.

To explore the behavior of the model and the spatial and non-spatial opinion patterns three different scenarios were run each having a different range of d values (drawn from a uniform distribution). For the first scenario each agent was assigned a d in the range of $[0.1, 03]$ indicating non-cooperative actors. The second scenario the d values varies between [0.3,0.6], and the third scenario d varies between [0.6,0.9] representing extremely cooperative actors. The parameters μ of equations 12 and 13 is kept constant for all scenarios.

5 Results

Figure 5 shows a map indicating the differences in the spatial distribuation of the opinions at time $t = 1$, and time $t = 25$, and $t - 50$ for the three scenarios. The maps show the difference between the range of opinions for each location in four classes. The white areas show locations having a low level of agreements amongst the agent i.e. the range of opinions is large while the dark areas have a high level of opinion i.e. the range of opinions is low. Looking at the patterns it shows that for the non-cooperative scenario ($d = [0.1, 0.3]$) many areas remain "non-negotiable" (the white area) while for the extreme cooperative ($d = [0.6, 0.9]$) scenario only few conflict areas remain after 50 iterations.

The differences of formulation of opinions can also be noted from the non spatial patterns presented in Fig. 6. The Figure shows the development of the

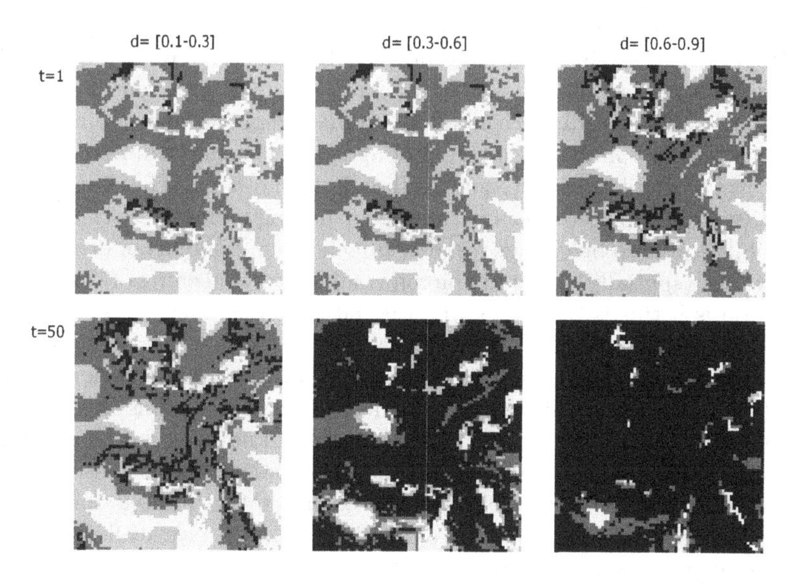

Fig. 5. Differences in spatial distribution opinions between the 3 agents for t= 1, and t=50 for various d in 4 classes of agreement (white = low agreement ... black = high agreement)

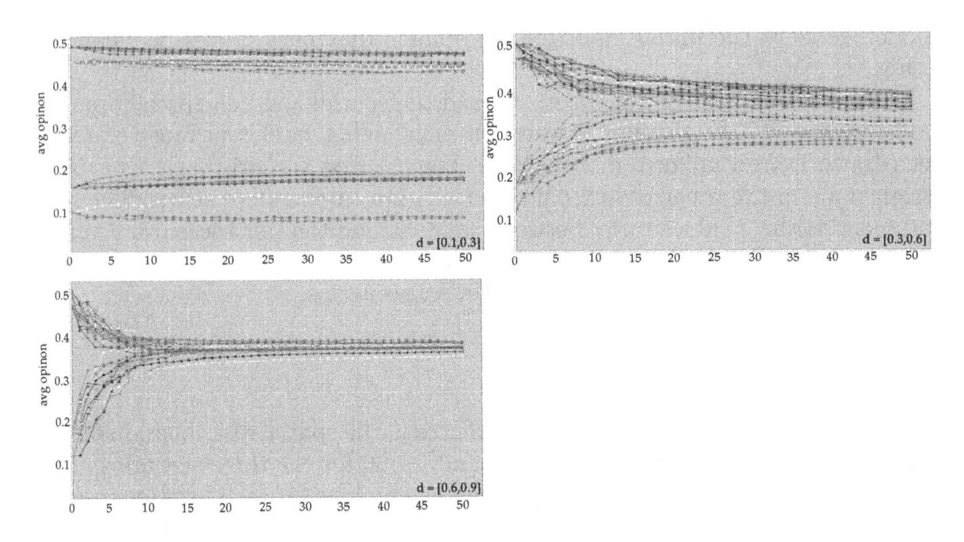

Fig. 6. Average change in opinion for each agent through time

average opinions through time for the three scenarios. Especially for the first scenario a clear segregation into two groups is shown.

6 Discussion

This paper only presents an initial attempt to apply an existing opinion dynamics model to a spatial situation. The original DW model was adapted to deal with spatial differences in opinions. Based on their initial preferences agents develop an opinion for specific locations. The influence of the spatial environment on the dynamics of opinions of a group of agents is currently limited to that of heterogeneity of opinions in the neighborhood of a location. The assumption is that relative small isolated areas with a agent having a clearly distinct opinion are more vulnerable to be changed under influence of a opinions of other agents. In other words, (spatially) isolated opinion results into weakening of this opinion. The question is if this is true? The opposite situation i.e. hardening of opinions in case of isolation would also be possible. Not much is know about the processes involved in spatial opinion dynamics. Except from vague notions such as NIMBY (not in my backyard) reactions of citizens on spatial plans affecting their immediate living environment [23, 24], not much is known about the effects of spatial factors such as distance, size, shape, and topological relations. The used model, based on the Deffuant-Weisburg approach is rather limited, and only allows for the representation of opinion dynamics at a aggregate level. Multi-actor spatial planning has a multi-goal and multi-issue characteristic, which increases the complexity of the opinion dynamics. The suitability of the DW approach to represent this complexity is not yet clear and needs further attention. Although in the presented model factors such as social distance (d) are made dynamic and depending on spatial and social influence of other actors, it is only a first step. Group dynamic processes such as cooperation, negotiation cannot easily be include in the model.

However, despite the above described limitations, the model generates plausible spatial patterns of opinions, consistent with what could be expected from the implemented scenarios. The model shows clearly the relation between level of cooperativeness of the agents and the resulting spatial opinion patterns. A society of agents with a higher willingness to cooperate leads to less differences, both spatially as well as socially.

Although the patterns resulting from the simulation seem plausible and logically consistent a proper sensitivity analysis will be essential to get more insight effects of the various parameters and input variables. Moreover, proper grounding in existing theories of opinion and innovation diffusion, and spatial planning is essential as well as a the development of methods and techniques to calibrate and validate these models.

References

1. Rogers, E.: Diffusion of innovations, 5th edn. Free Press, New York (2003)
2. Strang, D., Soule, S.A.: Diffusion in organizations and social movements: From hybrid corn to poison pills. Annual Review of Sociology 24, 265–290 (1998); ArticleType: research-article / Full publication date: 1998 / Copyright 1998 Annual Reviews
3. Kiesling, E., Günther, M., Stummer, C., Wakolbinger, L.: Agent-based simulation of innovation diffusion: a review. Central European Journal of Operations Research 20, 183–230 (2012)
4. Hagerstrand, T.: Innovation diffusion as a spatial process. [s.n.], Chicago (1967); Monograph Wageningen UR Library
5. Berger, T.: Agent-based spatial models applied to agriculture: A simulation tool for technology diffusion, resource use changes, and policy analysis. Agricultural Economics 25(2-3), 245–260 (2001)
6. Parker, C., Manson, S., Janssen, M., Hoffmann, M., Deadman, P.: Multi-agent system models for the simulation of land-use and land-cover change: A review. Annals of the Association of American Geographers 93, 316–340 (2003)
7. Matthews, R., Gilbert, N., Roach, A., Polhill, J., Gotts, N.: Agent-based land-use models: a review of applications. Landscape Ecology 22, 1447–1459 (2007), doi:10.1007/s10980-007-9135-1
8. Latane, B.: The psychology of social impact. American Psychologist 36, 343 (1981)
9. Deffuant, G., Neau, D., Amblard, F., Weisbuch, G.: Mixing beliefs among interacting agents. Advances in Complex Systems 3, 87–98 (2000)
10. Weisbuch, G., Deffuant, G., Amblard, F., Nadal, J.P.: Interacting Agents and Continuous Opinions Dynamics. In: Cowan, R., Jonard, N. (eds.) Heterogenous Agents, Interactions and Economic Performance. LNEMS, vol. 521, pp. 225–242. Springer, Heidelberg (2003)
11. Lorenz, J.: Continuous opinion dynamics under bounded confidence: A survey. International Journal of Modern Physics C 18, 1819–1838 (2007)
12. Lorenz, J.: Consensus strikes back in the hegselmann-krause model of continuous opinion dynamics under bounded confidence. Journal of Artificial Societies and Social Simulation 9 (2006)

13. Hegselmann, R., Krause, U.: Opinion dynamics driven by various ways of averaging. Computational Economics 25, 381–405 (2005)
14. Hegselmann, R., Krause, U.: Opinion dynamics and bounded confidence models, analysis, and simulation. Journal of Artificial Societies and Social Simulation 5 (2002)
15. Martins, A.: Continuous opinions and discrete actions in opinion dynamics problems. International Journal of Modern Physics C 19 (2007)
16. Martins, A.: Bayesian updating rules in continuous opinion dynamics models. Journal of Statistical Mechanics: Theory and Experiment 2009, P02017 (2009)
17. Delre, S.A., Jager, W., Bijmolt, T.H.A., Janssen, M.A.: Will it spread or not? the effects of social influences and network topology on innovation diffusion. Journal of Product Innovation Management 27, 267–282 (2010)
18. Salzarulo, L.: A continuous opinion dynamics model based on the principle of meta-contrast. Journal of Artificial Societies and Social Simulation 9 (2006)
19. Martins, A., Pereira, C., Vicente, R.: An opinion dynamics model for the diffusion of innovations. Physica A: Statistical Mechanics and its Applications 388, 3225–3232 (2009)
20. Ligtenberg, A., Wachowicz, M., Bregt, A.K., Beulens, A., Kettenis, D.: Design and application of a multi-agent system for simulation of multi-actor spatial planning. Journal of Environmental Management 71, 43–55 (2004)
21. Ligtenberg, A., Beulens, A., Kettenis, D., Bregt, A.K., Wachowicz, M.: Simulating knowledge sharing in spatial planning: an agent-based approach. Environment and Planning B: Planning and Design (2009) (advance online publication)
22. Leenders, R.T.A.J.: Modeling social influence through network autocorrelation: constructing the weight matrix. Social Networks 24, 21–47 (2002)
23. Van der Horst, D.: Nimby or not? exploring the relevance of location and the politics of voiced opinions in renewable energy siting controversies. Energy Policy 35, 2705–2714 (2007)
24. Dear, M., Taylor, S.M., Hall, G.B.: External effects of mental health facilities. Annals of the Association of American Geographers 70, 342–352 (1980), doi:10.1111/j.1467-8306.1980.tb01318.x

Agent-Based Evolving Societies

Loïs Vanhée[1,2], Jacques Ferber[1], and Frank Dignum[2]

[1] LIRMM, University of Montpellier II, France
lois.vanhee@gmail.com
[2] Utrecht Universiteit, The Netherlands
jacques.ferber@free.fr, f.p.m.dignum@uu.nl

Abstract. This paper describes a method to build artificial societies that are capable of expanding themselves from bottom-up in order to adapt to changes occurring in the environment. These changes trigger social issues at the individual level which are reported at the global level. Then, the society expands itself with an organization enforcing individual behavior that copes with the issue.

We apply this method to model the first stages of human societies. These stages are confronted with dramatic changes in the population size. These changes lead to the creation of social-control organizations to face the evolution from a familial tribe, to an autocratic chiefdom to finally reach a bureaucratic state.

Keywords: Methodologies for MABS, Social Simulation, Simulating Social Complexity.

1 Introduction

How can a society become aware of global problem and decide to solve it from bottom up? A simple solution consists in expanding itself with new organizations. Thus, firehouses are created to solve the problem of frequent fires. But, these firehouses do not raise from the ground by themselves: they result from a collective sense of the problem and a global decision to create this organization. This paper proposes a simple method to design agents that can collectively expand their society with new organizations in order to cope with social problems. In particular, we want organizations emerge from endogenous social interaction, without requiring external triggers.

We demonstrate how to use this method in modeling the first stages of human development inspired by social theories [2,3,7,8]. These theories describe describe how tribal societies evolve into chiefdoms and states. According to them, one of the keys of social development lies in the emergence of social control organizations. For instance, the transformation of a tribe into a chiefdom is initiated by the increase of violent conflicts within the society, resulting from looser family bonds due to population growth. These fights are prevented by the emergence of a social control organization: a police, which lead by a chief.

In Section 2, we present the method to design our extensible societies. Then, in Section 3, we detail the social science theories and computer science models we were inspired by. In Section 4, we present the model of our simulation and

B. Kamiński and G.Koloch (eds.), *Advances in Social Simulation*,
Advances in Intelligent Systems and Computing 229,
DOI: 10.1007/978-3-642-39829-2_25,

we test its dynamics against social science expectations in Section 5. Finally, we conclude this article with discussions about applications and future work.

2 A Method to Build Evolving Societies

We define a methodology to build societies which can dynamically expand themselves with new organizations in order to cope with social issues caused by changes in the environment or within the society itself. Moreover, we want these organizations to be created from bottom-up: to emerge from individual perceptions and actions and without exogenous trigger. In this section, we illustrate our method using the first transition of the model as an example. In this transition a tribal society is confronted with an increase in the number of fights. This triggers the creation of a police-like organization to prevent them.

Organizations are defined by the triple *(purpose, cost, effect)*. *purpose* determines the preconditions needed to create the organization. *cost* determines organizational costs. *effect* represents the influence of this organization on the society. In this bottom-up approach, one of the effects is the enforcement of individual behaviors b aiming at solving the purpose. Organizations can also be used as blackboxes in using wide range of effects (e.g. costs 10 units of resource and reduces the fight variable by 10%). Moreover, organizations may hold internal dynamics, in order to be able to tune their effect and their costs with regard to the environment and their purpose (e.g. an increase of the purpose leads to an increase of the organizational size).

The method relies on this sequence of steps: (1) A global problem which is observed by the individual variable $o \in [0, 1]$ affects individuals (step 1). o increases when the problem is frequently observed by an agent. Thus, o collectively increases when the problem affects the collectivity (e.g. fights become frequent). (2) This observation is reported to the rest of the society via a social merging mechanism $m \in [0, 1]$. m increases when o collectively increases. Various solutions exist to represent m, (e.g. a vote, a petition, a strike action). Simple computer-oriented solutions can also be selected (e.g. averaging o). (3) If the problem observed by m is important enough to meet the *purpose* of the organization O and the society can afford the *cost* of O (e.g. feeding policemen), then O is created (a police). The *effect* of O is applied, enforcing a problem-solving individual behavior b (e.g. the protection behavior). The indirect effect is the resolution of the problem and the cost to hire individuals performing b. (4) b should reduce the problem triggering o (e.g. violence decreases) and thus m (e.g. people feel safer). Note that organizations can rely on m as a performance indicator to determine if their effect should be amplified, reduced or kept stable while keeping the cost as minimal as possible. (5) Finally, O is removed if the problem is do not occur (*purpose* and *effect* are both low) or if the cost cannot be afforded. A diagram of the organization creation process is presented in Figure 1 using the MASQ formalism [9].

In this method, the emergence of an organization is endogenous: it is the consequence of agent-level parameters. There is no external trigger that directly creates the organization.

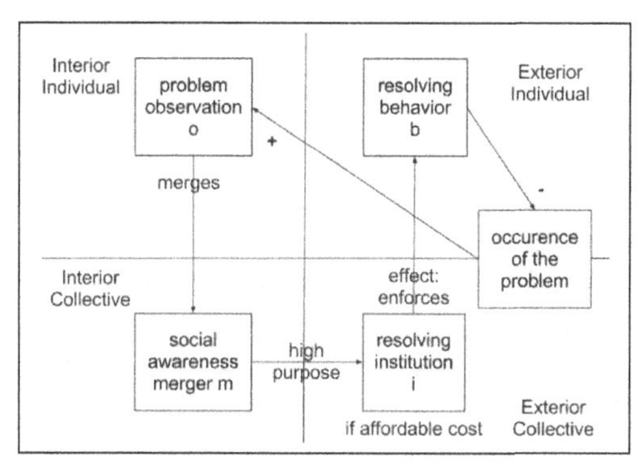

Fig. 1. Conceptual diagram of the organization creation process using the MASQ formalism [9]. Interior Individual represents the information process done within the agent, Interior Collective represents the shared beliefs amongst the society, Exterior Individual is what can be observed in the environment and Exterior Collective is what is globally observable from the simulation at the collective level.

3 Related Work

Evolving Societies: Social Sciences Perspective. Several authors [2,3,7,8] formulated theories on the long-term evolution of human societies. According to them, societies evolve through generic *stages*, each stage being characterized by the creation of a new population-control organization that tackles issues caused by shifts in population size. Thus, this theoretical framework fits the type of models we want to build with our method. Our model is mainly inspired by Diamond's theory because it proposes well-described transitions (problem and solutions) at the individual level. Nonetheless, depicted stages (band, tribe, chiefdom and state) are relatively similar across theories. As bands and tribes have similar organizational entities, with the difference that bands are smaller and nomadic our model starts with the tribe stage.

Each social stage corresponds to a social paradigm, which fits a given population size. For instance, a society with 100 members and another one with 10000 members cannot work the same way, for obvious practical reasons (e.g. resource management, decision making). Theories describe that the social stage changes when the population reaches the population-control limits of its society in creating a new population-control organization. Otherwise, if the population is too big for the social stage, social issues prevent the population to grow further. If the population is too small, some population-control organization become useless and are removed.

Tribes are groups from 80 to a few hundred people. The population is small enough to keep short the inter-individual familial distance. This kinship is a natural population-control mechanism favoring mutual exchanges and preventing fights to occur even in difficult conditions. There is no formal power centralizing information and decision which is reached through consensus.

When the population grows, the average familial distance between individuals increases. Consequently, the familial conflict resolution process fails to prevent fights. Moreover, reaching consensus on decision making and resource management becomes harder with numerous inhabitants. These limitations trigger the social transition from tribe to chiefdom.

Chiefdoms are groups up to thousands of individuals directed by a leader or a family of leaders. Leaders control power and information. Thus, this organization solves the scaling issue of the tribal consensual decision making. Moreover, leaders centralize and redistributes resources via a taxation system. This system allows leaders to feed specialists. These specialists can produce luxury goods for leader (inducing a cost of resource) but also control the population (policemen) preventing fights between unrelated individuals.

At the chiefdom stage, the management of critical information and decision making is made only by leaders. When the society grows (above 50.000 individuals), leaders become overwhelmed by information and decisions. They have to delegate a part of their power to subordinates. But, without tight control, subordinates tend to take over the leader.

To prevent this situation, the propagation of cultural values (e.g. law, norm, religion) that supports the leaders and their system have been proven successful. Such a culture generally contains rules and processes and decision and some ideology grouping members of the society within a moral circle. Thus, instead of obeying a leader, individuals act according to rules (decentralizing the decision process) for to defend the values carried on by the ideology. This culture is indirectly spread amongst the population via what we call Cultural Harmonizers (*CHs*) whose activity implies spreading the culture (e.g. priests, professors, itinerant storytellers). A society maintaining a strong culture with *CHs* is at the state level.

Table 1 provides a summary of the observed specificities for each social stage.

Table 1. Social patterns generally observed for each social stage (based on Diamond's theory [3])

	Tribe	Chiefdom	State
Number of people	Hundreds	Thousands	Over 50.000
Settlement pattern	1 village	1+ villages	Many villages
Decision making	"Egalitarian"	Centralized	Centralized
Leadership	Big man	Hereditary	Organizational role
Force and information monopoly	No	Yes	Yes
Conflict resolution	Informal	Centralized	Law, judges
Exchange	Reciprocal	Redistributive (tribute)	Redistributive (taxes)
Indigenous Literacy	No	No	Yes

Each stage is the consequence of the emergence of a new organization: a police at the chiefdom stage and a *CH* organization at the nation stage. Thus, these theories propose a framework that matches the purpose of our methodology.

Evolving: Computer Science Models. Previous research in computer science has also investigated the replication of emerging phenomena observed in social reality, while keeping low agents simple. The most famous ones being Sugarscape [5] and New Ties [6]. Sugarscape [5] introduces a very simple grid-like environment and simple agents with simple rules. But, in adding slight variations on top of this environment and agents (like reproduction, fight and trade), Epstein and Axtell were able to replicate a wide range of observed social phenomena (e.g. genetic selection, global wars, real-world market behaviors). New Ties [6] aims at building an artificial society with socially high mental and linguistic capabilities, but with simple agents and simple environment. The goal of New Ties is to build autonomous entities, living and adapting to the world due to learning. Moreover, these agents are capable to produce their own culture and share it with other agents. The research objectives of New Ties aim at a better understanding of culture formation and linguistics. In this article, our research differs from them in the sense that we aim at building societies capable of dynamically expanding themselves with explicit institutions, but still in their bottom-up approach.

4 A Model for Evolving Societies

Our model consists of two main features: the first concerns environmental aspects (e.g. food, reproduction) and the second is focused on social aspects (human interactions). Environmental aspects represent food collection and reproduction. Well fed human agents reproduce. Thus, the society grows until the food input matches the survival costs of humans. If the food input is lowered, agents starve to death and the society becomes smaller.

Social aspects concern human agents social interaction. In particular, it introduces social issues caused by population scale changes. When the population is bigger than a family, family-unrelated agents are likely to fight with each other when the food is sparse. In this case, a police organization is created when the population feels globally unsafe. This leads the society to the chiefdom stage. Similarly, when the population increases, the loyalty towards the leader decreases. When the loyalty is too low, a *CH* organization is created to keep high loyalty, leading the society to the state stage. As the chiefdom stage arises before the state stage, state is also chiefdoms. In the following, the model is presented in more detail using the SugarScape [5] formalism. This formalism splits the model in three sections: environment, agents and model rules.

4.1 Environment

The environment is similar to the Netlogo space: a bounded $\mathbb{R}^2$ space for agents on top of which food is represented by NetLogo patches (a $\mathbb{N}^2$ grid). Each patch has 3 states: $(sterile, harvested, full)$. A *sterile* patch never changes state. A $full$ one can be harvested by human agents, thus becoming *harvested*. A *harvested* cell can become $full$ each round ($p = 0.005$). In our experiments, patches in a radius of 8 around a village are initially $full$. Others are *sterile*.

All environments, even the simplest ones, introduce complex dynamics in the model (as shown in [4]). The most important environment-related aspect of this model to be aware of is that the food input is relatively constant each round and sublinear to the number of harvesters. As a rule, the gain for each additional harvester is marginal when the number of harvesters is above 20.

4.2 Agents

Human Agents. Human agents are entities populating the world. They are described by the following variables:

$hp \in [0, max\_hp]$: hit points or physical condition. This value decreases when the agent is attacked. If this value reaches 0, the agent dies. max_hp is set to 10 in the simulation.

$energy \in [0, max\_energy]$: remaining energy. If this value reaches 0, the agent dies of starvation and decreases by 1 every round. An agent with less than $max\_energy/2$ units of food is *hungry* (it cannot reproduce), otherwise it is $well\_fed$. max_energy is set to 400. Modifying max_energy impacts the sensibility of the population growth to external conditions: the lower this value is, the more the population fluctuates between periods of richness and extensive storage followed by overgrowth and starvation.

$bag \in \mathbb{N}$: amount of carried food. It is filled when agents perform the *harvest* action and its content can be *stored* in the village.

$culture \in [0, 1]^n$: a vector representing the cultural orientation of the agent (e.g. values and practices [7]), inspired by the Sugarscape cultural model [5]. Values of this vector are abstract, but can be used to compare inter-individual cultural similarity using Euclidean distance. This vector changes randomly with time (cultural innovation, change in opinion and divergence) and is be shared amongst individuals via the *harmonize_culture* action or inter-individual communication. Thus, inter-individual cultural distance evolves with time: it increases with the number of inhabitants and decreases with cultural sharing. We empirically set n to 50. n impacts on the variability of individual culture, but this impact is relatively low when $n > 20$. $culture$ defines the $loyalty \in [0, 1]$ towards the leader. The farther the distance of the agent's culture and the culture $culture_v$ of the village, the lower this agent is loyal. Formally, $loyalty = 1 - \alpha.distance(culture, culture_v)$ where $\alpha \in \mathbb{R}$ is an arbitrary constant to adjust the impact of cultural distance over loyalty. In our case α is 0.05.

job $\in$ {*harvest, protect, harmonize culture*}: the activity to be performed during the "perform job" phase. A *harvesting* human moves to the closest food patch, collects it and drops its bag in the village. A *protecting* human is a member of the police. It is inactive during the "perform job" phase but increases the probability of the villagers to be under control in the "meet other humans" phase. *Cultural Harmonizers (CHs)* unify the culture of the village in performing the *harmonize* action. This action sets the *culture* vector of 10 human agents to the village culture $culture_v$.

social_dissatisfaction $\in [0, 1]$: the dissatisfaction against the current society due to lack of physical protection. This value is reduced every round by 1% and increases by 0.2 (bounded by 1) when the agent is attacked.

Villages. A village agent accommodates human agents and serves as deposit for their food resources. This agent is *lead* if a special organization (police or *CH* office) requires it. The leader establishes the *social policy SP*. *SP* is a couple $(R_{police}, R_{CH}) \in [0, 1]^2$. Each member of SP determines the ratio of organizational jobs (police, *CH*) per inhabitant offered to the population. Thus, *SP* allows to tune the resource given to the organizations, in order to balance organizational costs and benefits.

food$\in \mathbb{N}$: the amount of food stored by humans. Human agents storing their *bag* increase the value of *food* by the amount of their *bag* and set *bag* to 0. When the village is lead, a portion (20%) of the stored food is spoiled by leaders (to buy luxury goods). With this mechanism, we simplify the burden of sharing resources via exchange, trade or redistributive economy (which can also be achieved with our method, the redistributive economy being a pre-requisite of an organization for instance). Thus, the salary of organizational workers is the ability to eat from the common pot.

$culture_v \in [0, 1]^n$: the culture of the village. Without a leader, it is the average *culture* of its inhabitants, otherwise it is the *culture* vector of the leader.

harvest_tech_level$\in \mathbb{N}$: the food production efficiency of the village (technological and environmental). When a human agent harvests some food, it puts *harvest_tech_level* units of food its bag. Empirically, the size of the society is correlated to the value of this variable. So, this variable is used in the experiments to determine the society size. Its values ranges between 10 and 300.

4.3 Rules

Human Rules. Figure 2 shows the main steps of the decision making process of a human agent (individual behaviour).

Update life constants: energy is decreased (starving agents are removed) and *culture* is randomly tilted (one of its items is set to a random value) with a probability of 0.2.

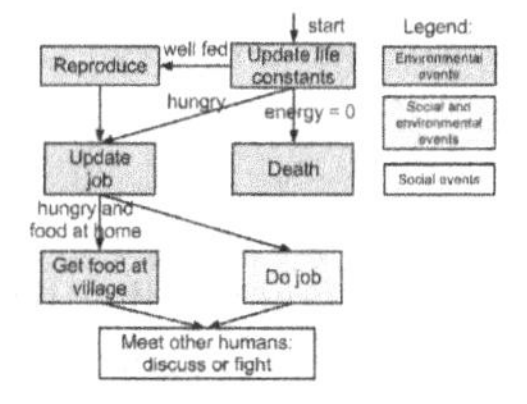

Fig. 2. The decision making process of a human agent

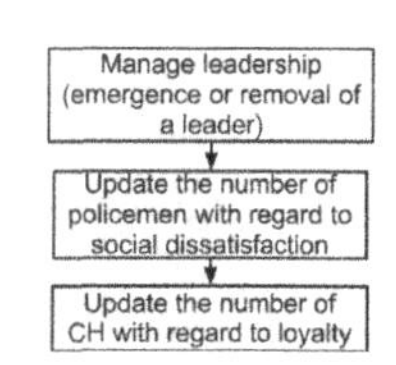

Fig. 3. Summary of the village rules

Reproduce: well-fed human agents can produce one child ($p = 0.01$). The energy value and culture are shared with the child. Other variables are set to default values.

Update job: Periodically, humans change their job by updating the *job* variable. Harvest can be selected by anyone without restrictions. Institutional jobs (protect and harmonize culture), can only be selected if offered by the social policy SP. The job selection is simple: the agent tries first to get an organizational job. If none is available, it selects harvest. To avoid erratic behaviors, an agent is committed to a job for multiple rounds (randomly between 20 and 70 rounds).

Get food at village: The agent moves towards its village. If the agent is at its village and the *food* value of the village is high enough (bigger than max_energy), the human agent performs an *eat* action.

Do job: The human agent performs its *job*.

Meet other humans: each human agent engages in some social interactions with another randomly selected human agent from the same village. If these two humans are from the same family or under control of a loyal policeman, they discuss with each other. They can mix their *culture* vector if their cultural distance is close enough (according to the Axelrod cultural model [1]). This mechanism is important for our purpose: more agents lead to a more important cultural divergence. If the food is sparse, the two agents come into conflict. One of them is hurt (*hp* is reduced by 1) and complains about the lack of physical protection (*social_dissatisfaction* is increased by 0.2).

Two randomly-drawn agents belong to the same family with a probability $s/|v|$ where s is the family size (set to 30) and $|v|$ is the village size. Two agents are under police control with a probability $P.L_v(police)/|v|$ where P is the number inhabitants a policeman can control (set to 10) and $L_v(police)$ is the sum of policemen *loyalty*. These mechanisms can of course be implemented in a different way. For instance, in a former version we represented explicitly family networks or policemen can patrol and defend their nearby space. The core property (whatever the complexity of its implementation) is that a population size increase triggers observable organizational failures (observed by o_1 and o_2).

Each round, each agent acts only once in the environment: during the "get food at village" or the "do job" phases. Available actions are: *move*: moves a distance of 1. *harvest*: sets the local *full* patch to *harvested* and fills the agent's *bag* with *harvest_tech_level* units of food. *harmonize*: merges 20 items of the *culture* of an agent to $culture_v$ (for sake of simplicity we did not forced agents to be nearby to make this exchange occur). *eat*: the agent removes $max\_energy - energy$ units of food from its village and sets *energy* to max_energy.

Village Rules. Village rules (Figure 3) represent orders given by chiefs (if any). They are split in 2 phases: organization creation and the social policy management (SP). If no police organization is present, agents evaluate their desire for physical protection *social_dissatisfaction*. Each round, a vote is performed to determine if the police is required: agents approve if *social_dissatisfaction*$>$ 0.7. If a majority is reached, then a police institution is created and a leader (the human

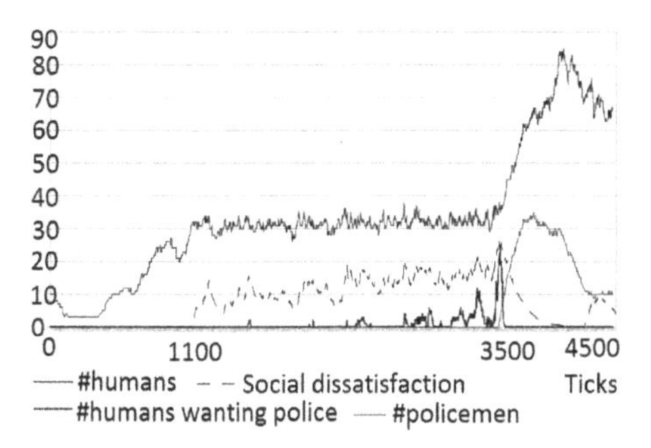

Fig. 4. The emergence of a police, leading the society to the chiefdom stage

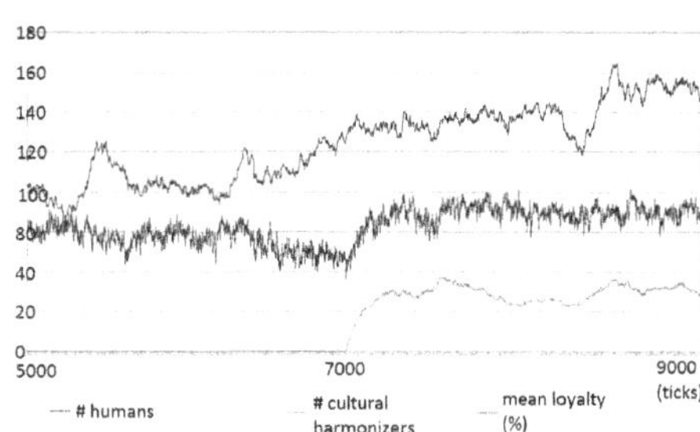

Fig. 5. The emergence of a *CH* office, leading the society to the state stage

agent who survived the more battles) rules the village. Then, one of the human agents is hired as a policeman to perform the "protect" behavior. Similarly with the *CHs office*: agents evaluate the village loyalty. If the average *loyalty* is below 0.4 then the CH office is created. One agent is hired as a *CH*, performing the "harmonize" behavior.

The policy management phase updates amount of resource allocated to each organization (their size) SP_v according to the needs of the village. Leaders update R_{police} to keep *social_dissatisfaction* below 0.5 and R_{CH} to keep the loyalty above 0.7. If R_{police} or R_{CH} is 0, the need for the corresponding organization is gone and thus this organization is disbanded.

Our method represents the police with organization O_1. $purpose_1$ is to prevent fights, $cost_1$ is the cost for feeding policemen and $effect_1$ is policemen action. The local observation o_1 is *social_dissatisfaction*, the social merging mechanism m_1 is the vote and the individual behavior and b_1 is the "protect" behavior. O_2 represents the CH office organization with $purpose_2$ to raise loyalty (o_2), $cost_2$ is the cost for hiring CHs and $effect_2$ is the action of *CHs*. m_2 is the average of o_2. b_2 is the "protect" behavior. Both costs and effects of O_1 and O_2 are tuned using m_1 and m_2.

5 Experimentation

Our model aims at illustrating our method on a real-case inspired setting. In addition, we internally validate our model in showing that macro and macro behaviors of the model are consistent with the ones from the theories. Of course, this validation is not a "proof" that theories are correct.

What Happens during the Transitions? In this experiment we describe how societies change stage and create new social control mechanisms. To this extent, we increase the population capacity of the society in exogenously improving life conditions with time (*harvest_tech_level* increases by 1 every 40 time units).

Figure 4 shows how the police organization is created, triggering the transition tribe-chiefdom. From time unit 0 to 1100, the population size is below the family size. Thus, even when the food is sparse, agents are not attacking each other. From time unit 1100 to 3000, some fights occur and are observed through o_1, but remains too sporadic to require the assistance of a police. During this period, life conditions improve, leading to more births and to more fights. Thus, o_1 increases during this period. The dissatisfaction reaches a climax from 3000 to 3500 where more and more agents are highly dissatisfied ($o_1 > 0.7$) leading to an increase of m_1. When m_1 reaches 50% of the population (tick 3500), the police organization (O_1) is created. Due to an initial high average dissatisfaction the organization is allocated an important amount of resources. So, many humans agents are hired to perform the protect behavior (b_1). This behavior reduces the number of fights o_1, reducing in turn the overall dissatisfaction m_1. Note that the organization becomes eventually tuned for its purpose: after an initial overgrowth, the ratio of policemen is reduced to keep the social dissatisfaction low without extra costs.

Figure 5 shows the emergence of the *CH* organization, leading the society to the state level. Due to a population increase, the traditional communication mechanisms are not sufficient for the population to globally agree leading in turn to a reduction of the average loyalty. The average distance between individual's *culture* and $culture_v$ tend to increase with the population size. The individual disagreement with the village culture is observed with o_2 and is collectively averaged with m_2. In time unit 7000, the society observes a global disloyalty ($m_2 < 40\%$) triggering the creation of a *CH* organization (O_2). O_2 hires *CHs*, who spread culture, reducing the average distance between individual's *culture* and $culture_v$ and thus raising the average *loyalty* to 0.7 with the "harmonize" behavior (b_2). These experiments show the adaptability of our model in dynamic environments: organizations can be created, kept and disbanded on the fly to fit in a dynamic environment and society.

As a general observation, when *harvest_tech_level* increases, society size increases as well as the pressure on the society to evolve to the next stage. The first transition can occur when *harvest_tech_level* reaches 80 and the second when *harvest_tech_level* reaches 150. Thus, changing *harvest_tech_level* update rate only influences the amount of time needed by the society to evolve.

In the current model, *harvest_tech_level* is modified exogenously, because we aim at observing the emergence of social control institution to illustrate our method of problem-tackling organizations. Nonetheless, we experimented more complex models (not presented here for space considerations) that include endogenous growth of *harvest_tech_level* . In these extensions, individuals perform research actions through a third organization dedicated to research (artists, as described in [3]). We obtained similar results as those obtained here (the same transitions) even if dynamics induced by research organizations influence other factors directly (e.g. the population growth speed) and indirectly (e.g. other organizations).

Some other observations made by theories at the macro level also appear in this simulation. For instance, a society in an "obsolete" stage cannot grow due to social issues. Thus, the emergence of a new stage generally triggers a demographic explosion. Similarly, the problem-solving organization is generally overemphasized during its first steps due to extremely high demand and is then reduced to a more reasonable size. Moreover, if the population is reduced, less fights and more loyalty are observed. If an organization achieves its purpose in spite of being empty $R_i = 0$, then this organization is disbanded.

Skipping Stages? Another important property of the human societies described in [2,3] is that even if life conditions are good enough, a stable state cannot be built directly from a tribe, while skipping the chiefdom stage. The chiefdom stage brings the necessary social control organization to prevent the society to collapse. They use this type of examples to describe why societies evolve through well-defined stages. We simulated this "what if" question with our model with the following hypothesis: *can a society in a wealthy environment expand to the nation stage (build a* CH *organization) and remain sustainable without reaching the chiefdom stage(build a police organization)?* To this extent, we run our model in setting *harvest_tech_level* to 300 (or any higher value) and preventing the police organization to be created (agents never vote "yes" for a leader in the village phase).

In these conditions, the society initially starts by growing, until reaching the point where the loyalty to the village becomes low. Then, the *CH* organization is created and the population keeps growing until the resource consumption meets the available resources. Then, individuals start fighting. Since no police exists to prevent combats, human agents fight until killing each other. Thus, the society collapses until reaching approximately the size of a family and is then limited at this point. This reduced size eventually triggers the removal of the *CH* organization, since the tribal communication mechanism is sufficient to keep the loyalty towards the village high enough. Thus, the society has returned to the tribal stage. Consequently, our model matches the predictions made by the theories at the collective level: each societies have to evolve one stage after the other and in the correct order in order to remain sustainable.

6 Conclusion and Future Work

In this paper, we presented a method to build societies capable of extending themselves with organizations solving collective issues. These organizations emerge from bottom up, with a mechanism which is internal to the society and based on the agent's local perceptions. Moreover, these extensions can be created, strengthened and removed on the fly, making the society capable of adapting to environmental and social changes. The method consists in merging the local observations of the issue amongst the society. This merging allows the society to become aware of the situation and then to build the appropriate organization in order to solve the issue.

We tested this method in modeling the first stages of human societies. These stages are characterized by the occurrences of two organizations, the police and the cultural harmonization (e.g. temples or schools) organizations, solving societal growth issues. The modeling and the implementation of the dynamic creation of organizations are straightforward from the method. Moreover, our artificial societies display similar behaviors to those of societies described by human development theories (e.g. skipping social development stages eventually leads to a social collapse).

Future work will focus on extending the model and the method. The model can be extended to include a more credible influence of the culture, as the one described by Hofsede et al. [7]. The model is also being refined in order to make the village emerge from bottom up on a grid. We expect that our methodology is sufficient to raise similar emerging organizations.

The method can also be refined into a logic, allowing to express dependencies for instance. Consider for example the need for bread in a society. Making bread requires flour, the flour requires weat and a mill. Thus, this logic would describe how to make an economy emerge from scratch and simplify its design. The outcome of this research has interests in various fields, like games (the previous example is drawn from Settlers II), simulation of complex societies and self-organizing systems. This logic would be capable to consider multiple organizational responses for the same issue, with different costs and benefits (e.g. official police, unofficial night brigade, civic education at school). So, this logic allows to build societies capable of growing several combinations of organizations in keeping into account its environment.

Acknowledgements. The first author wishes to thank Melania Borit for her feedback while writing this paper.

References

1. Axelrod, R.: The Dissemination of Culture: A Model with Local Convergence and Global Polarization. Journal of Conflict Resolution 41(2), 203–226 (1997)
2. Beck, D.E., Cowan, C.: Spiral Dynamics: Mastering Values, Leadership and Change. Wiley-Blackwell (1996)
3. Diamond, J.: Guns, Germs and Steel. Vintage Books (2005)
4. Dignum, F.P.M., Dignum, V., Sonenberg, L.: Exploring congruence between organizational structure and task performance: A simulation approach. In: Boissier, O., Padget, J., Dignum, V., Lindemann, G., Matson, E., Ossowski, S., Sichman, J.S., Vázquez-Salceda, J. (eds.) ANIREM 2005 and OOOP 2005. LNCS (LNAI), vol. 3913, pp. 213–230. Springer, Heidelberg (2006)
5. Epstein, J.M., Axtell, R.: Growing artificial societies. The MIT Press (1996)
6. Gilbert, N., Den Besten, M., Bontovics, A., Craenen, B.G.W., Divina, F., Eiben, A.E., Griffioen, R., Hévízi, G., Lõrincz, A., Paechter, B., Schuster, S., Schut, M.C., Tzolov, C., Vogt, P., Yang, L.: Emerging Artificial Societies Through Learning. Journal of Artificial Societies and Social Simulation 9(2), 9 (2006)
7. Hofstede, G., Hofstede, G.J., Minkov, M.: Cultures and Organizations: Software of the Mind, 3rd edn. (2010)
8. Johnson, A., Earle, T.: The evolution of human societies: from foraging group to agrarian state (2000)
9. Stratulat, T., Ferber, J., Tranier, J.: MASQ: towards an integral approach to interaction. Mind 2, 813–820 (2009)

Agent-Based Case Studies for Understanding of Social-Ecological Systems: Cooperation on Irrigation in Bali

Nanda Wijermans and Maja Schlüter

Stockholm Resilience Centre, University of Stockholm, Sweden
{nanda.wijermans,maja.schlueter}@stockholmresilience.su.se

Abstract. This paper describes the design phase of an ABM case study of Bali irrigation. The aim of the model is to explain the differences in the ability of rice paddy farmers to collectively adapt through cooperation. The model should allow exploring factors affecting self organisation within and between rice paddy farmer communities. The exercise of the ABM case study aims to move abstract models (theory) closer to real world phenomena, which requires contextualisation. This paper focuses on the first steps in model contextualisation: model selection and specification for the Bali irrigation case.

Keywords: ABM case studies, Bali Irrigation, cooperation, social-ecological systems, social dilemmas.

1 Introduction

We live in a complex world, where our actions affect and are affected by nature and other humans. Social-ecological systems (SES) research focuses on problems that involve both humans and the environment, which are tightly interconnected and in continuous interaction. For example, questions that address the impact of climate change, the spread of infectious diseases or the sustainable management of natural resources. One major challenge of SES research, with respect to sustainable use of natural resources, is to understand and manage social dilemmas, i.e., situations in which resources are shared and there is a need to individually restrain resource outtake to avoid collective over-exploitation of a common pool resource[1]. These social dilemmas can have tremendous impact on a large scale, such as, reduced water availability for food production, collapsing fish stocks or the inability to deal with climate

[1] Common pool resources are a type of resource where taking out resources (harvesting) by one user reduces the availability from a pool that is potentially available to others, i.e. subtractability, and it is difficult to exclude anybody from taking out resources, i.e., non-excludible [1].

B. Kaminski and G.Koloch (eds.), *Advances in Social Simulation*,
Advances in Intelligent Systems and Computing 229,
DOI: 10.1007/978-3-642-39829-2_26, © Springer-Verlag Berlin Heidelberg 2014

change. One of the main identified needs is to get basic information and understanding of SES dynamics to evaluate, learn and improve SES management [2]. Most work done in sustainability research/common pool research focuses either on rich descriptive case studies of real world cases or on highly abstract analytical models [3]. Both have their strengths by either having a strong relation to the real world phenomenon (case studies) or explorative power and clarity (generic models). Within this need of understanding we see a role for agent-based modelling: 1) to embed the complex adaptive nature of social-ecological systems; 2) to provide an insightful middle ground between (abstract) generic models and the (rich) real case studies; and 3) to provide a systematic way to discern between important general aspects of the social and ecological context of the social dilemma and case specific factors. We explore the contextualisation of concepts, mechanisms and interactions of a generic model and thereby to integrate general aspects of the social-ecological context that are relevant for explaining sustainable outcomes in complex SES. In the following we describe our first steps into an ABM case study: the contextualisation of an (abstract) model of cooperation in the case of Bali irrigation, see figure 1.

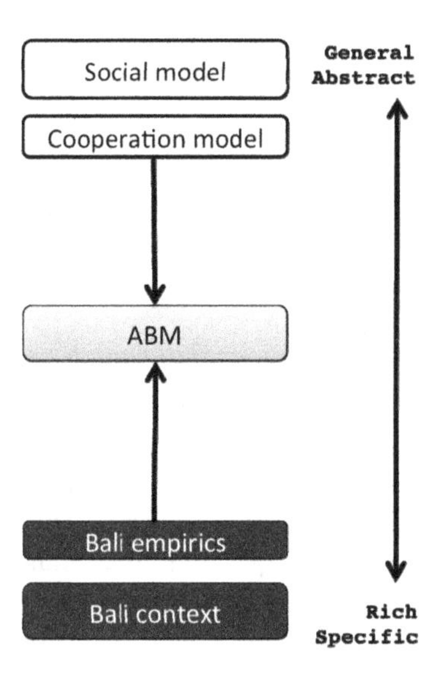

Fig. 1. The positioning of our approach in modelling an ABM case study

1.1 Bali Irrigation as an ABM Case Study

Bali irrigation is one of the well-know examples of a self-organised 'escape' of the 'tragedy of the commons' [4]. The tragedy of the commons describe the inevitable outcome of overexploiting shared resources in absence of a central authority [5]. Rice paddy farmers in Bali are part of a community-level organisation devoted to the management of the rice terraces, a subak[2], apart from being part of a village community. The rice paddy farmers in Bali demonstrate effective management of their rice fields in a bottom up manner by agreeing on an irrigation schedule *within* and *between* Subaks [6]. The irrigation schedule synchronises the cropping patterns to avoid pests and provides water for all the subaks. The success of the rice paddy farmers lie in their ability to adapt to the ecological circumstances, in other words, their capacity to engage in collective action. Any change in this ability of the farmers in a subak can potentially affect the whole system of subaks that are interconnected by the river. For instance, even short periods of lacking cooperation could have large impacts on the harvest on multiple scales. In addition, the damage that this lack of cooperation can produce takes time and effort to return to the state before cooperation was lacking, just imagine the devastating effects on ecology to recover from pests, or the long way a community needs go through in trusting each other again. The goal of our model will thus be to explore the vulnerability of a social group within the Bali context and explore on multiple scales the effect of such a system but also the role of our theories and assumptions on the understanding of such as system.

Recent work done by Lansing et al [7] indicate variation between subaks in their ability to adapt to environmental and social circumstances. . Lansing et al. [7] explored the age and demographic stability of subaks (their genetic diversity) in relation to their ability to adapt, i.e., to engage in collective action. Communities with lower adaptive capacity might potentially be more threatened by impacts of local or global change such as an influx of newcomers or expansion of tourism. These empirical hints of potential threats on to self organisation trigger questions: ' what affects the adaptive ability capacity of farmer communities?, 'what is the effect on different scales of varying capacity to adapt?' and in the larger frame of our project 'what does the understanding of this case mean for general models of social dilemmas?'. Inspired by these latest findings we focus on exploring potential threats on self-organising groups and thereby on the ability to maintain the necessary ability to adapt to ecological and social circumstances. The Bali irrigation case already proved its value for understanding both the Bali case[3] and more generic reflections on cooperation and ecological feedbacks [8].

This paper zooms in on the ABM design phase: the contextualisation of a generic model of cooperation placed within the context of Bali irrigation. Although we use

[2] The rice paddy farmers organise themselves in groups around a shared water resource, so-called subaks. These subaks are embedded in a nested network of temples matching the island's landscape: ranging from the main temple and lake on top of this steep volcanic island, via smaller temples at rivers, canals and weirs towards the rice paddy fields, where the water ends her journey finally in the sea.

[3] For instance the model developed by Lansing and Kremer demonstrated the necessity and power of bottom-up organisation that convinced the consultants and government to continue 'modernisation' of bali agriculture was a big drama, (for a detailled description of the 'green revolution, see chapter 1, [6]).

empirical data for our model design, our discussion restricts itself to model selection and specification; we do not focus on model calibration. We depart from an extensive body of work done on the Bali case, both descriptive ethnographical data and computational models. The models represent a realistic description of the important ecological functions, water availability and pest dynamics, e.g., [9-11]. The models of social-ecological dynamics in Bali focus on the level of coordination [9-11] and cooperation [6, 8] in irrigation schedules. Typically in these models the subaks are represented as the smallest dynamic entity. It is assumed that within a subak everyone sticks to the agreed rules, i.e., cooperates by executing a particular cropping plan, see figure 2. For the research aims of those Bali models this is a sensible simplification, however the recent insight of the variation in subak outcomes force us to zoom into the within subak dynamics itself, see figure 3. To represent the social dynamics, i.e., the ability of the rice paddy farmers to collectively adapt when environmental or social issues arise, guides the focus to research of cooperation. There exist a vast body of literature on cooperation, suggesting various mechanisms that lead to cooperation (that vary across the different social science fields, such as economics, social psychology and neuro-sciences). In light of our aim to connect move towards an intermediate level of the levels of abstraction, recall Figure 1. We selected to start with an abstract model of cooperation [12], an ABM reproduction of [13],. This paper will focus on illustrating some of the challenges and approaches we take to design a contextualised social-ecological model of cooperation in Bali.

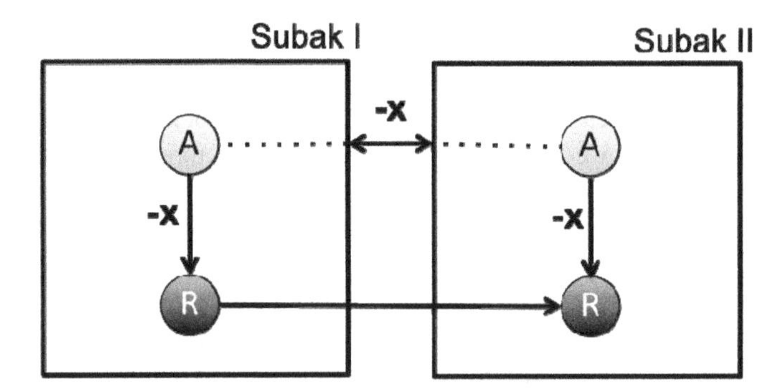

Fig. 2. Schematic overview of the focus on between subak dynamics of existing models of coordination and cooperation in Bali. The actors (A) in a subak take out x amount of water(R). The amount x is a result of the between subak coordination. It is a 'rule' that all actors know (dotted line) and are assumed to comply to. The subaks are interconnected via the network of rivers and canals of water (R). Where all actors have access to the water, but when actors from Subak 1 take out water (upstream) there is less water left for the actors in Subak 2, i.e., a typical common pool resource.

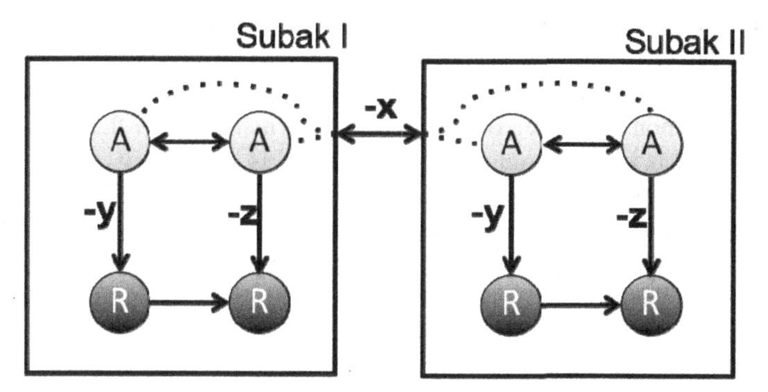

Fig. 3. Schematic overview of our focus on within and between subak dynamics in modelling cooperation of Bali irrigation. This is a variation on figure 2b, where the change in assumption is visualised. Actors in a subak all know about the agreed 'rule' (dotted line) to take out x but do not necessarily comply. The amount (y or z) taken out is influence by the interaction with other actors (social environment) too.

2 Modelling Subak Cooperation in Bali Irrigation

The aim of the model is to explain the differences in the ability of subak farmers to collectively adapt through cooperation. The model should allow exploring factors affecting self organisation within and between subaks. In parallel, the exercise of the ABM case study aims to move abstract models (theory) closer to real world phenomena, which requires contextualisation. Contextualisation describes a process of selecting a suitable model, specifying, sometimes calibrating, testing and starting a new design iteration: adapting a model, etc. This paper thus discusses the model design stage with a focus on contextualization. This section will discuss the first steps, model selection and specification for the Bali irrigation case.

2.1 Model Selection: Start of the Iterative Journey of Model Design

There exists a vast body of models explaining cooperation on a theoretical level [14-16]. At the same time, ethnographical observations and descriptions of subak life describe a richness of factors, actors and processes playing a role in the social cohesion of a subak [6]. Our process of developing an ABM taps from both sources. We choose to depart from a model of cooperation, the ostracism model [12, 13]. The reasons for choosing this model are as follows: a) context relevant, this model of cooperation is placed in the context of resource management, which matches the case context of a social dilemma; b) social driver for cooperation is, which we regard important to describe the adaptive capacity of a community; c) allows for comparison. The model is a replication of a theoretical model [12,13], which has advantages in comparing and reflecting on the case-based outcomes; d) ownership, it is a model created by one of us, which allows for short-links to interact about model specifics.

The Ostracism Model. This ABM model [12] replicates and further explores an analytical model [13]. Both ostracism models investigating the role of norms and social disapproval by norm followers for cooperation amongst a group of harvesters that share a common resource. If the group of norm followers that harvest the resource sustainably is large enough (i.e., social capital is high), they engage in ostracising over-harvesting norm violators. The system develops through agents imitating best performing strategies. When the initial number of norm followers and hence the social capital of the community is low, norm violators prevail and the resource is over-harvested. If the number of norm followers is high and defection of norm violators is not too large, a community of norm followers evolves. If the initial number of norm followers is high but the defection of the norm violators is large, coexistence emerges, where a small group of norm violators share the resource with a large group of norm followers. The ecological and social drivers of system dynamics balance out, e.g., the gain defectors get from higher resource levels due to high levels of cooperation is balanced with the social disapproval they experience through the community of norm followers.

The main concepts of the model are shown in Table 1. On the macro level, the emergent pattern (target) is the level of cooperation, e.g., the proportion of norm followers that harvest sustainably. In addition, the proportion of cooperators and defectors, the ostracism costs for defection and resource volume affect the agents in their aggregated form. On the micro level the agents are characterised by their behavioural options, to defect or to cooperate. They choose the behaviour that performs best, i.e., has the highest utility[4]. The environment (the meetingList) defines with whom the

Table 1. An overview of the main concepts of the (abstract) cooperation model we depart from filtered by main ABM dimensions

Macro Level	**Emergent pattern**	**Level of cooperation**
	Aggregated variables	Proportion of Cooperators, Defectors [%] Ostracism costs Resource volume
Micro level	Agent	Behaviours options: {Cooperate, Defect} Decision-making: behaviour = imitate if other has a better strategy
	Environment (topology)	MeetingList [random]
	Interaction	Physical environment: Receive utility() Social environment: Compare utility()

[4] In the model there is a distinction between payoff and utility. Payoff represents the 'crop output', whereas utility is the final gain (in money) that an agent can receive. Ostracism affects the step from payoff to utility, which can be illustrated by an agent being blocked access to the market.

agents interact, i.e., meet randomly, to learn about the best performing strategy by comparing the utility of its own behaviour with that of another agent. The interaction with the physical environment results in receiving utility as a consequence of the chosen behaviour.

The Challenge of Matching Theory and Case. The question how well suited the ostracism model is for the Bali case guides the iterative process of model design and model adaptation. To identify whether the ostracism model matches the irrigation context in Bali, the concepts of the ostracism model are placed in relation to what we know from irrigation context of Bali [6], see table 2. Without going into detail about each concept, we can identify 'easy' mappings' such as, resource is water and payoff is the rice harvest, however most concepts do not exist in an one-to-one correspondence with the real case. These notions are either more rich in the Bali context (light grey cells) or the data is not available (dark grey cells).

We see that the resource management context from the ostracism model matches the irrigation dilemma in Bali well in terms of: resource (dynamics), utility interlinked with the resource and the presents of social factors that could reflect the variation in adaptive capacity. More specifically, these social factors, relate to variables, e.g., sanctioning, norms, that are considered important to establish self organisation [17]. However, the Bali context also indicates aspects that are not addressed by the ostracism model. For instance, in simplistic representation of the physical environment: pests dynamics are not included, however play a crucial role in the Bali irrigation dilemma. Coupling the model to the existing models of Bali ecology is therefor the intended solution. Concerning the social environment (topology) is minimally existing on the micro-level, there could be good reasons to introduce a spatially dependent structure that affects how the agents meet. Another example would be (heterogeneous) agent attributes in which the role of caste could play a role in the way agents from different caste interact with each other.

When comparing these missing elements with other models of cooperation [14-16], we can identify some mechanisms that target some of the missing components addressed above to explain cooperation[5]. For instance, spatial explanations, such as network reciprocity, graph selection, or set selection, describe the influence of the network topology of an agent on the cooperation. Other explanations focus more on explaining cooperation based on the group that one belongs to, e.g., green beard, group selection or kin selection, which could be an option for representing the role of being heterogeneously part of a caste. Overall, most models of cooperation focus on one or a few mechanisms/drivers of cooperation. Probably the ABM model will result in a merger of theories/mechanisms. For now we consider the ostracism model good enough to continue, it is up to the next model specification phase to define what seems to be a good fit of model & context.

---

[5] Theories that describe the evolution of cooperation focus on identifying mechanisms.

Table 2. An overview of the main model concepts of the ostracism model the related available empirical data from the Bali context and some first ideas for the ABM of Bali irrigation

ABM dimensions	Abstract model	Bali context	Contextualised ABM
Emergent pattern	level of cooperation	Difference in ability to adapt to social and environmental challenges	Assume 'adaptive capacity' = level of cooperation
Aggregated variables	Resource volume f(constant)	Water availability (local) f(upstreamOuttake,rainfall)	Water availability
	Proportion of Cooperators, Defectors [%]	Number of cooperators	Number of cooperators
	OstracismCost OstracismCost = F(coopRatio, inequityEffect[6])	There is no empirical evidence for farmers knowing the overall cooperation/defection ratio (social capital) or the relative difference in size of defection. The age and demographic stability of the subak are indicators for effective sanctioning.	
Agent	Behaviour options {Cooperate, Defect}	Behaviours - Take out X water to farm land on time t - Perform rituals - Maintain canals - Perform Agricultural labour - Attend weekly subak meetings - other	
	Decision-making: imitate if other has a better strategy	This is a theoretical assumption of human decision-making. No empirical evidence about the imitating when others perform better.	

[6] The bigger the difference between the defector and cooperator payoff, the bigger the ostracism. Can be regarded as another type of gradual sanctioning: small offence, small punishment. Large offence, big punishment. In addition to the standard description of gradual sanctioning that describes an increase in punishment over time when the defection is repeated.

Table 2. *(continued)*

Environ-ment	MeetingList [random]	- Farmers are part of a Subak, they meet weekly in a Subak meeting. - Farmers own land (spatial location), and have neighbours - Farmers live some where (spatial location) and have neighbors - Farmers are part of a caste[7]	Options: - Network topology - fixed interaction group. - Local interaction with the same farmers neighbouring land - Maybe also interaction with farming neighbours of home-surrounding
Interaction	'Environment': Receive utility() Payoff = f(waterOuttake) Utility: f(payoff, ostracismCost)	- Rice harvest Harvest/payoff = f(waterAvailability, pestDamage,riceType) - Not aware of data on earnings.	Rice harvest - Including pest dynamics in the payoff function
	Other agents: compare utility()	Not aware of data on the knowledge of farmers on each other's harvest and income. Apart from utility, - There is a continuous tension between castes and the hierarchical position they belong - Sanction might affect utility comparison.	Option: - Find/Collect Data or include theories on other potential moderators on utility comparison (sanctioning, caste membership, ..)

[7] We have to investigate what this means for their interactions. However, when they interact, the caste determines the language. Low Balinese when talking to someone in a lower caste, high Balinese in a higher caste. The subak meetings are in this sense egalitarian, regardless of the caste everyone talks high Balinese to each other.

2.2 Model Specification[8]

Table 2 doesn't only give a first lead for model matching, it also forms a valuable source for model specification. In the last column (contextualised ABM) we indicated some first ideas of for the contextualised ABM. Particularly, the grey cells point out the focus for us as modellers to make decisions. The first type of decisions concern which factors to include and which not, since the context indicates more richness (light grey cells). It could also imply that these are the factors for manipulation in the simulation to explore their influence. The second form of decisions concern concepts in the theory (model) of which no data is available (dark grey cells). This is where we can choose to formulate assumptions or to collect data.

In a first version we will adopt the assumptions from the ostracism model, to have a baseline model to compare with while adapting the model gradually by adding contextual factors.

3 Conclusion

In this paper we highlight a story of contextualising models in which we typically start from a theory/model and relate it to empirical data to see which aspects of a social-ecological context matter and to derive focus points for model specification. Our aim is to open the discussion about the design stage of our model and share reflections among peers to increase the quality of (our) model(s) in a fundamental stage of modelling. Discussions could involve:

- Alternative options or suggestions for contextualising ABM (column 4)
- Assumptions in models
- Empirical data for model specification

The larger idea behind developing an ABM case study of Bali irrigation is to move the body of generic theoretical models closer to particular group of real world phenomena. These are phenomena where the collective interests of resource use, like water, are in conflict with individual interests, i.e., social dilemmas. Social-ecological research that is concerned with these dilemmas is provided with abstract theoretical models with strong analytical power, however little relation to real world social dilemmas. On the other hand, there exists an abundance of rich and descriptive case studies on real world social dilemmas that are case-specific. The use of ABM case studies is a way to discern between factors that are case specific and social dilemma specific. Enriching generalised models with these context sensitive social dilemma

[8] What we do here touches content wise with what [18] discuss. However, our focus lies on combining existing theoretical models in and contextualising it with empirical data. Where we agree with the communicated importance of empirical embeddedness, particularly also on the micro foundations of a model (model design). We focus on covering the first stages of model selection and model specification. Where Boero & Squazzioni [18] touches upon model specification the focus and research attention in general goes more to model calibration and output validation with empirical data.

factors allow for a stronger explanatory power, more realistic and more useful insights into the dynamics of this particular class of social-ecological systems.

References

1. McGinnis, M.D.: An introduction to IAD and the language of the Ostrom workshop: A simple guide to a complex framework. Policy Studies Journal 39, 169–183 (2011)
2. Carpenter, S.R., Mooney, H.A., Agard, J., Capistrano, D., DeFries, R.S., Diaz, S.: Science for managing ecosystem services: Beyond the Millennium Ecosystem Assessment. PNAS 106, 1305–1312 (2009)
3. Biggs, R., Schlüter, M., Biggs, D., Bohensky, E.L., BurnSilver, S., Cundill, G., Dakos, V., Daw, T.M., Evans, L.S., Kotschy, K., Leitch, A.M., Meek, C., Quinlan, A., Raudsepp-Hearne, C., Robards, M.D., Schoon, M.L., Schultz, L., West, P.C.: Toward Principles for Enhancing the Resilience of Ecosystem Services. Annual Review of Environment and Resources 37, 421–448 (2012)
4. Ostrom, E., Dietz, T., Dolsak, N., Stern, P.C., Stonich, S.: The Drama of the Commons. The National Academies Press (2002)
5. Hardin, G.: Garrett Hardin: The Tragedy of the Commons. Science 162, 1243–1248 (1968)
6. Lansing, J.S.: Perfect Order recognizing complexity in Bali. Princeton University Press (2006)
7. Lansing, J.S., Cheong, S.A., Chew, L.Y., Cox, M.P., Ho, M.-H.R., Wiguna, W.A.A.: Alternate stable states in a social-ecological system (in prep.)
8. Lansing, S., Miller, J.H.: Cooperation, Games, and Ecological Feedback: Some Insights from Bali. Current Anthropology 46, 328–334 (2005)
9. Lansing, J.S., Kremer, J.N.: Emergent properties of Balinese water temple networks: Coadaptation on a rugged fitness landscape. American Anthropologist 95, 97–114 (1993)
10. Lansing, J.S., Kremer, J.N., Smuts, B.B.: System-dependent Selection, Ecological Feedback and the Emergence of Functional Structure in Ecosystems. Journal of Theoretical Biology, 377–391 (1998)
11. Janssen, M.A.: Coordination in irrigation systems: An analysis of the Lansing–Kremer model of Bali. Agricultural Systems 93, 170–190 (2007)
12. Schlüter, M., Tavoni, A., Levin, S.: Robustness of cooperation in a commons dilemma to uncertain resource flows (in prep.)
13. Tavoni, A., Schlüter, M., Levin, S.: The survival of the conformist Social pressure and renewable resource management. Journal of Theoretical Biology, 1–10 (2011)
14. Dugatkin, L.A.: The Evolution of Cooperation. BioScience, 355–362 (1997)
15. Nowak, M.A.: Five Rules for the Evolution of Cooperation. Science, New Series 314, 1560–1563 (2006)
16. Zaggl, M.: Cooperation and reciprocity in two-sided principal-agent relations: an evolutionary perspective (2012)
17. Ostrom, E.: A General Framework for Analyzing Sustainability of Social-Ecological Systems. Science 325, 419–422 (2009)
18. Boero, R., Squazzoni, F.: Does Empirical Embeddedness Matter? Methodological Issues on Agent-Based Models for Analytical Social Science. Journal of Artificial Societies and Social Simulation, 1–31 (2005)

Narratives of a Drought: Exploring Resilience in Kenya's Drylands

Elizabeth Anne Carabine[1], John Wainwright[2], and Chasca Twyman[1]

[1] Department of Geography, University of Sheffield, Sheffield, United Kingdom
{e.carabine,c.twyman}@sheffield.ac.uk
[2] Department of Geography, Durham University, Durham, United Kingdom
john.wainwright@durham.ac.uk

Abstract. Drylands are complex, dynamic social-ecological systems under threat from potential climate change and land-use changes. Mixed qualitative methods, including participatory observation and informal and semi-structured interviews, have been used to capture narratives of livelihoods and environment in Kajiado District, Kenya, with particular reference to a severe drought event in 2009. Analyses have shown cultural changes to be important processes in responding to system shocks and stresses. A Beliefs-Desires-Intentions agent architecture has been employed in an agent-based model to explore the emerging narratives after the drought, offering a means of focussing on these key processes and their implications for the system. A series of iterative periods of fieldwork have allowed model strategy, parameters and assumptions to be tested and refined with the original research participants.

Keywords: agent-based model, qualitative, climate change, land-use change, pastoralism.

1 Introduction

The combined impacts of potential climate change and land-use change are likely to have significant effects on the sustainability of dryland ecosystems and the services they provide to the societies that depend on them [1]. Understanding the complex interactions between dryland ecosystems and local livelihood strategies is critical for exploring responses to shocks and stresses associated with climate change [2] and the implications for social-ecological resilience [3].

Employing an interdisciplinary approach, this research uses modelling as an integrating tool. An agent-based model (ABM) of pastoralist decision-making, derived from qualitative data, is combined with mechanistic vegetation modelling to explore the social-ecological dynamics of these globally important systems. With the inclusion of participatory research methods, the modelling approach attempts to combine actor-based and systems-based enquiry to explore narratives of livelihoods, land-use change and climate change in a study system in southern Kenya. In this paper we focus on the approach taken to develop the ABM.

B. Kaminski and G.Koloch (eds.), *Advances in Social Simulation*,
Advances in Intelligent Systems and Computing 229,
DOI: 10.1007/978-3-642-39829-2_27,

2 Modelling Approach

ABMs often take an abductive approach in that modelled systems give rise to self-organizing and emergent properties from which theories can be constructed, allowing for agents to act according to the meanings they ascribe to their systems [4]. The degree to which a model represents the system according to the perspectives of the agents (*emic*) or the researcher (*etic*) depends on the methods and analysis used [5]. The ABM methodology has parallels with grounded theory approaches which are subjective and interpretive, with meaning emerging through the analysis of qualitative data, and reflecting the beliefs of research participants [6:180].

Therefore, qualitative research methods can offer a valuable approach to developing ABMs [e.g. 4, 7, 8-10], generating detailed information about individual behaviours and group dynamics while increasing the opportunity for reflexive, inductive theories to emerge [11:142]. While qualitative data can be highly empirical and context-specific, ABM offers a means of identifying general patterns and processes that will have relevance in a general range of settings whilst operating within the parameters of the particular case study [12].

3 Understanding Social-Ecological Systems

The study system is a dryland system in the Kajiado District of southern Kenya. The area comprises two communally-managed Group Ranches, Eselenkei and Mbirikani, inhabited primarily by Maasai pastoralists (total population approximately 6,000). Rangelands in Kenya have been undergoing a process of land-use change but this section of Maasailand has proven relatively resistant to the pressures driving this change, maintaining livelihood strategies centred around communally managed resources and a pastoral economy [13]. The social structure of these communities is organized around clan and kin relationships and age-sets. Most notable here are the herder age-set (approximately 20-35 years of age) and the elder age-sets who generally no longer carry out herding duties and are customarily responsible for decision-making, although this is to some extent replaced by political actors and institutions.

In 2009, a drought event occurred which was particularly severe in its impact on this social-ecological system, resulting in the loss of approximately 85% of livestock and wild herbivores in a short period of time. This shock has been described by inhabitants as the "the worst drought ever" and some experts believe it may have brought the system to a catastrophic "tipping point" [14]. This drought event forms the focal point for exploring the responses of this system to climate and land-use changes and implications for its long term sustainability.

A total of 18 months' qualitative field research has been undertaken in the study system, over three visits since 2009. Qualitative data has been collected at each field site using a mixed methods approach [15] designed to address the multiple objectives of the research, including participant observation, informal discussion and semi-structured interviewing. The use of these mixed methods has allowed for a greater degree of triangulation of information and formal and informal data collection.

3.1 Participant Observation and Participatory Methods

Participant observation allows for detailed contextualization and analysis of both material and subjective processes which form the experience of actors within a system [10, 16]. At the early fieldwork stages, several months were spent in each of Mbirikani and Eselenkei Group Ranches, accompanying participants in their daily activities to gain an understanding of how users perceive the rangeland, the parameters of the system and the decision-making processes in utilizing the resources within it [17]. The information gained on these walks was triangulated through informal discussions with a range of different user groups including women, herders and elders in order to gain multiple perspectives.

3.2 Semi-structured Interviews and Focus Groups

Purposive sampling was used to identify interview respondents from households located throughout the study area. Sample size was greater than 30 in each of the Group Ranches [18] and included a wide range of interests and 'multiple realities' to develop a conceptual theory of decision-making processes in the system [19]. Topics included in interview schedules were livelihood activities, natural resource use over space and time, other actors and institutions, perceived socio-economic and environmental changes and the drivers of change, particularly with reference to the 2009 drought, as well as plans for the future.

A technique that was used during this work was discussions of scenarios and "what if?" situations. This approach helped to increase participation, depersonalize discussions, protect the privacy of other participants, and frame issues more effectively. Comprehensive notes were taken during interviews and field notes taken during participant observation and informal discussions. A field diary was kept on a daily basis also to ensure other information was captured.

4 Developing Behavioural Rules for Agents

4.1 Qualitative Data Analysis

Thematic analysis of transcripts and field notes was undertaken after the first and second field visits [20]. Particular care was taken to avoid biases, high-inference descriptors and jumping to conclusions too early [21] by sequentially coding the material by hand, starting first with descriptive codes and then more analytical coding to establish meaning as well as content [22].

NVivo software was used as a means of identifying broad-scale categories of information from secondary data such as policy documents and the wealth of information captured in previous studies undertaken in the area. In undertaking this analysis, particular emphasis was placed on reconstructing the perspectives of the research participants, incorporating *emic* interpretations of the system. Data were analyzed iteratively so that researcher and participant perspectives were appropriately integrated [23] and materials were re-read several times over a period of time to enable reflection on these issues.

4.2 Development of Theory

Through this analytical process, social and cultural shifts emerged as an important theme in the qualitative data and therefore form the basis of the theory of decision-making for the ABM of the system.

It is well documented that Kajiado, typical of many dryland pastoral systems, is undergoing a process of change driven by socio-economic, political and environmental factors such as population growth, globalisation, changes in land use and land tenure, shifting policy context, market access, technological innovations and climate change [24]. Analyses of pastoral societies have been focussed primarily on the socio-economic impacts of these factors in terms of assets, cash flows, livelihood incomes and natural resource availability. However, using the mixed methods described above, this fieldwork revealed shifting attitudes to livestock and pastoral livelihoods among the communities of the study area, which came out strongly in the coding of qualitative data. For example, when questioned soon after the drought, participants described how "people don't want to herd anymore". Young people "don't want to walk in the bush day after day", rather they "want to go to school and get jobs". Herding is thought to be "a tough life" compared to alternative employment opportunities, particularly in light of the lived experience of the drought [25].

Furthermore, differences emerged between the two communities within the study area. Although the inhabitants of Mbirikani and Eselenkei are from the same Maasai clans, share a broadly similar environment and pursue common livestock-based livelihoods, their differential socio-political experiences and exposure to commercial opportunities has led to differences in their broad worldviews. For example, Mbirikani herder age-groups claimed their counterparts in Eselenkei "are not learned, they are only interested in cows", they "buy many cows", whereas young men in Mbirikani believe "it is good to own land not cows", "cows are easily lost during a drought" and "you can keep cows once you have your land". Mbirikani participants described Eselenkei herders as "somehow innocent" due to the perception that they are interested in investing only in livestock. On the other hand, Eselenkei herders feel people in Mbirikani "should invest in cows" and they "don't know why they are moving away from livestock". Participants from Eselenkei claimed the need for diversification into horticulture on private plots of land purported by their age-mates in Mbirikani is "a big lie" and they are "just being greedy because they have seen others become rich with plots".

Further complexity is added by apparent shifts in worldview and livelihood strategies as a result of the lived experience of the drought. Herders from both areas were forced to travel further than at any time in the past 100 years with their livestock in search of forage and water for their family's herds. Most participants had never herded beyond their neighbouring areas and the different experiences they had on their journeys have led to different attitudes and behaviours upon their return. For example, some were exposed to alternative income opportunities, modifying their current and planned livelihood strategies [25].

These differences in norms were identified while working through the material, with initial descriptive codes such as “cows”, “herd”, “livestock”, “land” and “plots” applied, including the example statements above. Coding was carried out through several iterations and refined into more analytical codes such as “worldview”, “traditional perspective” and “commercial perspective”, which also included the above statements. In this way, and consistent with a grounded theoretical approach, a theory has emerged from the data that some Maasai favour more “traditional” livelihood strategies, investing in their herds as a means of coping with variable and uncertain conditions, while others tend to invest spare capital in diversifying their livelihoods with plots, employment and commercial activities. These worldviews are affected by differential economic and political conditions and individual experiences of shocks such as the 2009 drought. These findings indicate implications for the socio-cultural norms and values of the system which will interact in complex ways with dynamics of land-use change and climate change over time.

4.3 Agent Architecture

Based on the theory developed in the analysis of qualitative data, primary agents were identified at the herder level and share common attributes such as ethnicity (Maasai) and primary livelihood activity (pastoralism). Herder agents have a set of overarching goals, which were derived from the primary and secondary data. These goals are as follows, in order of importance:

1. Meet subsistence needs e.g. through pastoralism, agro-pastoralism or a mix of pastoralism with other livelihood activities;
2. Cope with stress and shocks e.g. maintain herd mobility, manage herd composition, herd-splitting, increase herd size; and
3. Maintain socio-cultural functions e.g. *osotua* (a practice by which Maasai will gift stock to each other when in need, in a relationship that is reciprocal over time), knowledge or information transfer through social networks, social capital between members of the same age-set, clan and kin groups.

While all herders share these common goals, the agents are differentiated in their motivations and options by criteria which include clan, relative wealth, opportunities for alternative strategies and worldview.

To capture the role of different worldviews identified in the analysis, a Beliefs-Desires-Intention (BDI) agent architecture has been developed [26]. This approach allows for agent actions to be mediated by their particular worldview or belief sets, which are derived from the environment, from their learned experiences and through their interactions with each other. The ABM is implemented in Netlogo and uses an existing BDI and communication library to manage beliefs and intentions and pass messages between agents [27].

Under this protocol, all agents have declared variables beliefs and intentions. Beliefs are composed of `belief-type`, e.g. `traditional` or `commercial`, and belief-content which specifies the conditions of that belief [28]. The library allows for multiple beliefs of the same type. A set of agent intentions are implemented as a stack and consist of intention-name and intention-done, which specifies the condition by which that intention is completed [28].

In the ABM, communication between agents is managed by a message passing library accompanying the BDI library for Netlogo [29]. In this way, social networks can be simulated which communicate information that influences decision-making and behaviour. For example, members of the same clan will communicate the condition of their current patches to each other to allow informed decisions about moving to new locations.

4.4 Building the ABM

In this initial version of the ABM, herder agents make decisions under certain conditions while pursuing their common goals. These decisions are implemented as sets of intentions and modified by their beliefs. According to the overarching goals identified in the previous section, herder agents execute the intention to subsist on a daily time-step, which includes more specific intentions to find-grass, to graze and to monitor the condition of their herd.

The second overarching goal is implemented as an intention to cope-with-stress. Under this intention, a more specific intention to assess-location is executed in particular months, as identified from the seasonal decisions identified in the qualitative data analysis. Agents score their current patches against the patches of ten random herders in their social network according to a set of identified criteria, including the relative condition of the 11 patches, the number of cattle in and neighbouring each of the patches, a preference to stay in their current locations and a preference to move to their home locations in the wet season months. When the score is received (intention-done), an agent will either remain in the current location or move towards a new one.

Decisions about coping with environmental stress depend on agent beliefs about rainfall. For example, in the dry season (March-June) herders will continue to execute the intention to cope-with-stress until it rains for a consecutive number of days (which is the intention-done condition) after which they will continue to subsist until the next intention is triggered off the stack. Strategies for coping with stress depend on agent beliefs. For example, more traditional agents will follow a set of intentions based on customary coping mechanisms including split-herd, reduce-herd, osotua and mix-species. If these strategies for coping with stress fail, beliefs may be modified and something else will be attempted. Agents that have pre-determined or adapted commercial beliefs will be more likely to sell-cows and to diversify their activities to cope-with-stress.

The goal to maintain socio-cultural functions is more complex and includes intentions such as maintaining social networks which play an important role in pastoralist societies, including the Maasai in Kajiado. Social networks are crucial in maintaining mobility, surviving shocks and stresses and gathering information about natural resources, cattle prices and so on [30]. For example, *osotua* is represented in the ABM as part of the coping with stress intention set. Many aspects of Maasai culture are adapted to support pastoral livelihoods, which is one reason why Maasai are relatively resistant to change [31]. However, differentiation driven by shifting worldviews and experiences may erode customary coping mechanisms based on social networks, leading to the loss and/or replacement of this source of resilience in the system.

4.5 Testing the Rules

A return visit to the field was carried out in 2013 to explore whether the interpretation of qualitative data and subsequent development of rules in the ABM hold meaning for participants [32]. The objectives for this fieldwork period were to test elements of the modelling strategy, model parameters and assumptions and to explore future scenarios for developing the model further. Interviews were held with subsets of the original participants to explore and reflect verbally on the findings of the research so far.

While the majority of participants were not literate or familiar with spatial diagrams, all of them had a degree of understanding of scientific research, having been exposed to many researchers from different disciplines over recent decades. Therefore, participants did not question a desire to understand their activities, implicitly accepting outsider interest in their culture which they encounter routinely through tourism, research and non-governmental organisations. Many participants, and certainly those who have been educated or exposed to outsider interest, through employment for example, have very good understandings of the issues in which researchers are interested such as natural resource use, livestock management and climate variability which provides researcher and participant with a general level of shared understanding. Furthermore, due to a relatively high level of research exposure, the local leaders and almost all participants expect researchers to feedback on the process in which significant time and assistance has been invested.

During interviews, a series of "if-then?" and "what-if?" questions were used to interrogate the model procedures and check explicit and implicit assumptions. In most instances, illustrative diagrams did not prove useful in structuring discussions with participants, who are unfamiliar with such media, and therefore were not generally used. Rather, questioning followed the model process step by step. In this way, the rules, process order and conditions in the ABM were tested with participants to see whether they seemed reasonable representations of their perspectives.

An example of how this final round of fieldwork altered the model is the adjustment of erroneous assumptions about seasonal decisions. When asked what they would do if it does not rain in the short rainy season (November-December), participants said they would carry on as normal rather take actions in anticipation of a drought, which had been the assumption in the model. Instead, a more realistic representation that agents will start to anticipate stress if the short rains fail and the subsequent long rains (March-June) are delayed was included. Returning two years after the drought, it appeared that while people had adopted very different livelihood strategies, they did not consider drought risk explicitly in their short-term decisions. As on participant explained it, "if you told me there was going to be a drought tomorrow

I would not believe you. There will be bad droughts again but only God knows what will happen in the future".

Also, participants were asked about future scenarios to guide questions for later versions of the model. For example, they were asked what they think will change in their area in the next 5-10 years, will there be more or less rain and will more or fewer people be farming when their children are herders, and so on. Based on the scenarios identified by the participants, future versions of the model should include a field of different agents which influence the motivations and options of the primary herder agents, incorporating important power dynamics and wider political and socio-economic processes. The factors affecting possible land-use changes, such as privatisation and sale of communal land, expansion of protected areas and spread of irrigation should also be explored.

A limitation of using qualitative methods to inform ABM development is that numerical thresholds and values often have to be applied, which can add rigidity beyond the evidence presented in that data [4]. What became evident in talking to participants was that decision-making processes are heuristic and qualitative. There are no set thresholds or criteria for particular decisions, rather daily or seasonal assessments based on the best available information and individual beliefs. For example, when asked about decisions of when to increase their herds, most participants said they generally prefer to buy stock just before the rains arrive when prices are low and forage not widely available, but this has to be weighed against the likelihood of a good rainy season to support extra stock and rising market prices after the rains arrive. In this sense, there is no set time or formula for decisions to go to market. For this reason, actual set figures and numbers have not been used in the ABM rather ranges to represent beliefs about rainfall for example, thus capturing the spaces within which decisions are made relative to other factors rather than linking decisions to specific variables like market prices.

5 Conclusions and Further Work

Qualitative methods have been used to capture narratives of livelihoods, climate change and land-use change in a dryland social-ecological system in southern Kenya. Building an ABM to explore these narratives offers a means of focussing on key processes and their implications for the system while incorporating the agency of actors to respond to change. Agent-based modelling also allows for the inclusion of material and normative aspects in system analysis. Using qualitative data to inform behavioural rules maximises this potential, especially when participants can be involved in model-building and interpretation to capture a more *emic* representation of the system which reflect the perspectives of the participants rather than the researcher. Repeated periods of fieldwork have allowed model strategy, parameters and assumptions to be tested and refined with the original research participants.

While qualitative data can be highly empirical, ABM also offers a means of identifying general patterns and processes that can have relevance in a general range of settings whilst operating within the parameters of the particular case study. Using sequential descriptive and analytical coding to develop theory and abstract rules for the ABM has proven a useful way of interrogating the narratives that are grounded in the qualitative data. Model-building acts as a further layer of analysis helping to interpret the data in different ways.

This analysis has identified worldview as an important factor in pastoralist decision-making processes. Agent architecture has been developed which reflects this finding, building agents with three overarching goals executed through sets of intentions which are modified by their belief sets (i.e. BDI agents).

Further versions of this ABM will be developed to explore scenarios of interest that have been identified by participants. Furthermore, the ABM will be fully coupled with a dynamic vegetation model for savannah ecosystems [33] which will represent the agents' shared environment and the potential impacts on and dynamics between socio-cultural changes and vegetation structure and function. The integrated social-ecological model will be tested and implemented at a range of spatial and temporal scales as appropriate to the system to investigate system resilience to climate and land-use changes.

Acknowledgments. This work was supported by the Economic and Social Research Council and the Natural Environment Research Council (grant number ES/I025758/1).

References

1. Galvin, K.A.: Transitions: Pastoralists Living with Change. Annual Review of Anthropology 38, 185–198 (2009)
2. Reynolds, J., Stafford Smith, D.M., Lambin, E., Turner II, B.L., Mortimore, M., Batterbury, S.P.J., Downing, T.E., Dowlatabadi, H., Fernandez, R.J., Herrick, J.E., Huber_Sannwald, E., Jiang, H., Leemans, R., Lynam, T., Maestre, F.T., Ayarza, A., Walker, B.: Global Desertification: Building a Science for Dryland Development. Science 316, 847–851 (2007)
3. Nelson, D., Adger, W.N., Brown, K.: Adaptation to Environmental Change: Contributions of a Resilience Framework. Annual Review of Environment and Resources 32, 395–419 (2007)
4. Yang, L., Gilbert, N.: Getting away from numbers: Using Qualitative Observation for Agent-Based Modelling. Advances in Complex Systems 11(2), 1–11 (2008)
5. Crane, T.A.: Of Models and Meanings: Cultural Resilience in Social-Ecological Systems. Ecology and Society 15(5), 19 (2010)
6. Mason, J.: Qualitative Researching. Sage, London (2002)
7. Huigen, M.: First Principles of the MameLuke Multi-Actor Modelling Framework for Land Use Change, Illustrated with a Phillipine Case Study. Journal of Environmental Management 72, 5–21 (2004)

8. Millington, J., Romero-Calcerrada, R., Wainwright, J., Perry, G.: An Agent-Based Model of Mediterranean Agricultural Land-Use/Cover Change for Examining Wildfire Risk. Journal of Artificial Societies and Social Simulation 11(4), 4 (2008)
9. Polhill, G., Sutherland, L., Gotts, N.: Using Qualitative Evidence to Enhance an Agent-Based ModellingSystem for Studying Land Use Change. Journal of Artificial Societies and Social Simulation 13(2), 10 (2010)
10. Smajgl, A., Brown, D.G., Valbuena, D., Huigen, M.A.: Empirical Characterisation of Agent Behaviours in Socio-Ecological Systems. Environmental Modelling and Software 26, 837–844 (2011)
11. Crang, M., Cook, I.: Doing Ethnographies. Sage, London (2007)
12. Janssen, M., Ostrom, E.: Empirically Based, Agent-Based Models. Ecology and Society 11(2), 37 (2006)
13. Spear, T., Waller, R. (eds.): Being Maasai. James Currey, London (1993)
14. Western, D.: The Worst Drought: Tipping or Turning Point? Swara 3, 16–22 (2010)
15. Barbour, R.: Mixing Qualitative Methods: Quality Assurance or Qualitative Quagmire? In: Bryman, A. (ed.) Mixed Methods. Sage, London (2006)
16. Drury, R., Homewood, K., Randall, S.: Less is More: the Potential of Qualitative Approaches in Conservation Research. Animal Conservation 14, 18–24 (2011)
17. Thomas, D., Twyman, C.: Good or Bad Rangeland? Hybrid Knowledge, Science, and Local Understandings of Vegetation Dynamics in the Kalahari. Land Degradation and Development 15, 215–231 (2004)
18. Voinov, A., Bousquet, F.: Modelling with Stakeholders. Environmental Modelling and Software 25, 1268–1281 (2010)
19. Baxter, J., Eyles, J.: Evaluating Qualitative Research in Social Geography: Establishing 'Rigour' in Interview Analysis. Transactions of the Institute of British Geographers 22, 505–525 (1997)
20. Slim, T., Thompson, A.: Listening for a Change: Oral Testimony and Development. Panos, London (1994)
21. Jackson, P.: Making Sense of Qualitative Data. In: Limb, M., Dwyer, C. (eds.) Qualitative Methodologies for Geographers. Arnold, London (2001)
22. Cope, M.: Coding Transcripts and Diaries. In: Clifford, N., Valentine, G. (eds.) Key Methods in Geography. Sage, London (2003)
23. Heckbert, S., Baynes, T., Reeson, A.: Agent-Based Modeling in Ecological Economics. Annals of the New York Academy of Sciences 1185, 39–53 (2010)
24. Catley, A., Lind, J., Scoones, I.: Development at the Margins: Pastoralism in the Horn of Africa. In: Catley, A., Lind, J., Scoones, I. (eds.) Pastoralism and Development in Africa: Dynamic Change at the Margins. Routledge, New York (2013)
25. Carabine, E.A., Twyman, C., Wainwright, J.: Trauma and Shock in Kajiado, Kenya: the 2009 Drought (manuscript in preparation, 2013)
26. Rao, A., Georgeff, M.: BDI Agents: from Theory to Practice. In: International Conference on Multi-agent Systems (ICMAS 1995), San Francisco (1995)
27. Sakellariou, I., Kefalas, P., Stamatopoulou, I.: Enhancing Netlogo to Simulate BDI Communicating Agents. In: 5th Hellenic Conference on AI. SETN, Syros (2008)
28. Sakellariou, I.: Agents with Beliefs and Intentions in Netlogo (2010) (unpublished)
29. Sakellariou, I.: An Attempt to Simulate FIPA ACL Message Pushing in Netlogo (2010) (unpublished)

30. Homewood, K., Rodgers, W.: Maasailand Ecology. Cambridge University Press, Cambridge (1991)
31. Spear, T.: Introduction. In: Spear, T., Waller, R. (eds.) Being Maasai. James Currey, London (1993)
32. Bousquet, F., Barreteau, O., d'Aquino, P., Etienne, E., Boisseau, S., Aubert, S., Le Page, C., Babin, D., Castella, J.C.: Multi-Agent Systems and Role Games: Collective Learning Processes for Ecosystem Management. In: Janssen, M. (ed.) Complexity and Ecosystem Management: The Theory and Practice of Multi-Agent Systems. Edward Elgar Publishers, Cheltenham (2002)
33. Scheiter, S., Higgins, S.I.: Impacts of Climate Change on the Vegetation of Africa: an Adaptive Dynamic Vegetation Modelling Approach. Global Change Biology 15(9), 2224–2246 (2009)

Towards a Context- and Scope-Sensitive Analysis for Specifying Agent Behaviour

Bruce Edmonds

Centre for Policy Modelling
Manchester Metropolitan University, Manchester, United Kingdom
bruce@edmonds.name

Abstract. A structure for analysing narrative data is suggested, one that distinguishes three parts: context, scope and narrative elements. This structure is first motivated and then illustrated with some simple examples taken from Sukaina Bhawani's thesis. It is hypothesised that such a structure might be helpful in preserving more of the natural meaning of such data, as well as being a good match to a context-dependent computational architecture. This structure could clearly be combined and improved by other methods, such as Grounded Theory. Finally some criteria for judging any such method are suggested.

1 Introduction

Agent-based modellers have always used a variety of sources to inform the design of their simulations. These have included: existing theories, tradition, expert opinion, intuition, experimental results, talking to stakeholders, summaries of other research, and narrative data. This paper considers the later of these, namely narrative data. This involves analysing transcribed accounts obtained from stakeholders and using these to inform the specification of the behavioural rules of corresponding agents within a simulation. This method goes beyond simply talking to stakeholders and using the understanding gained to inform ones modelling. It divides the process into stages: (a) conducting interviews, (b) transcribing these into text (the "narrative data"), (c) analysing this data, and finally (d) specifying the behavioural rules for a social simulation. As far as I am aware, Richard Taylor [17] and Sukaina Bharwani [4] were the first to do this.

The advantages of doing this analysis in this staged and formalised ways should be obvious. The data can be made available and examined by subsequent researchers who may spot aspects that the original observer has missed as well as seeing how the sense of the stakeholder might have been changed during the process. Unlike an informal approach, which does not work from transcribed text, a subsequent researcher can follow the chain of analysis and understand it better. This mirrors attempts to improve the documentation of simulation code and to make simulations easier to replicate and investigate. Although the process of analysing natural language data is never going to be a completely formal process and relies upon the understanding of the interviewer and/or analyst, formalising the process makes it more transparent and replicable.

B. Kaminski and G.Koloch (eds.), *Advances in Social Simulation*,
Advances in Intelligent Systems and Computing 229,
DOI: 10.1007/978-3-642-39829-2_28, © Springer-Verlag Berlin Heidelberg 2014

This paper aims to further formalise this approach by suggesting a structure for analysing such narrative data that (I argue) is particularly suitable for this task, but which also appropriate for specifying and writing the code that determines the behaviour of agents within a social simulation. This involves a three-stage analysis, into: context, scope, and narrative elements. This increased structure should make the process more transparent and systematic, but also (I hope) cause less distortion to the original sense and make the process easier!

In this paper I first discuss each of: context, scope and narrative steps into which narrative data can be analysed, before discussing how these might be brought together into a complete analysis, giving some examples. It then ends by discussing some aspects of the transformation into code, including a discussion about the kinds of computer architecture that might make this easier, and some ways forward.

2 Context

Context is a difficult word to use, for at least two reasons. Firstly it seems to be used to refer to a number of related, but different kinds of thing, and secondly it is not always clear that there is any *thing* that is "the context", but rather that the word is being used only as an abstraction to facilitate discussion. I am not going to go into a full discussion of the concept and definition of context here, as I have gone into this in previous publications, e.g. [5]. However, since context is central to this paper, a few words are necessary to make my intended meaning clear.

First there is the context in terms of the situation in which an event occurred. Thus the context for a certain conversation might be position: latitude N53:20:45, longitude W1:59:13 at 9.45 in the morning on December 22nd 2012. However this is not such a useful description since, although precise, it imparts very little useful information about what is relevant about that context. Even if one could retrieve what was around at that point in time and space, there is an indefinite number of potential factors that might be pertinent there. Thus it is almost universal to abstract from such precision and try to indicate the *kind* of situation one is referring to. For example, one might say "I was on the train on my way to work just before Christmas". This kind of context can be seen as the answer someone might give to the question "What was the context?" after having been told something they did not fully understand. In this sense it is what is needed to be known about the situation to sufficiently understand the specific utterance for a hearer's purpose. This kind of response is specific to the hearer's particular purpose; however, it is surprising how informative such a thin characterisation can be in that we can recognise a host of details that would normally accompany such a situation, filling out the details. In other words we are able to recognise the kind of situation being signalled by another person.

Many aspects of human cognition have been shown to be context-dependent, including: language, visual perception, choice making, memory, reasoning and emotion [18] [10]. Context-dependency seems to be built into our cognition at a fundamental level. There seems to be a good reason why this is so, as a mechanism for social coordination. That humans are able to co-recognise the same kinds of situation and thus bring to bear the same co-learnt norms, habits and other knowledge in the same kinds of situation. This is the importance of social context and why it can become so entrenched. The more recognisable a situation is as a kind of social context, the more special kinds of protocol, norms, terms, habits and infrastructure is developed for that context; the more a context is marked by differences of these kinds the more it is recognisable. Examples of such entrenched social contexts include: the lecture, a birthday party or an interview. We can all readily recognise such situations and can access a rich array of relevant expectations, knowledge and habits without apparent effort. Thus our cognition has a way of recognising the kind of exterior situation, in other words we have some kind of internal correlate of the relevant kind of situation, which I will call the cognitive context. It is this cognitive context that is most relevant to the process in this paper.

The very facility with which we recognise different social contexts, learn new knowledge with respect to them, and can bring to mind the relevant knowledge for them causes difficult for us when trying to identify the cognitive contexts with respect to which people act. This is due to the richness of the information that seems to be used in such recognition, and also the fact that its recognition is done largely unconsciously. It is this that makes context such a difficult thing to handle. However social context, due to the fact that it is often entrenched in our common practices and institutions, is easier to identify, and it is social contexts that are of the greatest interest here. Also although we are frequently unconscious of which context we are assuming, we are very sensitive to violations of context, or when the wrong context has been assumed.

3 Scope

Scope is the label I have given to the consideration of what is and is not possible within any particular situation. Thus when entering the context of a lecture one may assess where it is possible to sit – it may be very crowded and it might not be possible to slip in at the back and one might have to sit on the steps rather than a seat. Thus scope is something one assesses within a context, since its assessment will often depend upon context-dependent information (such as what is socially acceptable in a lecture context, which may rule out vaulting over the seats).

Working out what is and is not possible in any situation is computationally onerous. This is why the "frame problem" of AI [12] is so hard. This is the reason that humans only do such calculations rarely, for the most part assuming past assessments of scope to frame decision making rather than checking these frequently. Indeed, it seems likely that there are a number of heuristics at play to make assessments of scope feasible, including: not working it out explicitly but relying on a learning process of trial and error, or only attending to scope when meeting new kinds of situation or when something has gone wrong in an unexpected manner. If Luhman [11] is right, a major function of social institutions is to simplify calculations like these for its participants, and it is very plausible that social contexts evolve so that they ensure that scope is more stable inside a context, so that people can learn it and then not have to revaluate it. Thus reckoning about what is possible it central in court case, but the rules, norms and habits that are pertinent in this context serve to limit what kinds of possibility there are at each stage and to ensure that possibilities are increasingly constrained as one approaches the verdict.

To sum up, there may be some quite complex reasoning about scope by subjects, especially when encountering new contexts or when something fundamentally surprising has occurred, but that between these events scope may be quite stable and is something that can be elicited.

4 Narrative Elements

Once context and scope have been established, this leaves the everyday kind of reasoning that people use to reason about action. As Simon observed [16] within very regular and constrained environments (office situations) the reasoning of people is better characterised as 'procedural' rather than approximating any kind of ideal (what he called 'substantive') rationality. That is they have a set of interconnected but fairly simple set of rules that they use if circumstances are as they expected them to be. This was also a lesson from the field of "Expert Systems" which attempted to elucidate rules for how domain experts behaved. The rules they discovered where surprisingly simple but, of course, this was only within a settled kind of situation with fairly constrained goals. In a similar vein Giegernezer [8] has identified a host of fairly simple heuristics that are effective in our daily lives.

The simplest and most common structure for expressing these procedural rules is a simple conditional: "if *this* (some condition) then *this* (some consequence, action or calculation)". Such conditionals can either be used to represent a decision point by the agent or local causation within the situation. However there are others that are maybe so obvious that one tends to overlook them, such as: sequencing ("*this* follows *that*", or conjunction ("*this* … together with *this*"), alternatives ("*this* or else *that*"), or simple expectations about the situation ("*this*" or "not *that*"). Such local rules form a kind of logic[1], which together determine what a person might do, although on the whole these are not manipulated in very complex ways (except if a revaluation of possibilities are required, for example in debugging code). Rather than to rely on complex reasoning, humans often seem to prefer other tactics such as habit, imitation, or trial and error learning. For example, when faced with 1000s of food choices in a supermarket we do not make extensive price-utility comparisons, but constrain our own choices via habit and use alternatives to comparison such as social imitation. Only in very limited situations (e.g. we have decided to buy baked beans and now wish to work out which has less sugar in them) do we resort to detailed comparison.

The best (alternatively most natural) way of representing these narrative elements is not entirely clear. Maybe the argumentation framework of Toulmin [19] would be helpful here, or maybe we need more of a narrative structure such as that of Able [1].

---

[1] These are not a formal logic but rather an expression of common sense relationships and beliefs. However the fact that these elements are situated within context and scope makes them easier to relate to a more formal representation to program the behavior of an agent.

This is fruitful area for future research. This paper focuses only on the distinction between context, scope and narrative elements – its necessity and usefulness and hopes that others will help improve the method for coding these elements.

5 CSNE Analysis

Thus, at the first level of approximation the proposal is that narrative data be analysed for the following kinds of element/referent: context, scope, and narrative elements (CSNE). These correspond to different aspects of the narrative, namely: relevance, applicability, and the detail of the narrative steps. The idea is that the analysis of narrative data should seek first for clues as to context, then to indications of scope (given the hypotheses as to the relevant context) and finally to base narrative element pairs (given the contexts and scopes identified): antecedent and consequent. These are summarised in Table 1.

Table 1. The three main aspects in the CSNE framework and their corresponding properites

CSNE Aspect	Corresponding Property
Context	Relevance
Scope	Applicability
Narrative Element	Local: cause-effect pairs, decision points, sequences, alternatives etc.

The hypothesis is that this kind of structure will (a) be a better 'fit' to natural accounts of occurrences and so distort them less as well as revealing more useful information about behaviour; (b) be a structure that is readily amenable to programming agent behaviour (given a suitable architecture); and (c) thus make more transparent the assumptions used since these can be associated with the level at which they are made (context, scope or narrative element). I have motivated the approach above, and give a few worked examples below, but ultimately these are pragmatic hypotheses – either they will help extract usable specifications for agent behaviour from narrative data or they wont.

Identifying context is tricky due to the fact that normally we do this unconsciously. However we do seem to have an innate ability to recognise the appropriate contexts where these are (a) social and (b) part of a society we have sufficient experience of. Moreover since people are particularly sensitive to misidentified context, this suggests an approach where contexts are first coded at based on the intuitions of the analyst and then repeatedly re-confronted with the data and the opinions of others (e.g. the sources of the data).

Once the implicit contexts for chunks of text are coded, the data should next be scanned for indications as to judgements of possibility with respect to the context. It should be born in mind that some of the constraints concerning what is possible will be implicit in the context (e.g. some things are not acceptable to say when greeting a stranger and can only be said once the parties have got to know each other). Such judgements might be accompanied by some complex reasoning, but in other cases might come out of previous experience, or simple assumption.

Finally, with the identified contexts and scope judgements in mind, the narrative elements can be identified and coded. This is the more straightforward stage, since these are those 'foreground' elements that people commonly talk about – those that they consciously bring to mind. There might be a whole menu of constructions that an analyst might look for, but not all narrative elements will fit these and thus a flexibility in terms of allowing new constructions should be maintained.

Any imposed structure or theory can limit and/or bias subsequent analysis, but on the other hand being completely theory/structure free is impossible. I am suggesting that the CSNE structure is relatively 'benign' in this respect to the degree that it is rooted in how humans think. However this also suggests the sort of approach as championed by Grounded Theory [20] since this seeks to avoid constraining analysis by high theory (where possible) but rather letting the evidence lead. As in GT, I would expect that the emerging analysis will need to be repeatedly confronted with the data to check and refine it. For example, if the analysis assuming a particular kind of social context for some text seems to result in a bad 'fit' for several elements of the identified scope and narrative elements then this may suggest revisiting and reconsidering the context identification (in contrast to a single 'oddity' which might indicate a lower level of misfit).

6 An Illustrative Example

The examples of narrative data are lifted from [3] – they are a series of excerpts from an interview with a particular farmer on the various farming decisions he is faced with. I don't take them in the same order as in [3] and I number them in the order I chose. Obviously I am selecting them to illustrate different aspects of the CSNE approach.

The first considers the effect of people switching from rice to wheat and thus an emerging change in the world market for wheat.

> *(Quote 1) "The one conundrum here is that there are more people in the East who want to get away from rice and gradually upgrade to more wheat allied products, that may alter the value of the end product to us. You see the worst thing that has happened to us worldwide is the collapse of the Eastern economy... over the past 4 or 5 years, but it is coming back again now and that actually may help us again. It is a great shame because we were getting into the Eastern markets and it was beginning to grow and suddenly it collapsed." ([3] p.113)*

From reading the complete selection of quotes in [3], I would guess at the following cognitive contexts by imagining the sort of situation that the farmer is talking from. Clearly these are merely hypotheses and could be wrong, but they are a starting point.

- "**survival**" – things are continually getting worse and the primary goal is to keep in farming, battle against nature

- "**comfort**" – conditions are comfortable with no immediate survival threat, one could stop worrying so much and take things a little easy
- "**entrepreneur**" – one is looking for big profit, taking risks if necessary

Clearly, most of the quotes in [3] come from a "survival" perspective, with the other contexts being briefly imagined, or used as a contrast to "survival".

The above quote is firmly from the survival context – in other contexts people moving from rice to wheat would have been interpreted as an opportunity implying higher prices, but here it is interpreted just as a trap, citing the East European example. The quote acknowledges implicitly that prices *could* increase due to this change in consumption in the East, but then anticipates that this might be followed by a collapse (and thus pose a risk if resources were committed on the basis of rising prices). The survival context implies several facts about what possibilities for action there might be, namely: (a) one is not going to be able to discover a "killer" profit, (b) to survive one has to focus on margins, looking for small improvements that will allow one to break even. Once the context and scope has been stripped away this quote seems to boil down to two observations (1) prices for wheat might increase (2) increases in price can be followed by a sudden collapse.

Table 2. CSNE Analysis of Quote 1

Context	Scope	Narrative Elements
Survial	*No "killer" profit available*	• Prices for wheat may increase in near future • Price increases can be followed by a sudden collapse

The second quote imagines the situation under possible climate change.

(Quote 2) "I, as a farmer would imagine that if the summers were warmer and the autumns were wetter you would have an earlier harvest, and therefore all that would happen is that the harvest would come early and your drilling [cultivation of ground for sowing seeds] would come early so that you would still be able to establish your winter crops before the rain really started. If the rains were really early then we would have to resort to spring sown varieties... The net effect would be that you would be drilling as soon as you possibly could which may be later than normal, but because the weather is warmer that would make up for lost time, so harvest would still be about the same time I would suspect.... If the autumn was continuously wet, wet, wet and we were under water, I mean that is a serious change - that is like this year, every year. If it was like this year every year, then yes there could be a problem." ([3] p.112)

The last sentence implies the survival context again, the possible changes in weather do not indicate any positive possibilities. It documents in detail some of the complex reasoning about scope. The scope is considered that under the conditions that "*if the summers were warmer and the autumns were wetter*" with refinements of this, adding: "*If the rains were really early*" and then "*If the autumn was continuously wet, wet, wet and we were under water*". Each of these implies different narrative elements.

Table 3. CSNE Analysis of Quote 2

Context	Scope	Narrative Elements
Survival	*Summers warmer and autumns wetter*	• Harvest comes early • Therefore drilling needs to be early
	Summers warmer and autumns wetter + rains really early	• Need spring sown variety • Therefore drilling as soon as possible • Probably harvest at the same time due to warmer weather
	Summers warmer and autumns wetter + autumn was continuously wet and we were under water	• If wet like this every year then there is a serious problem

In the next excerpt the farmer reflects upon his own and other's attitudes, the text implies that the farmer has just been asked whether the aim is to maximise profit.

> *(Quote 3) "Well not necessarily some people would say, now you see we have often had this conversation around this table. Some people don't want to maximize profit.... They are happier to take a slightly easier, lower level approach and have an easier life, and not make quite so much money.... And I can relate to that... But because I'm a tenant I don't own my own land... Everything we farm is rented and therefore we have an immediate cost, the first cost we meet is to our landlord and that tends to go up. ([3] p.127)*

In the piece the farmer imagines how another farmer might feel, one who owns their own land and hence is more secure, inhabiting a comfort context. However this is not possible for him, being a tenant. This implies he has to maximise profit to survive and meet the immediate cost of possibly rising rent. The theoretical possibility of the "entrepreneur" situation is implicitly ruled out.

Table 4. CSNE Analysis of Quote 3

Context	Scope	Narrative Elements
Comfort	*Does not have to maximise profit to survive*	• Can take life easier • Does not make quite so much money
Survival	*Has immediate cost (rent) which tends to go up*	• Has to maximise profit to survive

The next quote I consider is where the farmer considers the future's market for fixing the price he would get for his crops ahead of time.

(Quote 4) "Trading in Futures is always an option but I think it is almost a different ball game that is not really farming as such, but... it's hedging your bets on guaranteeing a price ahead. So if you think you can make money on a particular price level and you can see that on a Future's price, say next April, next May, next June, and so on, you could hook on to that and guarantee you are going to get that price but if the market goes up of course you lose out, course if the market goes down you've won. So you know it's 50/50 one way or the other. Not many farmers do it - there was a period when a lot of us tried it and... it wasn't a vast success. I think I have tried it three times now. This is on Options - three times I've done it, and I think twice we have missed out and once we won out, so there's not much in it." ([3] p.145)

Note that this is not considered from a comfort point of view (then not having to worry about the final price) or the entrepreneur point of view (making a profit via speculation) but, again, the survival viewpoint. If this had been within an entrepreneur context it might have led to quite a different analysis.

Table 5. CSNE Analysis of Quote 4

Context	Scope	Narrative Elements
Survival	*Fix price on Futures Market*	• If market drops over the year you gain • If market rises over the year you lose • No gain in long term

The final quote I consider is about other farmers.

(Quote 5) "We used to grow onions here years ago and I have been wondering whether we should go back into onions, but so is every other Tom, Dick and Harry, and I think that will be the next thing that is over done... So, got to be careful there." ([3] p. 120)

Being in the pessimistic survival context, he expects that if there is any opportunity everyone else will also see it and hence the price advantage disappear (although this will be more relevant to a risky and smaller-demand crop like onions). Growing onions is a possibility because he already has experience in growing it.

Table 6. CSNE Analysis of Quote 4

Context	Scope	Narrative Elements
Survival	*Could try growing onions*	• If it is good to grow onions, maybe other farmers think so to and the price would be poor, thus needs to be make with caution (avoiding excessive risk)

All these quotes are from a farmer reflecting on his (he is reported as male) everyday practice, and thus is not situated within the various seasons of the year. If one was simulating a month-by-month process it may be that the seasons also help determine the immediate context as well as the general attitude of the farmer, but in this case the quotes do not indicate this, but a year-long view taking into account the whole cycle. Indeed many of the quotes involve consideration of the year taken as a whole.

Looking at the above, very sketchy, analysis I hope it is starting to be clear how a simulation of many farmers might be programmed. The narrative elements clearly talk about different stages of the yearly cycle and their timing, so at least a monthly granularity needs to be considered. The farmer obviously considers the response of other farmers to opportunities, so that a multi-agent model makes sense with each looking to the innovations and decisions of others as part of their decision making. Different kinds of situation relating to kinds of risk dominate the thinking and might form the backbone of the model, involving some rich but "fuzzy" mechanism determining whether and when farmers switch between survival, comfort and entrepreurial contexts, for example it may be there is evidence that switching to survival is fairly rapid, but it takes many good years to switch to comfort and more if you are a tenant farmer. The quotes do not present any evidence the farmer has been in an entrepreneurial context, though it is possible that either newcomers to farming or very comfortable farmers might. Clearly the choices made by a farmer are perceived as being tightly constrained by what is possible, so this is an important part of behaviour determination, and probably should be part of a simulation.

The simulation in [3] took a particular framework (Bennett's theory of adaptive dynamics [2]) and a knowledge engineering approach, looking at the factors that seem to be significant in each decision (primarily how much of each crop to grow). There were two kinds of agent: adaptive and sceptic (sceptic of climate change and hence effectively non-adaptive. Given these kinds of agent there is essentially a complex risk-benefit analysis depending on climate, capital etc. of the farm. In contrast the CSNE analysis suggests an extension of this where farmers might change their

Thus, although the above illustration clearly has depended on the excellent standard of elicitation and selection in [3] making my task easier, it suggests an enhanced version of the model developed there.

7 From CSAR Analysis to Program Code

The overall aim is to make the relationship between the program code that defines agent behaviour within a simulation and the original narrative data as close as possible. To this end we proposed above a structure that we hypothesise will facilitate this by brining the coding analysis closer to how humans and human language work. Another step to facilitate the same goal is to bring the computer code closer to the analysis. That is to make the structure of the coding as coherent with the analysis as possible. Almost no agent-based coding schemes provide ready structures for context recognition/use and very few have anything that might help with reasoning about scope. Thus part of the development of a CSNE analysis should be to provide these structures, narrowing the 'jump' from analysis to code, just as we are trying to narrow the 'jump' from data to analysis.

Just as with the CSNE analysis of natural language, this involves some compromises. Using high-level structures in programming eases the task of program construction, but it also can make the understanding and verification of code harder, because the micro-level steps are not directly specified and the high-level structures might encode assumptions that programmers do not fully understand. However, to a large extent it should be the computer science that adapts to suit the kind of accounts that people make of their behaviour rather than the other way around. Work is under way to produce such architectures building on existing architectures (BDI) [14] and past experience in developing architectures that suit the task of social simulation [7] [13]. An early attempt to build-in context into an agent is in [6].

8 Concluding Discussion

The central claim of this paper is that a CSNE structure (or something similar) could play a useful role in mediating between narrative data and the code that determines agent behaviour within a simulation. Clearly however, there needs to be a synthesis of methods to develop the most effective technique possible, drawing on a variety of approaches, such as Grounded Theory [20], Conversational Analysis [15] or KnETs [4]. This should be a creative and appropriate mix, which will not occur if each approach holds inflexibly to its own traditions and shibboleths. The criteria for judging any such method should be that it:

- Preserves as much of the meaning in the original data as possible;
- Introduces as few distortions as possible;
- Is as transparent as possible, that is that when assumptions are used/added they are clear from the report of the procedure and not implicit/hidden;
- Is practical as a process and not demanding of impossible or infeasible steps;
- Is as systematic as possible, so that others can attempt to retrace a reported analysis;
- Is as honest as possible, in that it does not fudge results appearing to do more than it can deliver.

A method that goes some way to meeting the above criteria could have a transformative effect on social science in general by bringing together the worlds of qualitative and quantitative evidence in a principled way via agent-based simulation. Further enriching our simulations with information gained from narrative accounts might change how agent-based social simulations look like, since they might include some of the "mess" and complexity inherent in the social world we observed.

Acknowledgements. This research was partially supported by the Engineering and Physical Sciences Research Council, grant number EP/H02171X/1. Many thanks to Emma Norling, Cathy Urquhart, Richard Taylor and all the participants of the Informal Workshop on the topic held in Manchester (http://cfpm.org/qual2rule/informal-workshop.html) with whom I have discussed these ideas.

References

[1] Abell, P.: The role of rational choice and narrative action theories in sociological theory the legacy of cole- 6 Figure 5: events and goals. man's foundations. Revue Franaisede Sociologie 44(2), 255–273 (2003)
[2] Bennett, J.W.: The Ecological Transition: Cultural Anthropology and Human Adaptation. Pergamon Press Inc., New York (1976)
[3] Bhawani, S.: Adaptive Knowledge Dynamics and Emergent Artificial Societies: Ethnographically Based Multi-Agent Simulations of Behavioural Adaptation in Agro-Climatic Systems. Doctoral Thesis, University of Kent, Canterbury, UK (2004), `http://cfpm.org/qual2rule/Sukaina%20Bharwani%20Thesis.pdf`
[4] Bharwani, S.: Understanding complex behaviour and decision making using ethnographic knowledge elicitation games (Kn ETs). Social Science Computer Review 24(1), 78–105 (2006)
[5] Edmonds, B.: The Pragmatic Roots of Context. In: Bouquet, P., Serafini, L., Brézillon, P., Benercetti, M., Castellani, F. (eds.) CONTEXT 1999. LNCS (LNAI), vol. 1688, pp. 119–132. Springer, Heidelberg (1999)
[6] Edmonds, B., Norling, E.: Integrating Learning and Inference in Multi-Agent Systems Using Cognitive Context. In: Antunes, L., Takadama, K. (eds.) MABS 2006. LNCS (LNAI), vol. 4442, pp. 142–155. Springer, Heidelberg (2007)
[7] Edmonds, B.: Towards an Ideal Social Simulation Language. In: Sichman, J.S., Bousquet, F., Davidsson, P. (eds.) MABS 2002. LNCS (LNAI), vol. 2581, pp. 105–124. Springer, Heidelberg (2003)
[8] Gigerenzer, G., Goldstein, D.G.: Reasoning the fast and frugal way: models of bounded rationality. Psychological Review 103(4), 650 (1996)
[9] Kemp-Benedict, E.J., Bharwani, S., Fischer, M.D.: Methods for linking social and physical analysis for sustainability planning. Ecology and Society 15(3), 4 (2010), `http://www.ecologyandsociety.org/vol15/iss3/art4/`
[10] Kokinov, B., Grinberg, M.: Simulating Context Effects in Problem Solving with AMBR. In: Akman, V., Bouquet, P., Thomason, R.H., Young, R.A. (eds.) CONTEXT 2001. LNCS (LNAI), vol. 2116, pp. 221–234. Springer, Heidelberg (2001)
[11] Luhmann, N.: SozialeSysteme. GrundrisseinerallgemeinenTheorie, Frankfurt/M (1984); Engl.: Social Systems. Stanford University Press (1995)
[12] McCarthy, J., Hayes, P.J.: Some Philosophical Problems from the Standpoint of Artificial Intelligence. Readings in Planning, 393 (1990)
[13] Moss, S., Gaylard, H., Wallis, S., Edmonds, B.: SDML: A Multi-Agent Language for Organizational Modelling. Computational and Mathematical Organization Theory 4, 43–69 (1998)
[14] Norling, E.: Folk psychology for human modelling: Extending the BDI paradigm. In: Proceedings of the Third International Joint Conference on Autonomous Agents and Multiagent Systems, vol. 1, pp. 202–209. IEEE Computer Society (July 2004)
[15] ten Have, P.: Doing Conversation Analysis: A Practical Guide (Introducing Qualitative Methods). SAGE Publications (1999)
[16] Simon, H.A.: Administrative behaviour, a Study of decision-making processes in Administrative Organization. Macmillan (1947)

[17] Taylor, R.I.: Agent-Based Modelling Incorporating Qualitative and Quantitative Methods: A Case Study Investigating the Impact of E-commerce upon the Value Chain. Doctoral Thesis, Manchester Metropolitan University, Manchester, UK (2003), `http://cfpm.org/cpmrep137.html`
[18] Tomasello, M.: The cultural origins of human cognition. Harvard University Press (1999)
[19] Toulmin, S.: The uses of argument. Cambridge University Press (2003)
[20] Urquhart, C.: Grounded Theory for Qualitative Research: A Practical Guide. SAGE Publications Limited (2012)

Participatory Policy Making in Practice: Simulating Boundary Work in Water Governance

Nicolas Gailliard, Olivier Barreteau, and Audrey Richard-Ferroudji

IRSTEA, UMR G-EAU, Montpellier, France
nicolas.gailliard@irstea.fr

Abstract. Concerted and participative management has emerged in recent years in the French water policy to respond to the current projections of climate change, and an increasing demand for water in response to population growth. Therefore water management needs to move towards more sustainable practices.

In France decentralization caused legislative changes. The Water Act of 1992 and the establishment of SAGE (Local Water Management Plan) and river contracts have generated the need for people facilitating them. We consider here a new category of people named boundary worker which will be part of what some authors call intermediary people (Mauz, Granjou, Billaud, Moss, and Medd).

This new approach to public policies is not completely stabilized. Its implementations on the ground are very diverse. Little is known on their efficiency. Our work aims at providing means to improve the assessment of this aspect of participatory governance for public policies.

In this paper we propose a model to represent consequences of the involvement of a boundary worker in river basin governance, taking in account the context (social, institutional, physical) of this involvement. To achieve this we will particularly relate and articulate an analysis of several interviews with the boundaries worker.

Final aim of this model is exploration of various conditions of involvement of boundary workers and consequences on the evolution of socio-hydrosystems they are attached to.

Keywords: Agent-based modeling, qualitative data, boundary worker.

1 Introduction

Water resources management is increasingly requiring interfaces between the various users and resources. The aim is to facilitate an evolution towards more sustainable practices of the socio-hydrosystem. One of the interfaces mobilized to facilitate this implementation is the intervention of boundary workers, such as a river basin manager or a basin institution facilitator. In recent years there has been an increasing number of boundary work from different background in matter of water management. All these new people take with them their own scientific and political knowledge and also a personal vision of the situation. How the dynamic of socio-hydrosystem evolve with the introduction of these people into the collaborative management process?

B. Kaminski and G.Koloch (eds.), *Advances in Social Simulation*,
Advances in Intelligent Systems and Computing 229,
DOI: 10.1007/978-3-642-39829-2_29, © Springer-Verlag Berlin Heidelberg 2014

These people facilitate dialogue processes that end up as generators of innovation processes in the sense described by Villani and Serra [1]. For example, they endorse a role in the networking between various stakeholders. This networking generates a process of interaction and then novelty and thus innovation within the socio-hydrosystem. Stringer and Dougill show that a range of participatory mechanisms can be employed at different stages of the adaptive cycle, and can work together to create conditions for social learning and favorable outcomes for diverse water users[2]. Since boundary workers are supposed to generate and increase level of participation, they increase "social learning" which, in turn, generates new knowledge that is considered of an innovation.

ABM is an appropriate technique to represent the dynamics of heterogeneous entities, based on assumptions on how people can communicate with each other and with their environment [3]. ABM has been increasingly used to simulate the dynamics of water use and management [4-7] [8-11] either to understand or to support these collective action processes.

In order to better understand the dynamics of socio-hydrological systems featuring involvement of dedicated boundary workers, we seek to analyze the impact of these new people on the behavior of heterogeneous actors interacting in the same socio-hydrosystem. On the follow up of previous modeling and simulation efforts to represent the dynamics of socio-hydrological systems, we use agent based modeling and simulation to investigate pathways taken by socio-hydrosystems according to various scenarios of boundary work. This work is in progress. We are currently testing the feasibility of designing a suitable ABM on the basis of empirical knowledge, so that it encapsulates main drivers at hand for a river basin manager. In a first stage we focus on water availability issues as they appear in basins with irrigation as a major land and water use activity.

This paper is made of three parts. The first part presents the implementation of participatory devices in water governance and more precisely the role and the actions of boundary workers. A second part will present the outcomes of interviews with a set of boundary workers and our articulation of qualitative data and ABM. And in the third part we will detail the first modeling choice and a structure of the model.

2 Boundary Work as Participatory Interface

There is no agreed definition for "intermediary work"[12] while it has become increasingly popular. In the follow up of this paper we will name "boundary worker" the category of people endorsing activities considered as boundary or intermediate work. It is necessary that it appears as a distinct class of people so that they are "not captured by the all-encompassing concept of stakeholder" [12]. The technocrats in basin institutions for example belong to this class. These are between water users and policy maker. They are environmental specialists and must possess important relational capacity. They need to harness three types of competences: scientific/technical/ legal.

They are advisers on technical issues and as such participate in project development. They have also to facilitate working groups in order to relay initiatives on a specific study area. They are promoting "good water management" by bringing out a multi-agent system across the watershed. One of their promoting possible roles is to strengthen relations with local stakeholders. They serve as a relay and "buffer zone" between stakeholders. It is common to involve all water users who use the same water resource but in a different way and with various representations. Therefore a major role of boundary workers is to act as translators between users, policymakers and experts. These river basin manager can be considered as key people in the implementation of new water management modalities [13]. This is a new form of cooperation in public space [14]. Consultation and negotiation are central features of actions where the issue of mediation is central. It is a policy of negotiated management environment designed to be implemented.

In public space, the role of boundary worker can be endorsed by different people. It can be someone, like broker, science impact coordinators, and local authorities.

3 Combining Qualitative Data and ABM

3.1 "An Ethnographic Seduction"

I relied on the paper : "An ethnographic seduction": how qualitative research and Agent-based models can benefit each other [15]. It discusses how qualitative research and Agent-Based Modeling can be combined to help each other. The authors proposed in this paper an analytical framework informed by empirical data in order to design an ABM, and feedback on theory and empirical knowledge. They structure the flow of knowledge from field observation to theory and simulation in the same kind of cycle process as companion modeling [16-17]. This method could facilitate an adequate representation of social behavior in an ABM.

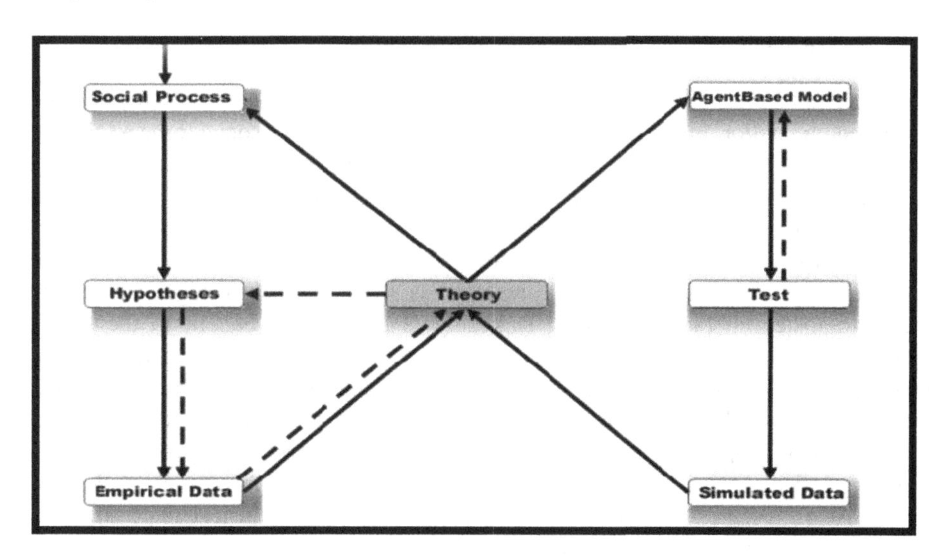

Fig. 1. Analytical framework

Qualitative research provides detailed descriptions of social phenomena. It enables then to feed representations of several behaviors and attitudes of people, for example in a socio-ecological system. The authors explore the promises of bringing together ethnography and ABM. Indeed one of the objectives of ABM is to evaluate the effects of micro-behavior (interactions) within a social system. They argue that ABMs are particularly adapted for instantiation of qualitative data and it will be explored in this article the advantages of the combination of the two could offer.

The compatibility between ABM and the qualitative and quantitative data has driven some theorists to consider ABM as the "third way" in social science research [18]. The model is able to reproduce the facts already observed and must also have a predictive role in envisioning a future state of the system.

Once associated with sociological methods, ABMs enable to fill some gaps usually associated with the use of qualitative data as the construction of forecasting. With ABM, replications are possible to evaluate the sensitivity to variations of controlled characteristics of the population. ABMs enable to describe a variety of groups of people, to emphasize similarities and expanding the field of observation. It is important to note that ABM could be used as tools for the design of policy, simulating the possible forms of intervention before the field tests.

ABMs entail to achieve more details of the description of social process. To ensure that the model reproduces these social processes as accurately as possible, the design rules of agent behavior must be informed by detailed micro-data that include information on the types of relationships and interactions between agents and the conditions they occur.

We are conducting our investigation on the consequences of boundary work on evolution of socio-hydrosystems along a process similar to the one formally described above. In the follow up of the paper we describe our implementation of first stages.

3.2 Interviews

Starting point in this analytical framework is a given social process, which is a stake for research and/or management reasons. In our case, this social process is the influence of boundary workers in the evolution of socio-hydrosystems. Due to lack of comprehensive knowledge on the practices of these actors, we started with twelve interviews from various watersheds in France. The hypothesis (step 2 in the Tubaro and Casilli's framework) that we could elicit from these interviews is that the boundary workers' activity can be represented as a sequence of management actions meant to foster interactions among stakeholders in order to benefit to the socio-hydrosystem. It comes out that boundary workers have a list of "recipes", dedicated to foster interactions and suited to specific conditions and objectives. We assume then that boundary workers arbitrate the implementation of these recipes under budget (funding, social capital...) constraints. With this assumption in mind, we came back to the empirical material collected through recorded interviews. We shaped then an initial list of "recipes" following the interpretation of the empirical material. We provide below examples of these recipes together with the empirical material supporting them.

Organize a Thematic Workshop

Extract of interview:

- *"We have organized a thematic workshop in order to inform a large number of water users in our projects".*
- *"We proposed thematic workshop to gather requirements and questions of watershed water users and also to submit our. We will be able to make some decisions that will be validated in decision meetings".*

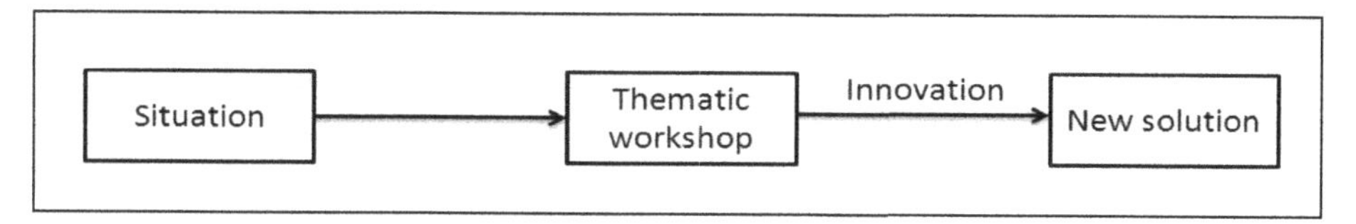

Fig. 2. Thematic workshop

The boundary worker organizes this workshop when he finds a weak stakeholder engagement. The thematic workshop includes some watershed water users to discuss a particular topic. This makes it possible to gather and raise awareness these people and gets their opinion. Following this workshop the participants will increase their knowledge of the rules of the watershed.

Organize an Information Meeting

Extract of interview:

- *"We organized this kind of meeting that the water users can position themselves giving the keys to understanding and not to try to pass something. There are somehow meetings diagnostic. There is no immediate concern".*

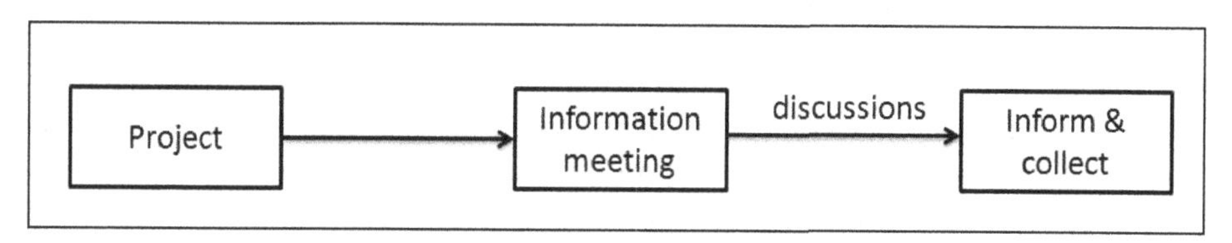

Fig. 3. Information meeting

When the boundary worker wants to present projects he invites the stakeholder to this meeting. This kind of meeting serves to inform water users and to obtain their views on the possibility of work or actions. These meetings will make a diagnosis on a given situation from different angles. The boundary workers will be able to determine the stakeholders who wish to engage in a project presented.

Organize a Meeting on the Field

Extract of interview:

- *"In a particular situation we try to go into the field and escort the water users to find a result. This leads to new solutions not considered in more formal meetings."*

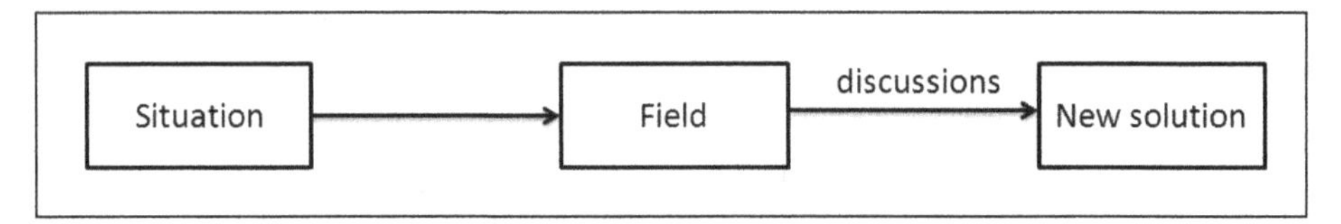

Fig. 4. Field meeting

When the boundary worker notes a disagreement on a topic he organized a meeting on the field. New ideas can emerge after discussions that have hatched in the new space. Situations that seemed blocked at the meeting can reach a favorable outcome. These are moments in the informal water management that can be characterized as a kind of black box [19]. Different users will feel freer to speak, the context is different and will facilitate exchanges between the actors. The work of the boundary workers is easier, this informal meeting is a time of dialogue and negotiation can lead to beneficial adjustments.

As stated earlier the "recipes" have emerged as a result of interviews with a sample of boundary workers. These recipes represent different tools at their hand to set up different formats of interaction. Depending on the context and purpose he chose in his library of recipes the one suited better a given situation. The combination of recipes is a key element in the successful implementation of a process of interaction.

4 Structure of the Model

We describe in this section the structure of the agent based model following the field work investigation. With this model we aim at understanding the possible paths taken by a socio-hydrosystem with a boundary worker.

4.1 Agents Description

In this initial prototype, we designed the model with two types of agents: water users and boundary worker. All agents know each other and can interact. Each class is composed of several attributes (figure 5). For example, every user has a confidence index towards the other users and the boundary workers. This index will change according to the interactions format proposed by the boundary workers. The knowledge index will evolve in the same way. The water users have also an access and can modify the status of their pumps.

On the other hand, one of the boundary worker attributes is a recipe list. They can, for example, observe the river flow and perform one of the recipes. They have access to the institution including various hydrological thresholds.

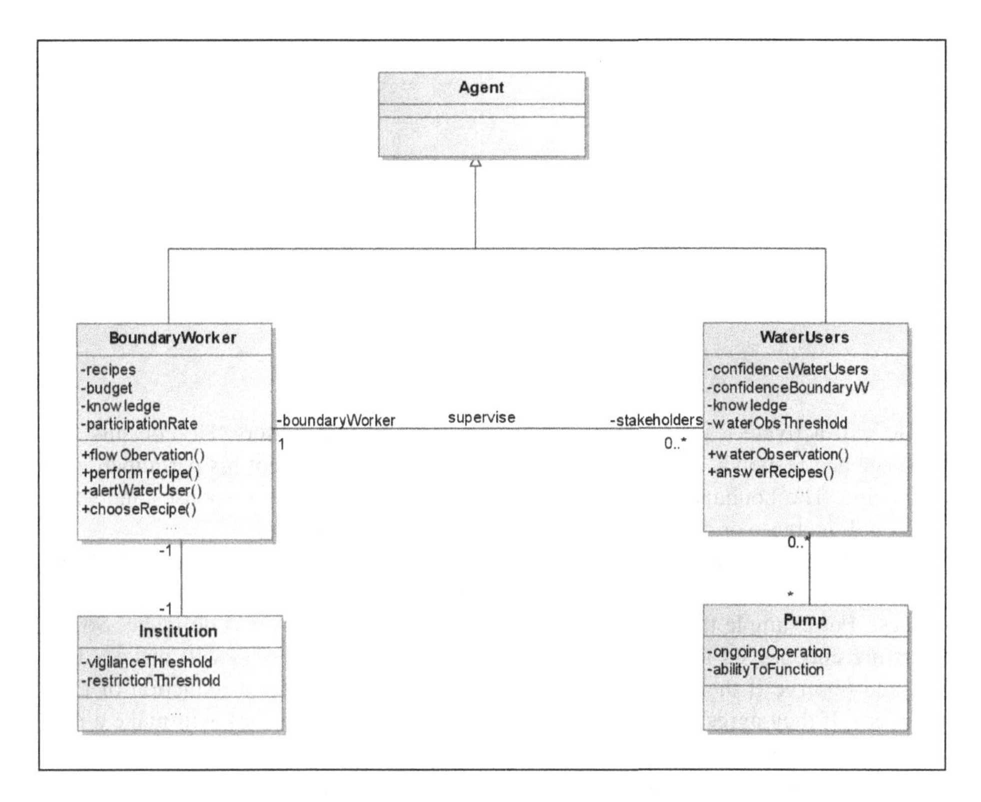

Fig. 5. Class diagram

4.2 Changes Induced by Interactions

We want to portray the capacity of perception of people, including showing their ability to perceive their surroundings. The water users can perceive their environment with sensors that will provide numerical measures, such as the water level in a river. The environmental perception can also be evaluated directly by the boundary worker. In this process, the boundary worker has his own assessment of the environment. In case he believes that there is a problem (when the water level is lower as a defined threshold by the institution) several recipes may be considered.

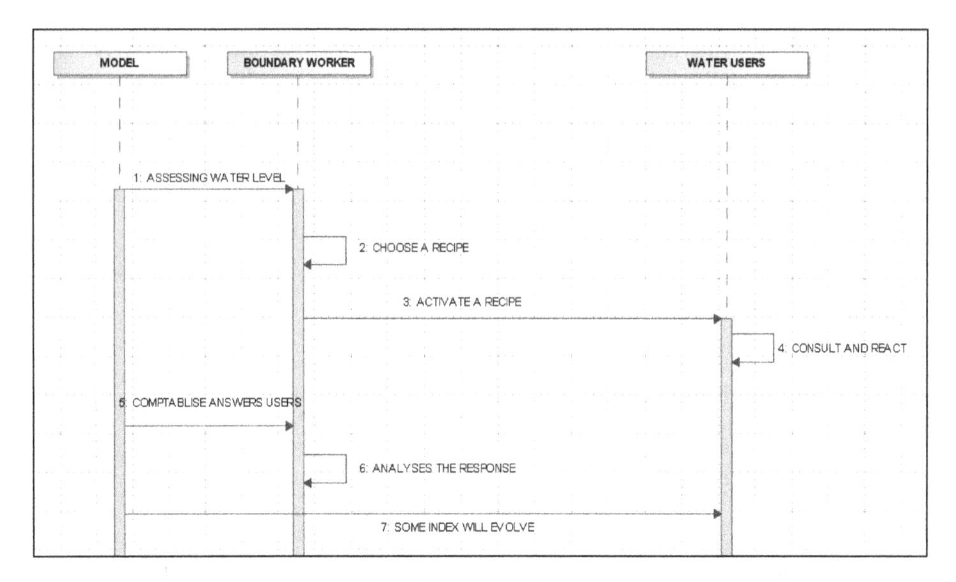

Fig. 6. Sequence diagram

He will activate a recipe and send the message to water users. He consults this message and responds to the recipe sent indicating whether or not his participation to the recipe. The boundary worker will take into account the user responses and some index will increase or decrease depending on what the recipe implied.

The user answer to a recipe sent by the boundary worker is subjected to various tests. To begin, the user will observe if he has received in his "mailbox" a recipe request. For example the recipe is that the water level in the river is deficient. So the users are encouraged to reduce their water withdrawals. The users will make a first test and observe if their own assessment of water level requires a reduction of their practices. If they agree they reduce their activity, but otherwise they will make a new test. Each user will test his confidence index on the boundary worker. If this index exceeding a determined threshold the users will accept the recipe otherwise he will not change anything.

5 Conclusion and Perspective

This paper discusses the creation of an ABM which is based on empirical data. Qualitative data are involved in the construction phase of the model and will describe research questions, rules of behavior, interactions between agents and some scenarios to simulate. This empirical work has resulted in the development of a recipe list which has been validated by a boundary worker.

This framework will also help to prepare a return to the field and to conduct a revision of the ABM. In our case study it seems clear that following the first simulations new questions appear and force us to return to the field to ask for some boundary workers. The return phase seems necessary to calibrate the model as finely as possible. A model based on an articulation of recipes may not be enough to quantify the impact of the boundary workers. Thus it is essential to include at this model several indicators that show the effects of these recipes on the socio-hydrosystem. The model is still in development phase. We will explore indicators of resilience of socio-hydrosystem as: the level of the resources, the index of knowledge and the confidence of stakeholders.

References

1. Serra, R., Villani, M., Lane, D.: Modeling Innovation. Complexity Perspectives in Innovation and Social Change, 361–388 (2009)
2. Stringer, L.C., Dougill, A.J., Fraser, E., Hubacek, K., Prell, C., Reed, M.S.: Unpacking "participation" in the adaptive management of social–ecological systems: a critical review. Ecology and Society 11, 39 (2006)
3. Janssen, M.: Complexity and ecosystem management: the theory and practice of multi-agent systems. Edward Elgar Publishing (2002)
4. Barreteau, O., Bousquet, F.: SHADOC: a Multi-Agent Model to tackle viability of irrigated systems. Annals of Operations Research 94, 139–162 (2000)
5. Becu, N., Perez, P., Walker, A., Barreteau, O., Le Page, C.: Agent based simulation of a small catchment water management in northern Thailand: Description of the CATCHSCAPE model. Ecological Modelling 170, 319–331 (2003)
6. Lansing, J.S.: Anti-chaos, common property and the emergence of cooperation. In: Kohler, T., Gumerman, G. (eds.) Dynamics in Human and Primate Societies, pp. 207–223. Oxford University Press (1999)
7. Pahl-Wostl, C., Hare, M.: Processes of social learning in integrated resources management. Journal of Community and Applied Social Psychology 14, 193–206 (2004)
8. van Oel, P.R., Krol, M.S., Hoekstra, A.Y., Taddei, R.R.: Feedback mechanisms between water availability and water use in a semi-arid river basin: A spatially explicit multi-agent simulation approach. Environmental Modelling & Software 25, 433–443 (2010)
9. Barreteau, O., Bousquet, F.: SHADOC: a multi-agent model to tackle viability of irrigated systems. Annals of Operations Research 94, 139–162 (2000)
10. Le Bars, M., Le Grusse, P.: Participative modelling of agricultural water demand at regional scale: an example in central Tunisia (2008)
11. Hoanh, C., Le Page, C., Barreteau, O., Trébuil, G., Bousquet, F., Cernesson, F., Barnaud, C., Gurung, T., Promburom, P., Naivinit, W.: Agent-based modeling to facilitate resilient water management in Southeast and South Asia. In: International Forum on Water and Food, pp. 10–14 (2008)
12. Moss, T., Medd, W., Guy, S., Marvin, S.: Organising water: The hidden role of intermediary work. Water Alternatives 2, 16–33 (2009)
13. Barreteau, O., Richard-Ferroudji, A., Garin, P.: Des outils et méthodes en appui à la gestion de l'eau par bassin versant. La Houille Blanche 6, 48–55 (2008)
14. Billaud, J.-P.: Ce que faciliter veut dire. A propos d'un retour d'expérience de chargés de mission Natura 2000. In: Remy, J., Brives, H., Lemery, B. (eds.) Conseillers en Agriculture. Educagri/INRA (2006)
15. Tubaro, P., Casilli, A.A.: An Ethnographic Seduction: How Qualitative Research and Agent-based Models can Benefit Each Other. Bulletin de Méthodologie Sociologique 106, 59–74 (2010)

16. Bousquet, F., Barreteau, O., Le Page, C., Mullon, C., Weber, J.: An environmental modelling approach: the use of multi-agent simulations. Advances in Environmental and Ecological Modelling 113, 122 (1999)
17. Étienne, M.: Companion modelling: a participatory approach to support sustainable development. La modélisation d'accompagnement: une démarche participative en appui au développement durable (2010)
18. Gilbert, N.: Quality, quantity and the third way. In: Holland, J., Campbell, J. (eds.) Methods in Development Research: Combining Qualitative and Quantitative Approaches. ITDG Publications, London (2004)
19. Richard-Ferroudji, A.: Limites du modèle délibératif: composer avec différents formats de participation. Politix, 161–181 (2012)

An Adaptation of the Ethnographic Decision Tree Modeling Methodology for Developing Evidence-Driven Agent-Based Models

Pablo Lucas

University College Dublin, Geary Institute, Ireland
Maastricht School of Management, The Netherlands
pablo.lucas@ucd.ie, lucas@msm.nl

Abstract. This paper introduces the integration of the Ethnographic Decision Tree Modelling methodology into an evidence-driven lifecycle for developing agent-based social simulations. The manuscript also highlights the development advantages of using an Ethnographic Decision Tree Model to promote accountable validation and detailed justification of how agent-based models are built. The result from this methodology is a hierarchical, tree-like structure that represents the branching and possible outcomes of the decision-making process, which can then be implemented in an agent-based model. The original methodology grounds the representation of decision-making solely on ethnographic data, yet the discussed adaptation hereby furthers that by allowing the use of survey data. As a result, the final model is a composite based on a richly descriptive dataset containing observations and reported behaviour of individuals engaged in the same activity and context. This in turn is demonstrated to serve as a useful guide for the implementation of behaviour in an social simulations and also serve as a baseline for testing.

Keywords: methodology, evidence, qualitative, development, validation.

1 Introduction

Unless the purpose of an agent-based model is purely theoretical, e.g. strictly regarded only as thought experiment or demonstration, it is most likely that the modeller will have to deal with qualitative data to some extent. Despite noteworthy progress in the development of evidence-driven and participatory modelling methodologies[1] [Moss, 2008] [Edmonds et al., 2013], agent-based developers are still generally struggling with fundamental aspects as to how should quantitative and qualitative data be used to inform model building and how experimental data, obtained from simulation experiments, ought to be appropriately analysed.

The overall flexibility to develop agent-based social simulations (henceforth ABSS) particularly highlights the need for modellers to adopt rigour throughout the specification and implementation of simulation assumptions and processes.

[1] I.e., involving stakeholders throughout modelling, implementation and validation.

B. Kamiński and G.Koloch (eds.), *Advances in Social Simulation*,
Advances in Intelligent Systems and Computing 229,
DOI: 10.1007/978-3-642-39829-2_30,

As a consequence of that, one can find various examples of models developed that are either mainly theory- or evidence-driven. The earliest approaches to implement ABSS are undeniably theory-driven, as that is how this research area begun: by providing a computational alternative for testing –or demonstrating– theories and hypotheses that could not be tested in real settings. In an attempt to improve the *status quo*, the current trend in the ABSS community is rather skewed towards the preference for implementing evidence-driven models. Due to the nature of social phenomena, ABSS modellers confront the inherent subjectiveness that accompanies the task of representing –unambiguously, as it has to be implemented computationally– all relevant social structures and processes.

To code these into an ABSS, one shall adhere to a non-discursive and precise notation (i.e. a computer language). Experience acquired in using ABSS development frameworks raise one's awareness that computational experiments must be clearly designed, specified and correctly implemented so that the tested hypotheses indeed shed light on the actual phenomena and not only on models themselves. Thus ABSS modellers are in a position which requires them to ensure that observed simulation results are due to correct model specifications, and not artefacts resulting from particular implementations or errors –such as those discussed in [Galán and Izquierdo, 2005]. However, there are still various open issues on how should evidence –particularly those qualitative in kind– be used to inform the development of an ABSS with regards to a social phenomena.

A systematic way of incorporating qualitative data is put forward in this paper, which is by means of integrating an adapted version of the Ethnographic Decision Tree Modelling (henceforth EDTM) methodology [Gladwin, 1989] in an evidence-driven approach to develop ABSS (introduced in [Lucas, 2011]). This methodological proposal is aimed at strengthening the cross-validation process [Moss and Edmonds, 2005], from model design to implementation, by grounding the representation of decision-making processes in a composite model of how individuals engaged in the same activity and context actually behave. The argument is that developing an ABSS based on an EDTM promotes a more accountable validation and detailed justification of how modelling decisions are made. That is because the result of an EDTM is a hierarchical, tree-like structure that represents the branching and possible outcomes of the decision-making process based on data collected directly from the decision-makers. The original EDTM methodology grounds the representation of decision-making solely on ethnographic data, yet the discussed adaptation hereby furthers that by allowing the use of survey data to account for when subjects are not directly accessible.

2 An Evidence-Driven Approach to Modelling (EDAM)

Both Within and beyond the ABSS, it is generally accepted that the available development platforms[2] are useful for building models that allow accurate demonstration and testing of hypothetical scenarios. Notwithstanding the steady

[2] Including the following popular ones, each providing their own particular features: SWARM, NetLogo, Repast, CORMAS, Ascape, SeSAm, SDML, MASON and M4A.

technical progress in the provision of better tools for developing and analysing ABSS, there are still rather difficult open methodological issues with regards to the life-cycle development of these models. As a consequence, to date there is no *de facto* standard to evaluate and validate how an ABSS is developed, including:

- (a) how available qualitative evidence can systematically be used in ABSS;
- and (b) how the modeller decide the implementation of qualitative evidence.

Thus provided there are enough resources and time, modellers could effectively experiment with different implementation methods in search of the most successful for carrying out the task of correctly representing a social phenomenon in an ABSS. However this is rather unrealistic, resource- and time-wise, so it is likely that ad-hoc, particular techniques will be used by the modeller. And that is probably the one which the researcher has most experience with. Generally, a different coding technique is tested only –if at all– when the model is replicated. These aspects are important in the modelling life cycle of ABSS because challenges recur within the processes of data collection, analysis, and particularly during the process of deciding the roles of both qualitative and quantitative data in the simulation experiment [Polhill et al., 2010] [Geller et al., 2010].

The next page contains an explanation of the Evidence-Driven Approach to Modelling (henceforth EDAM) which has been integrated with the adapted EDTM that is introduced in the next section. Considering the process illustrated in Figure 1, in case the ABSS consistently diverge from what has been observed in reality about the social phenomena, it is likely that something in the model has either been misrepresented, implemented incorrectly or that parameters were set unrealistically [Lucas, 2011]. Both within the academic ABSS community and practitioners beyond it, the common understanding is that models should at least generate results that are plausible in light of the existing (qualitative and quantitative) evidence [Moss and Edmonds, 2005] [Edmonds et al., 2013].

Furthemore, it is recommended that modellers test technicalities of the ABSS well before reaching the 5th step, described in the next page. A full validation of obtained results is only possible once the model has been scrutinised and deemed plausible, both by academics and stakeholders. From this milestone onwards modellers can attempt to mediate the development of knowledge about phenomena –not the model itself – via the interpretation of simulation results.

The lifecycle for developing ABSS models is described in the following steps:

1. The target system is the social phenomena itself, from which evidence is collected and analysed. This might require, for instance: administration of questionnaires, survey socio-economic circumstances or setup of an automated strategy for collecting social data[3] from the social phenomenon participants.
2. With analysed evidence, modellers proceed to discuss the plausibility of observations and assumptions with stakeholders. Here there is a potential loop as researchers and domain experts must reach an understanding of what has been analysed and whether hypotheses are based on realistic assumptions.

[3] Larger datasets have greater chances of providing richer, more useful findings.

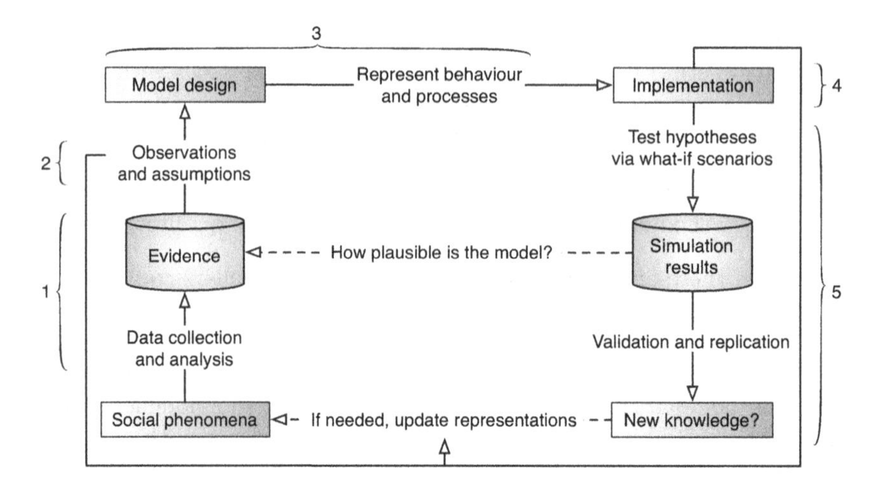

Fig. 1. Evidence-Driven Approach to Modelling (EDAM) [Lucas, 2011]

3. To design an ABSS based on the scrutinised evidence, which at this stage has been verified by stakeholders[4], one shall then differentiate what is essential to be in the model from what is contextual data about the phenomena. The latter comprises much more than the former, thus caution must be observed.
4. Deciding how to implement agent behaviour and processes is entirely up to the modellers, as no evidence favours any particular computational paradigm. Most approaches nowadays use object-orientation due to the relative ease of representing behavioural features in simplified, self-contained threads.
5. Having a feasible model built with guidance of evidence, modellers can proceed to test hypotheses using scenarios that resemble observations. Results are then compared to evidence and, to strengthen validation and knowledge development, findings should ideally be discussed with domain experts.

3 Adapting the Ethnographic Decision Tree Model

An Ethnographic Decision Tree Model (EDTM) is the result of a systematic, mainly qualitative, methodology for building a model of decisions that are taken by a specific cohort –where the research process is driven by the data collected from the surveyed subjects themselves, rather than the researcher hypotheses [Gladwin, 1989]. An EDTM requires therefore an approach to development that begins with the researcher selecting the decision-making context. As it can be observed in the figure below, the original EDTM combines the ethnographic research cycle with a linear plan for hypothesis testing. It also does suggest the setup an ethnographic database, which shall contain both quantitative and qualitative information. The intention is for these resources so to become the

[4] I.e., the data can serve both to verify and justify how the ABSS has been built.

reference material for consultation throughout the process of the EDTM development, which is argued hereby that it can also serve for the ABSS development.

In the original methodology there is a rather arbitrary suggestion that the qualitative cycle would need to be done at least 10 times first, and then proceed with the EDTM development [Beck, 2000]. Another rather arbitrary suggestion from the original proposal is that an EDTM is successful if it can predict between 85% and 90% of the tested individual choices. This criterion may have been inspired in the concept of statistical confidence intervals, yet it remains unconfirmed if that is the case and whether the related literature would be applicable to EDTMs. Given the likelihood of EDTMs being based on relatively small samples, there is a significant risk of statistical errors in testing hypotheses because normality tests would –most likely– not detect non-normality in such limited datasets.

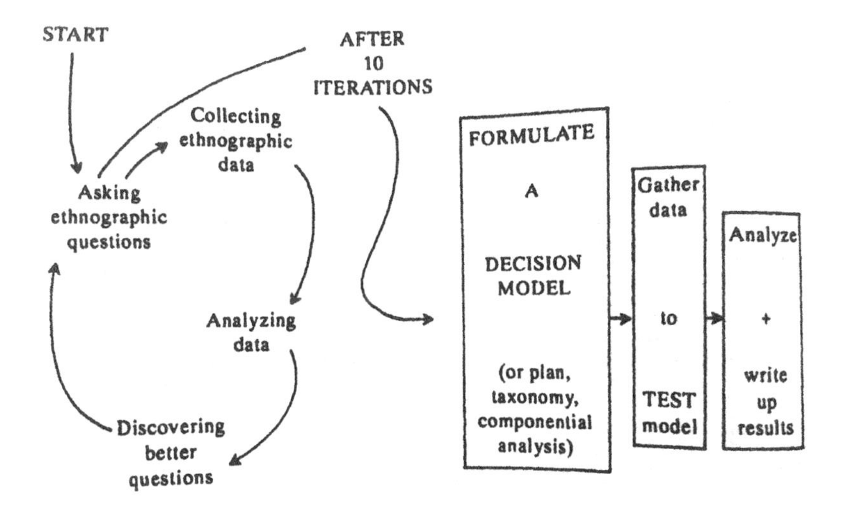

Fig. 2. The original EDTM development cycle proposed by [Gladwin, 1989]

For the purpose of integrating this approach with the previously discussed EDAM, The proposed adaptations of the original EDTM include the following aspects:

– To allow the process to begin with surveys, which shall –at a later stage– be complemented with an ethnographic approach. This is to provide flexibly in terms of reaching a larger population and circumvent the common difficulty of modellers not being physically able to access the decision-makers *in situ*. This means that the first phase can hold a series of surveys, with the intention to improve the understanding of the decision-making process in question.

– Use the developed EDTM as the behavioural reference for the ABSS implementation. This is pertinent as the EDTM is originally suggested to be tested by administering the decision tree with the decision-makers. Thus this adaptation proposes testing of the EDTM also at the ABSS level. The advantage of this is to leverage the qualitative validation at the micro-level, whilst contributing to cross-validate results at the macro-level. This validation approach is similar with the proposal discussed in [Moss and Edmonds, 2005].
– As normality may not be an important issue in large datasets for forecasting purposes [Maindonald and Braun, 2010], the testing of the developed EDTM as an ABSS implementation is proposed to remain within the bounds of whether the model results are deemed plausible by the stakeholders –without necessarily providing predictions. Yet the model may still serve such purpose.

Figure 3 below depicts how the aforementioned adaptations interrelate in practice with the the EDAM development, discussed in the previous section.

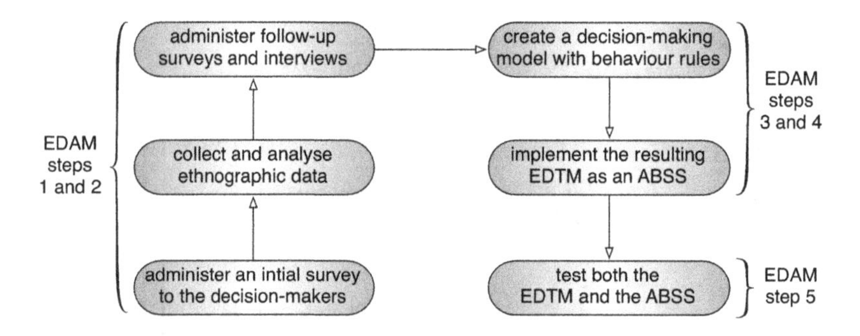

Fig. 3. The adapted EDTM development cycle, following the EDAM

Once the surveys have been administered, one shall consider the ethnographic approach –which can be done via semi-structured interviews and field observation. In terms of the ethnographic database, perhaps the most important aspect is to make notes about the data as soon as it has been collected, preferably on the spot, otherwise analyses can be adversely impacted by the increasing likelihood of the qualitative data being misinterpreted, misremembered or forgotten [Murchison, 2010]. Such detailed qualitative notes can be helpful in constructing a narrative of what has been researched and also does help with fine tuning the EDTM. One cannot record absolutely every detail, yet the modeller should have enough information as to be able a clear understanding of the decision-making process represented in an EDTM. As this is done from an ethnographic perspective, it may include –whenever appropriate and consented– recorded audio and video as part of the database which grounds every step of the EDTM.

Due to the qualitative nature of the process, irrespective of whether the original methodology or the adapted one is chosen for data collection, the usage of an EDTM require modellers to take the following aspects into careful consideration:

- framing effects (i.e. how the initial exposure to the phenomenon and presentation of the decision-making process can influence the process of modelling);
- the criteria for choosing subjects, design of data collection and techniques for data analysis (i.e. how are the instruments of surveys and interviews going to be actually administered in the field where the decision-makers are located);
- and testing of the EDTM, which includes the qualitative administration of the hierarchical model to decision-makers and implementation in an ABSS.

These are important because the model assumes the creation of a tree-like, hierarchical structure to represent decision-making that is culturally-tuned by surveying a specific groups of individuals [Gladwin, 1989] [Beck, 2000]. That is the reason for arguing that an EDTM can be systematically tested via an ABSS, where agents are equipped with the behavioural rules. As the EDTM is based on qualitative and quantitative evidence, modellers can use this resource to systematically justify implementation decisions that would otherwise remain unaccounted. Thus the main contribution by guiding the development of ABSS with an EDTM is the integration of a methodology that better equip modellers to deal with qualitative data. Due to the hierarchical nature of an EDTM, it can be straightforwardly implemented in an ABSS as a series of IF-THENS or using a system of inference techniques –such as forward chaining in JESS [Doorenbos, 2001] and backward chaining in Prolog [Roy and Haridi, 2004].

4 Final Remarks

This paper is an introduction to the integration of the EDTM qualitative methodology, designed to elicit what stakeholders actually do when taking decisions, to facilitate the bottom-up implementation of behavioural rules in an ABSS. The result of applying this approach includes a culturally grounded database, containing richly descriptive qualitative and quantitative data, about observations and reported behaviour of individuals engaged in the same activity and context. This in turn serves both for building an EDTM and for improving validation of evidence-driven ABSS. This is because modellers can use EDTM to guide and justify the implementation of behavioural rules, whilst still being able to use the model as a baseline for testing. Discussing a complete example of this development approach would have been beyond the scope of this paper; yet for an ABSS that has been built based on the EDTM, refer to [Lucas and Edmonds, 2013].

Acknowledgments. I would like to express my gratitude for the useful discussions held with Bruce Edmonds and Cathy Urquhart; plus institutional support from the University College Dublin and the Maastricht School of Management. I would also like to thanks the comments and suggestions from the two anonymous reviewers, which are being taken into consideration to develop this paper.

References

[Beck, 2000] Beck, K.A.: A decision making model of child abuse reporting. PhD thesis, Doctor of Philosophy (PhD) Thesis, University of British Columbia (2000)

[Doorenbos, 2001] Doorenbos, R.B.: Production matching for large learning systems. Technical report, Carnegie Mellon University, Pittsburgh, United States (2001)

[Edmonds et al., 2013] Edmonds, B., Lucas, P., Rouchier, J., Taylor, R.: Human Societies: Understanding Observed Social Phenomena. In: Simulating Social Complexity: A Handbook. Understanding Complex Systems. Springer (2013)

[Galán and Izquierdo, 2005] Galán, J.M., Izquierdo, L.R.: Appearances can be deceiving: Lessons learned re-implementing axelrod's 'evolutionary approach to norms'. Journal of Artificial Societies and Social Simulation 8(3), 2 (2005)

[Geller et al., 2010] Geller, A., Goolsby, R., Hoffer, L.D.: On qualitative data in agent-based models. In: 3rd World Congress on Social Simulation –Scientific Advances in Understanding Societal Processes and Dynamics (2010) (working paper)

[Gladwin, 1989] Gladwin, C.H. (ed.): Ethnographic Decision Tree Modeling (Qualitative Research Methods). SAGE Publications, Inc. (1989)

[Lucas, 2011] Lucas, P.: Usefulness of simulating social phenomena: evidence. Artificial Intelligence and Society, 1–8 (2011), 10.1007/s00146-010-0315-1

[Lucas and Edmonds, 2013] Lucas, P., Edmonds, B.: Modelling endorsements amongst microfinance clients using an ethnographic decision tree model. University College Dublin, Manchester Metropolitan University (working paper, 2013)

[Maindonald and Braun, 2010] Maindonald, J., Braun, W.J.: Data Analysis and Graphics Using R: An Example-Based Approach, 3rd edn. Cambridge Series in Statistical and Probabilistic Mathematics. Cambridge University Press (2010)

[Moss, 2008] Moss, S.: Alternative approaches to the empirical validation of agent-based models. Journal of Artificial Societies and Social Simulation 11(1), 5 (2008)

[Moss and Edmonds, 2005] Moss, S., Edmonds, B.: Sociology and simulation: Statistical and qualitative cross-validation. American Journal of Sociology 110, 1095–1131 (2005)

[Murchison, 2010] Murchison, J.: Ethnography Essentials: Designing, Conducting, and Presenting Your Research (Research Methods for the Social Sciences), 1st edn. Jossey-Bass (2010)

[Polhill et al., 2010] Polhill, J.G., Sutherland, L.-A., Gotts, N.M.: Using qualitative evidence to enhance an agent-based modelling system for studying land use change. Journal of Artificial Societies and Social Simulation 13(2), 10 (2010)

[Roy and Haridi, 2004] Roy, P.V., Haridi, S.: Concepts, Techniques, and Models of Computer Programming. MIT Press, Cambridge (2004)

Grounded Simulation

Martin Neumann

University of Koblenz, Koblenz, Germany
maneumann@uni-koblenz.de

Abstract. This paper investigates the contribution of evidence based modelling to Grounded Theory. It is argued that evidence based modelling provides additional sources to truly arrive at a theory in the inductive process of a Grounded Theory approach. This is shown at two examples. One example concerns the development of an ontology of extortion racket systems. The other example is a model of escalation of ethno-nationalist conflicts. The first example concerns early to medium stages of the research process. The development of an ontology provides a tool for the process of theoretical coding in a Grounded Theory approach. The second example shows stylised facts resulting from a simulation. Stylised facts are broad patterns which are characteristic for certain domain fields. Thus they generalise over a particular case. This provides credibility for the claim to inductively generate a theory, i.e. to overcome a purely descriptive level.

Keywords: Grounded Theory, Evidence Based Modeling, Theoretical Coding, Ontologies, Theoretical Sensitivity, Stylized Facts.

1 Introduction

In the past years, research in computational social science has reached a certain stage of scientific maturation [1,2]. The purpose of this paper is to contribute to the research area of a cross-fertilisation of simulation and the standard methods of empirical social research [2]. This paper aims at exploring additional sources of empirical credibility that evidence based modelling approaches can provide to the qualitative account of Grounded Theory. The objective of this paper is not to examine the diverging variants of Grounded Theory approaches [3]. Rather, the research question of this paper is to investigate the question of what is a *theory* in the Grounded Theory approach.

This paper will enfold two main arguments: *first*, it will be shown that the research process of evidence based modelling has a number of parallels with the Grounded Theory approach. This suggests to take experiences and guidelines of Grounded Theory approaches into account explicitly in a research process of evidence based modelling. This thesis addresses primarily ICT specialist working in the field of evidence based modelling. *Second,* a more surprising result may be that evidence based simulation provides methodological tools as well as terminological concepts to strengthen the theoretical element of a Grounded Theory. The objective of this argument is to inform specialists in the field of Grounded Theory about the possibilities of a methodological cross-fertilisation of Grounded Theory and evidence based simulation.

B. Kaminski and G.Koloch (eds.), *Advances in Social Simulation*,
Advances in Intelligent Systems and Computing 229,
DOI: 10.1007/978-3-642-39829-2_31, © Springer-Verlag Berlin Heidelberg 2014

The paper is organised as follows: first, an overview of the methodology of the Grounded Theory approach is provided. Particular emphasis is put on the terms theoretical coding and theoretical sensitivity. Subsequently, a brief sketch is provided of how evidence based modelling parallels a Grounded Theory approach. Third, at two examples it is demonstrated, how the process of theoretical coding gains from the development of a software ontology and how stylised facts provide additional evidence to theoretical sensitivity. Finally, the paper ends with concluding remarks.

2 Grounded Theory

The terminology Grounded Theory is slightly misleading since it is not a classical 'grand' theory or a middle range theory of a certain phenomenon. Rather, it denotes a certain methodological advice to *generate* theories. Thus Grounded Theory is an inductive approach to study social phenomena. This is done in an iterative process in which data collection and analysis are in a reciprocal relationship. The analysis of data should stimulate new questions posed to the data that stimulates new collection of data. This process should be iterated until a stage of theoretical saturation is reached [4]. This parallels companion or participatory modelling approaches [5] in which the process of model development is constantly informed by the expertise of stakeholders and vice versa this process aims to inform also the stakeholders by unfolding tacit knowledge of the domain. This *iterative* account parallels the concept of theoretical saturation.

2.1 Theoretical Coding

Grounded Theory aims at inductively reaching a theory. In this process, the notion of theoretical coding is of central relevance. The concrete instructions shall not be reported here [6]. However, in broad terms the notion of theoretical coding describes the process of building categories from key terms and relations. This is a process of increasing abstraction from a detailed description of the phenomenon to discover categories and their properties. Categories denote not the individual phenomena, but relate certain groups of phenomena into a single concept, i.e. they denote a set. Furthermore, the set of categories that enfold a picture of the target system are embedded in a web of relations which describe the properties of the categories in various dimensions. Briefly, this is the process of how Grounded Theory aims to enfold a *theory* of a phenomenon, rather than merely describing it. Nevertheless, the relation between theory and description remains ambiguous. The objective of the first of the following two examples is to demonstrate how software ontologies enable to assist this process.

2.2 Theoretical Sensitivity

The terms 'theoretical saturation' and 'theoretical sensitivity' provide quality criteria of the theory development in the Grounded Theory approach. The objective of the second example is to show how simulation contributes in particular to the quality criteria of theoretical sensitivity. Briefly, theoretical saturation is the criteria to stop the iterative process of data collection and analysis. This is indicated if no additional categories or properties can be found anymore. Theoretical sensitivity indicates the meaningfulness of the results. It is claimed that Grounded Theory cannot be reduced to a routine application of certain methods. However, in the Grounded Theory literature [7-9] this is specified as the credibility of the researcher. For instance, the imagination and creativity of the researcher is highlighted [7]. This is a very personal conception [10] and lacks a more objectifiable criterion. The assessment of the creativity of a researcher depends to a large degree on the person undertaking the assessment. The second example will demonstrate that the notion of stylised facts, developed for the investigation of simulation results [11] can provide a means to develop criteria to evaluate the quality of a Grounded Theory. Admittedly, this is not the original conception of theoretical sensitivity [7,9]. Nevertheless, it will be argued that stylised facts provide a source of evidence that the inductive research process generated distinctive theoretical insights rather than merely a description of a phenomenon. Insofar it is a criterion for theoretical sensitivity as it indicates that the data revealed meaningful theoretical insights.

3 Evidence Based Modelling

Before turning to the concrete examples, a brief overview of core principles of evidence based modelling will be provided in order to show the key similarities between both research methodologies. Evidence based modelling is an umbrella term for a number of approaches which evolved in the past decade. They aim to include qualitative, descriptive sources of evidence in the model assumptions [5],[12,13]. [13] coined the term KIDS principle (Keep it descriptive, stupid). Partly this modelling approach arises from the fact that agent-based modelling allows for a rule based modelling approach [13]. This enables an implementation of a detailed description of individual decisions and actions on a social micro-level.

The modelling process in the evidence based account is a bottom-up process, meaning that it is an iterative process, cycling between modelling and field work [5]. Thus the research process does not conform with the distinction between logic of discovery and logic of confirmation. Evidence based modelling as well as Grounded Theory describes an inductive research methodology of an iterative process of constant comparison. Both methodologies consist of a process of increasing abstraction to gain a consistent and coherent representation of the most salient features of the target system. As it is the purpose of Grounded Theory, this approach enables to discover inductively a (middle range) theory of the field of investigation, starting from an idiographic description.

4 Two Examples

In the following, two examples will be consulted to demonstrate at concrete cases how simulation technologies can be utilised in a Grounded Theory research process. Both examples will highlight the role of simulation technologies at different stages of the research process. The first example will draw attention to the opportunities provided by software ontologies for the process of theoretical coding. The second example will draw attention to the role of simulation results for reaching a theory in a Grounded Theory process. Here the notion of stylised facts will be consulted.

4.1 Software Ontologies: An Example of the Contribution to Theoretical Coding

Ontologies are used in information systems and knowledge engineering for purposes of communication, automated reasoning and representation of knowledge. For this reason an ontology is defined as an explicit specification of a conceptualisation [14]. Following the formalisation of [15], this can be clarified using intentional logic. Without going into technical details, a *conceptualisation* can be defined as an intensional relational structure, consisting of a universe of discourse D, a set of possible worlds W and a set of conceptual relations R on the domain space <D,W>. This means that a number of objects D exist in the domain with certain relations R between them. Broadly speaking, the set of possible worlds is introduced in order that the domain does not only consist of the relations that have actually been realised but includes also possible relations. This is a first step of abstraction in reaching a theory from a description of the domain. However, the question of what are possible relations is not unequivocal. For this reason they have to be restricted to the intended models of the domain, i.e. to those relations that should be considered as relevant for the domain. Others need to be left out. The relevant relations need to be defined in a formal language L with a vocabulary V. This restriction of the scope of application is denoted as ontological commitment K. Again without outlining the technical details, an ontology can be defined as follows:

Let C be a conceptualisation, L a language with vocabulary V and ontological commitment K. An ontology O_K for C with vocabulary V and ontological commitment K is a logical theory consisting of a set of formulas of L, designed so that the set of its models approximates as well as possible the set of intended models of L according to K [15: 11].

Thus the formulation of an ontology enables *first* a formal precision of the description of domain of study. *Second*, intensional logic is used to achieve a theoretical generalization of the description. In the following it will be shown at an example how this contributes to the process of theoretical coding. It is drawn from the ongoing FP 7 research project GLODERS, which aims to investigate the global dynamics of extortion racket systems (ERS). The project is collaborative work of different project partners. The purpose of the GLODERS project is to develop an computational model for understanding the dynamics of ERS. The model development is informed in a first instance by empirical evidence based on detailed analysis of the Sicilian Cosa Nostra [16].

The history of the Sicilian Mafia goes back to the mid-19th century. The Mafia offered private protection and established a monopoly of violence in case of a weak state authority and a lack of civil society [17]. This triggered a view on the Mafia as a state-like organization [18]. However, in the 1970s and 1980s the Mafia radically enlarged its business segment with drug trafficking and other illegal economic activities. Economic enrichment has become a dominant motive [19– 22]. This went along with two Mafia wars in the 1960s and from the mid-1970s to 1980s [20]. This suggests to integrate the phenomenon in the discussion of transnational crime taking advantage of globalized markets [23]. However, evidence remains mixed. While factually the Mafia migrated to other territories, often this was not voluntarily and changed the structure and the course of action [24]. Extortion remains a local phenomenon. There exist evidence that Mafia is a territorially based phenomenon that is not easy to reproduce in a foreign territory [25,26]. Thus empirical evidence indicates that Mafia operates differently within different social environments. This can be summarized into the distinction between Mafia in a *traditional* environment, undertaking *systematic* extortion within a territory, and Mafia in an non-traditional environment, using opportunities of various types of legal and illegal economic activities and only occasionally undertaking *predatory* extortion. In contrast to non-traditional types, the Mafia in a traditional environment is perceived as a *legitimate* authority.

A first step in the development of a simulation model is the development of a shared domain ontology which provides the key terms and relations. Thus we need to define a space of discourse D and a set of relations R of our conceptualisation of ERS. A first sketch will be presented at a different talk at this conference. For the question of what ontology development contributes to the process of theoretical coding it need to be emphasised that this is an inductive and iterative process of refinement of the code, starting from data analysis to ontology development and going back to the data. Ontology development enables to identify gaps in the data basis, which suggests to gather new data. In the following, some questions will be outlined which so far arose in the ongoing process of ontology development to demonstrate how disentangling a verbal description into a formal and explicit specification of an ontology facilitates the process of *theoretical* coding:

- A result of a first investigation, undertaken by Sicilian project partners, is that it is a distinctive feature of the Mafia in a traditional environment that it has been perceived to some degree as a legitimate ruling authority over the Sicilian territory. However, beside the fact that the question of legitimacy is a highly normative question, it is also empirically not easy to disentangle the verbal concept of legitimacy into behavioural terms and relations between interacting individuals. Currently lab experiments are undertaken to investigate the underlying socio-psychological mechanisms. Thereby the ontology development enables a triangulation between different empirical research methods.
- A result of the first empirical field work is a distinction between a traditional and non-traditional environment which are differentiated by different normative systems. However, it need to be clarified how these normative systems are characterised. The development of an ontology requires to disentangle the multiple dimensions of the concept of a traditional and non-traditional environment and to stimulate further empirical research.
- One important dimension of the normative system of a traditional environment is that on the one hand, Mafiosi are well-known, while on the other hand Mafia is nevertheless a secret organisation. This indicates that the concept of reputation is required in an explicit micro-specification of the notion of a traditional environment. Mechanisms of gathering reputation need to be investigated in further empirical field work.
- Beginning with the second Mafia war in the 1970s and 1980s Anti-Mafia movements emerged in Sicily. This is a cultural change, indicating a loss of legitimacy of the Mafia authority. An ontology development requires to spell out such a cultural change in terms of micro mechanisms to identify the core dimensions.

Interviews will be undertaken to disentangle the cognitive components of cultural change. This will lead to a refinement of the code which will then be adjusted to textual data about (the change of) Mafia action until a stage of theoretical saturation is reached.
- Finally, the mechanisms of transformation of Mafia type organisations and their adaptation to different environments need to be specified. To explicitly specify the mechanisms of Mafia migration or the conditions which foster the emergence of Mafia type organisation, data from different regions such as the Netherlands and Germany is under investigation. This will then lead to further refinement of the code.

These examples demonstrate how the development of an ontology facilitates the process of *theoretical* coding. First empirical fieldwork enables the identification of core concepts of the domain. The explicit specification of an ontology allows to identify open questions. This stimulates an iterative process of gradually refining verbal concepts into a formal explicit specification of micro-mechanisms. Moreover, the ontology provides a platform for a triangulation of various empirical methods and data sources.

4.2 Stylised Facts: An Example of the Contribution to Theoretical Sensitivity

The second example will shed light on how simulation experiments contribute to the formulation of theoretical statements, resulting from simulation results. Here emphasis will be put on the concept of stylised facts.

The term stylised facts had been coined by [27] in macroeconomic growth theory. [11] demonstrated that it can be applied beyond economic analysis to the evaluation of simulation results. The central tenet of stylised facts is "to offer a way to identify and communicate key observations that demanded scientific explanation" [11:2.2]. For this purpose 'stylised facts' denote stable patterns that can be found throughout many contexts. Thus details of concrete empirical cases are left out in favour of a broad description of tendencies that have been proven as robust patterns that can be discerned in a certain class of phenomena. This can be used for a cross-validation to check whether the micro assumptions put in a simulation model reveal broad patterns in the simulation results that are characteristic for the field of investigation [28]. With regard to the question of what is a theory in a Grounded Theory approach, this provides an additional source of credibility: If the transformation of the description of a certain field case into a source code of a simulation model generates simulation results which are consistent with broad patterns that can be found in a number of contexts, then this would be an indicator of a theoretic insight generated by the inductive process. In the following this will be illustrated at a concrete example.

The example is a model that investigates the escalation of ethno-nationalist tensions into open violence. The evidence has been drawn from the case of the former Yugoslavia. The puzzling question is how and why neighbourhood relations between people with different ethnic background changed from good and peaceful relations to traumatic violence. A simulation model has been developed to study the dynamics of the escalation of tensions into violence. Model assumptions had been based on the empirical evidence of historical narratives of this much studied case [29–32]. The evidence suggests a two-level design of the model, namely to specify the mechanisms of the escalation dynamics as a recursive feedback between political actors and a socio-cultural dynamics at the population level. While a focus purely on the population level [33] mask the responsibility of political actors, explanations that focus purely on the political level [31] need to explain why certain politics had been successful. Integrating both accounts generates a self-organised feedback cycle of political actors and attitudes. On the one hand politicians mobilise value orientations in the population to get public support. On the other hand, politicians appeal to the most popular value orientations in order to maximise the support. In abstract terms, the feedback relation can be described as a recursive function.

The simulation model cannot be explained here [34]. Simulation experiments had been undertaken with the Bayesian assumption of complete ignorance about the empirical distribution of the political attitude of the citizens and the political agenda of the politicians. Initially both are determined by chance. This is a theoretical experiment to study the pure effect of the feedback cycle. The model is only calibrated at the population census of 1991. Thus differences in the dynamics are due to differential population distribution. The results reveal theoretical insights by generating stylised facts of two basic mechanisms of the escalation dynamics. The first mechanism concerns political processes, the second mechanisms concerns micro processes of neighbourhood relations. The inter-penetration of the processes reveals a sequential ordering:

- *First,* visibility of the political appeals plays an essential role for radicalisation. Initial radicalisation attempts stimulate counter-radicalisation. This is driven by the political level. Its effectiveness accounts for a rather homogeneous population. Ethnically mixed populations such as in Bosnia-Herzegovina provide more power of resistance against political radicalisation prior to the outbreak of actual violence. This casts doubt on the widespread scepticism about the political stability of multi-ethnic republics such as Bosnia-Herzegovina.
- *Second,* refugees and rumours play an essential role to reach a stage of self-perpetuating radicalisation. This holds for the later radicalisation in Bosnia-Herzegovina. Here, radicalisation is imported from outside. In contrast to the early radicalisation, the radicalisation is driven predominantly by the population level. In this process, dense networks increase the likelihood of the spreading of radicalisation. A comparative analysis shows that once a stage of self-perpetuating micro-level radicalisation is reached, multi-ethnic societies are in danger of much more intensive violence than homogeneous societies.

These results are not limited to an idiographic description of the particular case of the former Yugoslavia but may also hold for other cases. Thus a simulation based on the empirically grounded evidence of a certain case enables to generate stylised facts which reveal a middle range theory of mechanisms of ethno-nationalist radicalisation.

5 Conclusion

The paper argues that the theoretical element in a Grounded Theory approach can be strengthened by supplementing the methodology of Grounded Theory with evidence based simulation. This is demonstrated at two examples: first, an example of an ongoing research project is provided. It is shown that the development of an ontology refines the process of theoretical coding by providing a means to detect gaps in the data and to specify the intended model, i.e. the scope of the domain. This facilitates the recursive iteration between data gathering and coding. Second, at the example of the escalation of ethno-nationalist conflicts it is shown how simulation results of an evidence based model generates stylised facts. If a model based on the evidence of an empirical case generates broad patterns that are characteristic for the field, then this shows that a process starting with an idiographic description succeeded to generate a theory. This can be done with the help of simulation experiments. This contributes to a clarification to the precarious relation between a mere description and a truly theoretical Grounded Theory.

Acknowledgment. This work has been undertaken as part of the Project GLODERS, in the funding scheme FP-7 ICT, Grant no. 315874.

References

1. Lorscheid, I., Heine, B.O., Meyer, M.: Opening the 'black box' of simulations: increased transparency and effective communication through the systematic design of experiments. Comp. and Math. Org. Theory 18(1), 22–62 (2012)
2. Squazzoni, F.: Agent Based Computational Sociology. Wiley, Chichester (2012)
3. Kelle, U.: "Emergence" vs. "Forcing" of Empirical Data? A Crucial Problem of "Grounded Theory" Reconsidered. Forum Qual. So. 6(2), Art. 27 (2005), http://www.qualitative-research.net/index.php/fqs/article/view/467/1000
4. Strübing, J.: Zur sozialtheoretischen und epistemologischen Fundierung des Verfahrens der empirisch begründeten Theoriebildung. VS Verlag, Wiesbaden (2004)
5. Strauss, A., Corbin, J.: Basics of qualitative research: Techniques and procedures for developing grounded theory. Sage, Thousand Oaks (1998)
6. Barreteau, O.: Our compagion modelling approach. J. Art. Soc. and Soc. Simulation 6(1) (2003), http://jasss.soc.surrey.ac.uk/6/2/1.html
7. Flick, U.: Qualitative Sozialforschung. Eine Einführung. Rowolth, Hamburg (2002)
8. Glaser, B., Strauss, A.: The discovery of Grounded Theory. Strategies for qualitative research. Aldine, Chicago (1967)
9. Glaser, B.: Theoretical sensitivity. Advances in the methodology of Grounded Theory. Sociology Press, Mill Valley (1978)
10. Birks, M., Mills, J.: Grounded Theory a practical guide. Sage, London (2011)
11. Heine, B., Meyer, M. Strangfeld, O.: Stylized facts and the contribution of simulation to the economic analysis of budgeting. J. Art. Soc. and Soc. Simulation 8(4) (2005), http://jasss.soc.surrey.ac.uk/8/4/4.html
12. Edmonds, B., Moss, S.: From KISS to KIDS: An anti-simplistic modelling approach. In: Davidsson, P., Logan, B., Takadama, K. (eds.) MABS 2004. LNCS (LNAI), vol. 3415, pp. 130–144. Springer, Heidelberg (2005)
13. Yang, L., Gilbert, N.: Getting away from numbers: Using qualitative observation for agent-based modelling. Adv. in Complex Syst. 11(2), 175–185 (2008)

14. Studer, R., Benjamins, R., Fensel, D.: Knowledge engineering: Principles and methods. Data & Knowledge Eng. 25(1-2), 161–198 (1998)
15. Guarino, N., Oberle, D., Staab, S.: What is an ontology? In: Staab, S., Studer, R. (eds.) Handbook on Ontologies, pp. 1–17. Springer, Heidelberg (2009)
16. Scaglione, A.: Reti Mafiosi. Cosa Nostra e Cammorra. Organizziazoni criminali a confronto. FrancoAngeli, Milano (2011)
17. Franchetti, L.: Condizioni politiche ed administrative delle Sicilia. Vallecchi, Florence (1876)
18. Nozick, R.: Anarchy, state and utopia. Basil Blackwell, Oxford (1974)
19. Gambetta, D.: The Sicilian Mafia. The business of private protection. Harvard University Press, Cambridge (1993)
20. Dickie, J.: Cosa Nostra. Die Geschichte der Mafia. Fischer, Frankfurt a.M. (2006)
21. Gambetta, D., Reuter, P.: Conspiracy among the many. The Mafia in legitimate industry. In: Fiorentini, G., Peltzman, S. (eds.) The Economics of Organized Crime. Cambridge University Press, Cambridge (1995)
22. Paoli, L.: Mafia Brotherhoods. Organized crime Italian style. Oxford University Press, Oxford (2003)
23. Shelley, L.: The Globalization of Crime and Terrorism. EJournal USA, 42–45 (February 2006)
24. Varese, F.: Mafias on the Move. The Globalization of Organized Crime. Princeton University Press, Princeton (2011)
25. Campana, P.: Eavesdropping on the Mob: the functional diversification of Mafia activities across territories. Europ. J. of Crim. 8(3), 213–228 (2011)
26. Varese, F.: How Mafia take advantage of globalization. The Russian Mafia in Italy. Brit. J. of Crim. 52, 235–253 (2012)
27. Kaldor, N.: Capital Accumulation and Economic Growth. In: Lutz, F., Hague, D. (eds.) The Theory of Capital, pp. 177–222. St. Martin's, London (1961)
28. Moss, S., Edmonds, B.: Sociology and Simulation: Statistical and qualitative cross-validation. Am. J. of Soc. 110(4), 1095–1131 (2005)
29. Bringa, T.: Being muslim the Bosnian way: Identity and community in a central Bosnian village. Harvard University Press, Cambridge (1995)
30. Silber, L., Little, A.: Yugoslavia: death of a nation. Penguin, New York (1997)
31. Gagnon, V.: The myth of ethnic war. Serbia and Croatia in the 1990s. Cornell University Press, London (2004)
32. Calic, M.: Der Krieg in Bosnien-Herzegowina. Surkamp, Frankfurt a. M. (1995)
33. Horowitz, D.: The Deadly Ethnic Riot. University of California Press, Berkeley (2001)
34. Markisic, S., Neumann, M., Lotzmann, U.: Simulation of ethnic conflicts in former Yugoslavia. In: Troitzsch, K., Möhring, M., Lotzmann, U. (eds.) Proceedings of the 26th European Simulation and Modelling Conference 2012, Koblenz (2012)

Generating an Agent Based Model from Interviews and Observations: Procedures and Challenges

Tilman A. Schenk

University of Leipzig, Department of Geography, Germany
tschenk@rz.uni-leipzig.de

Abstract. In the course of an increasing impetus to connect agent based simulations to empirical data, also the potentials of qualitative social science methods to inspire such models are explored. In this work, qualitative interviews, participating observation, and document analysis are combined to analyze a political process that relies heavily on the communication and the collaboration of stakeholders to serve as a text based data source for an agent based model of the process. The simulation outcomes are also produced in text format so that they are easily understandable to stakeholders and other users. The simulation reproduces the discussions among the stakeholders and their subsequent decisions, is able to react on changes in their general settings, and can be used to explore different sets of rules for the decision processes and their results.

Keywords: Empirical modeling, qualitative reasoning, communication, cooperation, narratives.

1 Introduction

During the past two decades, the adoption of communicative and participatory practices in the planning discourse has made substantial impact on regional policy strategies that are consequently increasingly formulated as stakeholder oriented learning and governance processes [1], [2]. Mostly resulting rather in qualitative objective statements than measurable objective parameters, their analysis and assessment thus suffers from a methodological dilemma: While quantitative approaches often struggle with low data availability or lacking measurability of the processes in question, the results of qualitative inquiries are hard to generalize and are often – at least outside the academia – perceived as imprecise.

This leads to the postulation to develop sets of methods that will be able to integrate these complexities and at the same time lead to precise evaluation statements. Agent based models prove advantageous in that respect as they do not require to exclude either one of the approaches for purely technical reasons. Furthermore, they are able to represent communication and learning processes among the modeled stakeholders from their own point of view without general simplifying assumptions. The specifications of the underlying model do not have to rely on quantifications, but can be formulated on text basis. This makes them accessible to the results of a rich variety of well-established qualitative empirical methods in the social sciences [3].

B. Kaminski and G.Koloch (eds.), *Advances in Social Simulation*,
Advances in Intelligent Systems and Computing 229,
DOI: 10.1007/978-3-642-39829-2_32, © Springer-Verlag Berlin Heidelberg 2014

Thus, a combination of qualitative methods and social simulation could prove to be an ideal tool for the analysis and evaluation of communicative planning processes. However, there is no broad and general understanding of how to transfer qualitative research results into agent rules [4], [5]. This paper aims at contributing to fill this gap by combining three qualitative methods to produce evidence of a political process, and then demonstrating how this qualitative evidence can be turned into action rules for agents of a simulation. The simulation should not only be able to reconstruct past discussions, but also provide a tool to explore different institutional settings in which these discussions take place.

The remainder of this paper is organized as follows. The next section will describe the setting and the details of the political process analyzed. A methodology section will then focus on the selection of the stakeholders interviewed and the coding of interviews and observation results. A larger section 4 will describe in detail how the findings from the empirical phase were transferred into action rules for the agents. A short discussion of the results will conclude the article.

2 The Setting

Current development programs for rural areas increasingly emphasize the collaboration of regional stakeholders in an attempt to turn planning into integrative and participatory processes that have been widely discussed and termed as governance approaches [1], [2]. This has made substantial impact on regional policies for rural areas in the European Union such as in the LEADER (Liaison entre actions du développement de l'économie rurale) programs.

The idea of the LEADER program is to bring together stakeholders of different sections (political, economical and social) to define and pursue development goals for their rural region. Cooperation between the sectors is strongly encouraged in designing projects that will support the achievement of the defined goals. The project ideas have to be submitted to a regional assembly (RA) consisting of representatives of all the three sectors, who discuss the projects and decide about the spending of funds to support them. Although European, the local implementation can vary due to differing national regulations as well as more informal understandings of planning in general. Furthermore, the social practices performed within the external regulations differ from one another and will be described in more detail in the following sections.
Two study areas were chosen in Eastern Germany and Northern Sweden that both implement EU-funded, communication based development programs for rural areas in the LEADER framework. The two study areas are comparable as they are both rural and sparsely populated and containing a regional centre, even though these characteristics appear on different scales in the two countries.

The meetings are organized by the "regional management", an institution that consists of usually one or two persons who consult stakeholders in designing projects, send out invitations to the meetings and supervise the discussion and decision process.

The social processes relevant for this study are described in more detail in the following chapter, after a brief introduction to the empirical methods used for their analysis.

3 Data Collection and Analysis

3.1 Methods

While there are techniques to include the stakeholders involved in the studied processes in the model building [6], [7], methods requiring less stakeholder involvement were applied. The intention was to focus on more widely used methods such as interviews and observation, in order to explore the extent to which they were suitable to generate evidence for an agent simulation. Qualitative methods in general, and the ones applied here in particular produce narratives, i.e. text based information about human actions and mental concepts of their environs. Three such methods were combined for this study:

1. Participating observation was conducted by a researcher attending several meetings of the regional assembly and recording the course of the discussions about the submitted projects. The analysis focused mainly on the arguments used in the discussions and identifying the features of submitted projects that were regarded important by the discussants. The observation method is termed as 'participating', because it seems fair to assume that the presence of the researcher cannot be unnoticed in a group of 20-25 persons. However, the researcher did not actively take part in the discussions, in order to minimize the influence of the observer on the processes taking place.
2. A number of qualitative interviews with members of the regional assembly served as the second empirical source. The interviewees were chosen equally from the three stakeholder subgroups. The interviews focused on issues that were not apparent from the observations of meetings, such as the arrangements of networks between the stakeholders (emergence and maintenance of relations), their ways of collaboration in the assembly (course of the meetings, their personal appraisal, and decision processes), and finally their own development goals for the region. For the analysis, the audio recordings of the interviews were transcribed into written text.
3. Because observations rely to a considerable extent on the interpretations of the researcher, and the interviews focused on those of the assembly members, both groups of findings were controlled and underpinned by analysis of the official documentation of the work of the regional assembly.

While observations served to grasp the regularities in the stakeholders' actions, the interviews were capable of generating information about the motivations behind those actions. Another advantage of this combination of methods is that it helps to identify discrepancies between stakeholders' self reflections as expressed in the interviews and their overt behavior. However, due to the object oriented programming of agent models, these contradictions may well be a part of the resulting simulation model.

3.2 Observation Results Concerning Assembly Meetings

Using the observation protocols and partly also interview transcripts, the process of discussions and decisions could be captured. Figure 1 illustrates the prevalent

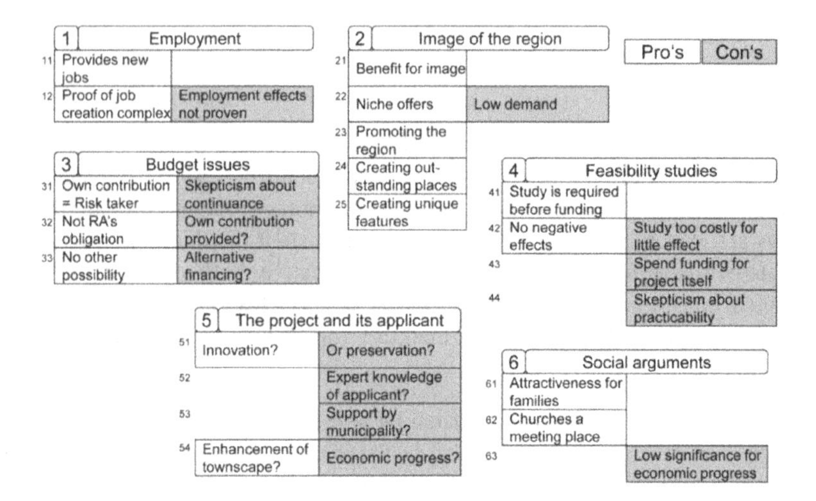

Fig. 1. Categories of discussion arguments

arguments used in the discussion (exemplary for the German region), organized into six categories.

Within these categories, discussion arguments can furthermore be subdivided into "pro's" and "con's". For clarification, consider category 3 ("financing"): Oftentimes, skepticism about the survival of a project is expressed. Some members suspected that funding may be invested in a project that will not achieve the desired effects and will then be abandoned. In opposition to this argument is the notion that due to the own contribution of the applicant, she/he participates in both success and failure and hence has an own interest in the survival of the project (line 31 in figure 1). Not all categories equally provide "pro" and "con" arguments. For instance, when talking about the "image of the region", the assembly members argue widely positively.

A similar scheme of discussion arguments and their categories has been deployed for the Swedish case study. It also accounts for the fact that the discussions are structured quite differently from the German case. While in the latter, the entire assembly exchanges arguments and then makes a decision, in Swedish assembly first discusses the projects in the three subgroups (political, economic, social) of members and each of those suggests a vote. A following joint discussion then prepares for the final vote. A further result of the observation was that the types of arguments that appeared in the discussion were dependant on the category of the project discussed. For instance, in the German case road improvement projects were even approved without discussion. For the Swedish stakeholders however, road improvements were not included in their understanding of regional development, which is why such projects are not even admitted to the discussion.

3.3 Results of the Qualitative Stakeholder Interviews

Because functioning and trust based networks are considered important in communicative planning processes, the interviews focused on the construction, the development and the current quality of the existing networks. Three sorts of networks have been identified by the stakeholders (in both case studies): The first group of contacts results from ties before the existence of the LEADER region. Especially in rural areas, this sort of strong local involvement is typical, but is often rather a means of identity building than of actual significance for the LEADER process. The second group of contacts does play an important role in this, but did not originate from it. Bindings of political stakeholders often fall in this category, as they have further means of communication and opportunities for exchange apart from the LEADER institutions, which are often perceived as additional structures. In case those ties are stronger than the relatively fresh ones in the LEADER context, incentives for new bindings are low, remain arbitrary and do not lead to stronger networks. Finally, a strikingly high number of stakeholders denoted themselves as "outsiders" or "newcomers" to the process, leading to a suspicion of being generally cut off from procedures.

Another important aspect of the interviews was to address the development goals of the stakeholders for their region. These were connected to the aspects of the submitted projects that assembly members regarded important for the region and that made significant contributions to convince them in their decision. They were complimented from the observation results of the meetings and assigned to categories. Again, the German case serves as an example for figure 2:

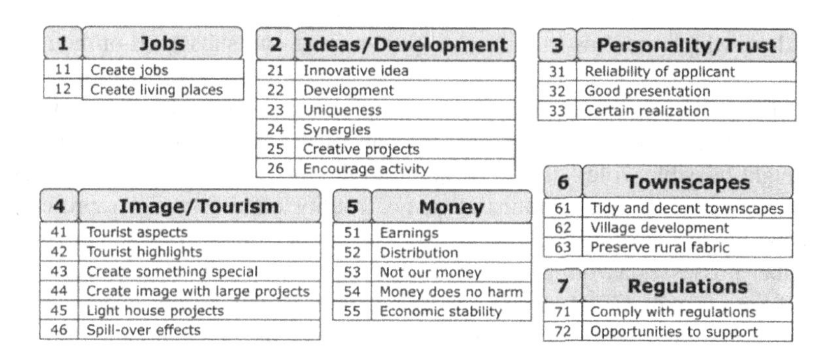

Fig. 2. Categories of stakeholders' development goals

The relative importance of the categories "ideas/development" (2) and "image/tourism" (4) highlight the emphasis on innovative and highly visible projects in the LEADER approach. Two of the categories may deserve further elaboration: "personality/trust" (3) refers to the confidence of the decision makers that a submitted project will be realized, depending to some extent also on a personal impression of the applicant. The category "money" (5) summarizes some quite differing economical arguments: While "earnings" (51) and "economic stability" (55) refer to the survival chances of a project, "distribution" (52) and "not our money" (53) reveal a desire to allocate funds as quickly as possible.

Just like the others, the category "townscapes" (6) appeared in interviews in both study regions, this one however in quite contrary sense. While it was an important aspect of LEADER funding in Germany to preserve the rural settlements ("tidy and decent townscapes" (61)), the majority of the stakeholders in the Swedish region were frustrated with spending funds for renovating and modernizing rural building structures simply to replace owners' commitment to their real estate, or that soon afterwards decayed.

4 Transferring the Analyzed Communicative and Behavioral Regularities into Activity Rules for an Agent Model

This section will now focus on the derivation of agent rules for the simulation from the results of the qualitative analysis of the communication structures and regularities in the stakeholders' actions. The procedure follows an ontology developed by [8] that requires defining 'endurances' (physical and non-physical objects and constructs existing invariantly throughout the time span of the analysis) and perdurances (processes and constructs changing over time such as actions or events). In this case, endurances doubtless include the submitted projects and the involved stakeholders. Furthermore, also the development goals of stakeholders and their discussion arguments fall into this category, as empirical evidence from the observation of the discussions suggested that they remain relatively stable over the time of the analysis. Which of the available arguments were applied in the discussions however, is a result of the decisions of the stakeholders, which are thus to be termed perdurances and subsumed in their activity rules.

As endurances, two agent classes[1] were implemented, labeled as 'regional managament' and 'regional stakeholder'. Each agent of the latter class adopts three of a total of eight possible 'roles' that correspond to the involved stakeholder groups (political, economical, social, administrative i.e. county administration representatives supervising the legal consistency of the process) and also indicate whether an assembly member possesses a voting right, which is regulated by the internal rules of procedure of the assembly. This also allows agents to assume multiple roles; in one case for instance, a hotel owner, officially a representative of the economic sector in the assembly, is also the mayor of her municipality, and hence inherits activity rules for both roles. The remaining endurances (projects, discussion arguments and development goals) are implemented as 'resources' classes. 'Resources' in the terminology of the used software refers to objects possessing state variables but no activities.

The process of transferring the empirically identified perdurances into agent activity rules follows a three-step procedure of empirical evidence, its interpretation and derived agent rules, which will be presented in table format as suggested by [4]. This will be performed separately for the issues contacts/meetings/networks of stakeholders, discussions and decisions in the assembly, and project design/stakeholder cooperation.

---

[1] The simulation is implemented in the software „SeSAm" developed by the Artifical Intelligence working group at the University of Würzburg, Germany. More information can be obtained at www.simsesam.de.

4.1 Contacts, Meetings, and Networks of Stakeholders

Networks are termed endurances in the ontology by [8] as they are non-physical entities that persist over the time span of the analysis, although their shape and intensities may vary. For this study, it is presumed that networks evolve through contacts that may either be physical by meeting other agents at assembly sessions or non-physical by phone or mail. One implication from interview statements is that network ties become more intensive with higher contact frequencies (table 1). The frequencies of contacts can be derived from interview statements as well (table 2).[2] From these rules, average contact frequencies were deduced and converted into daily contact probabilities.

Every agent saves a list of other agents met, with those met most recently being tossed to the top of the list. The position that a particular agent holds in a list of a fellow agent may hence be interpreted as the intensity of the tie between these two agents. Note that they are also directed, i.e. one agent may have a strong tie to another agent, while the latter has stronger ties to other agents than to the former. This procedure is applied for all types of contacts, within and outside of the regional assembly meetings.

Table 1. Selected interview statements concerning network ties and their transformation into agent rules

Interview statement	Interpretation	Agent rules
"With stakeholders from the neighboring municipalities, I have only few contacts. I'm a man in the economy, not represented in the municipality. I did make contacts here and there, but it's hard to keep them. I feel like an outsider sometimes." (economic stakeholder) "We mayors often meet in advance and talk about these things. We mayors have strong ties due to the many other opportunities where we meet." (political stakeholder) "But just for the meetings every four or six weeks, I don't have the time and the desire to build up relations with these people." (political stakeholder)	The frequency of contacts has an influence on the intensity of network ties.	Ties will be stronger for agents meeting more often.

[2] The interviews were conducted in German or Swedish, the original statements were translated into English for this paper by the author.

Table 2. Selected interview statements concerning contact frequencies and their transformation into agent rules

Interview statement	Interpretation	Agent rules
"I only meet the others once a month when we have the assembly" (economic stakeholder) "So I'm also in the county council, that brings us together every two weeks for different purposes" (political stakeholder) "There is not a single day without contact to one of the other mayors" (political stakeholder)	• Political stakeholders meet each other more often than the other groups • Time intervals between meetings range from daily to once a month	Meeting events are simulated in different frequencies for each agent group

Assembly meetings are organized by the Regional Management. Analysis of the documents revealed that they were always summoned for the second Wednesday of a month, in case a sufficient number of projects were ready to be decided upon. The Regional Management agent sends out invitations to the assembly member agents by inserting the date in their schedules, whereupon they decide whether to attend. Again, from the assembly documents it was evident that its members chose not to attend at an average rate of 20% of the meetings, which was used as a general no-show probability. Thus, the present assembly members are defined and their network lists are updated using the above mentioned mechanism.

4.2 Discussions and Decisions in the Regional Assembly

In the simulated discussions, the present assembly member agents bring forward arguments from the pool of available arguments (fig. 1). The course of the discussion is logged by the regional management agent. Because a clear assignment of argument categories to groups of agents is not empirically supported, the starting point of discussions is chosen at random. However, depending on the category of the project in question, arguments are preselected in a way that is consistent with the meetings' observation results. Figure 3 shows a simulated course of discussion as an example.

When the discussion has come to an end, the assembly members entitled to vote decide about each project application. A project is approved if the number of supporting votes is greater than that of dissenting votes. In this case, the application is removed from the list of pending projects and its budget subtracted from the total budget available.

<table>
<tr><td>Project ID</td><td colspan="2">20101114</td><td colspan="2">Project category</td><td>Feasibility study</td></tr>
<tr><td>Initiator</td><td colspan="2">P6</td><td colspan="2">Partner</td><td>P2</td></tr>
<tr><td colspan="6">Course of Discussion in Regional Assembly</td></tr>
<tr><td colspan="6">• P8 says: Study is required before funding (41+)
o E1 replies: Study too costly for little effect (42-)
• S2 says: Study has no negative effects (42+)
o S5 replies: Study has no negative effects (42+)
o A4 replies: Skepticism about practicability (44-)
• S3 says: Study is required before funding (41+)
o S4 says: Study is required before funding (41+)
o A2 replies: Skepticism about practicability (44-)
• P8 says: Spend funding for project itself (43-)</td></tr>
<tr><td>Date</td><td>08-12-2010</td><td>Supporting votes</td><td>13</td><td>Dissenting votes</td><td>3</td></tr>
</table>

Fig. 3. Simulated course of discussion in the regional assembly (anonymized). Key to stakeholder symbols: P=political, E=economic, S=social, A=administrative. The numbers indicate the position of the argument in fig. 1; '+' and '-' refer to positive and negative arguments respectively.

In the assembly meetings observed, projects were seldom turned down. Asked to give a reason for that, the interviewees stated that so far the budget was far from being exhausted, which is why the representatives felt no urge to prioritize projects and direct funding. However, from some of the interview statements it became evident that the stakeholders were conscious of a need to prioritize in the future:

- "If we had more applicants, so that the budget will become tight, we will have to talk about how to establish priorities." (Political stakeholder)
- "But when we are told that this is your budget, and there isn't any more, then we will have to trade off which of the projects we can support and which we cannot." (Social stakeholder)

So for the simulation it appeared worthwhile to assume dynamic decisions influenced by the course of discussion beforehand. That way, the decision results in the simulation are opened for the design of alternative discussion and decision scenarios, which aim at supporting decision mechanisms in situations where the available budget is shrinking and projects will have to be prioritized. An impact of the discussions on the decisions is indicated in several of the interviews. In order to maintain a consistency of decisions and discussions inside the simulation, the agents decide predominantly according to the proportion of the own 'pro' and 'con' arguments in the discussion to support or dissent a project. Agents that had been silent during the discussion direct their vote according to the total numbers of 'pro' and 'con' arguments placed, thus being 'convinced' in the course of discussion.

4.3 Project Design and Stakeholder Cooperation

The idea of the LEADER process is to link stakeholders and activities to initiate cooperation among those involved, so that they will generate new ideas resulting in projects, evaluate their own agendas, and draw their own conclusions from that, resulting in what could be termed a learning cycle (fig. 4).

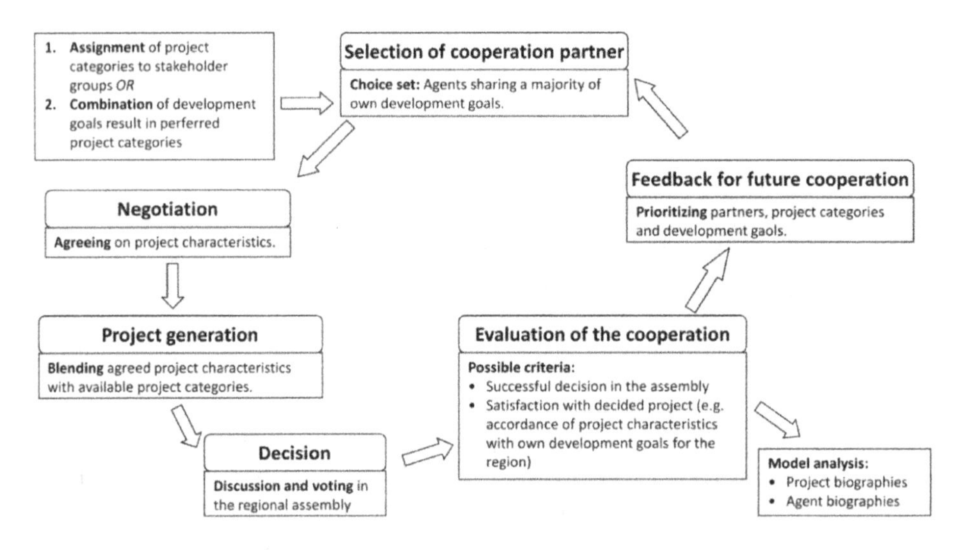

Fig. 4. Learning cycle concerning agent cooperation

Unlike the previous, this simulation process is not a replication of past events, but rather an idealized sketch of possible future developments. Some of the elements of the learning cycle have been supported by empirical evidence, such as the project discussions and decisions (fig. 3) or the development goals of the stakeholders that form the basis for cooperation (fig. 2).

Although numerous simulations of negotiating processes between agents exist, the social science research about the design of such models is in an early state [9]. In many cases, such simulations start with a number of assumptions that are successively brought in line with empirical evidence. One of the challenges in this process results from the finding that some of the influences on negotiation processes such as emotions, personal affection, social control, and norms are harder to connect to simulation rules than overt behavior. This led to a similar strategy in this case: using a few plausible assumptions, the feasibility of simulating this learning cycle is shown, and subsequently opportunities for empirically inspired improvements are identified.

Since a distinct assignment of development goals to specific stakeholders or groups of stakeholders was not empirically evident, each agent received a random sample of one third of the available goals (fig. 2). Also, the extent to which stakeholders will have to agree on these goals to trigger cooperation could not be identified empirically. Again, this is replaced by a plausible rationale: Agents agreeing in none of the development goals will not cooperate. Since interview statements about development goals always referred to more than one category in figure 2, agreement in one category is quite likely and thus not a reasonable basis for cooperation. The lowest possible requirement thus would be an agreement in at least two of the goal categories.

After the selection of possible cooperation partners, the agents send cooperation queries to their possible partners in distinct time intervals that are replied positively or negatively depending on their degree of agreement concerning the development goals. Agents with higher shares of positive replies are selected for cooperation. The final step is to have the agents design a project together. It is assumed that agents intersect their goals and subsequently select a project category that matches that intersection best. At the end of this process, the agents may evaluate their cooperation e.g. by the number of projects that they designed with a particular partner or by the share of their own development goals that the designed projects achieved.

5 Discussion

The objective of this paper was to investigate and model a political process that involves a number of stakeholders with their respective viewpoints, and which relies fundamentally on the communication and collaboration among them (a detailed description of the results may be found in [10]). The achieved simulation results should be validated, i.e. be compared to the empirical evidence from the social phenomena investigated [11].

It became evident that the text based representations of actions of the stakeholders can in principle be transferred into rules of an agent based simulation model, in part requiring substantial empirical effort. This may be illustrated by means of some of the model elements: Empirical evidence suggested that political stakeholders had closer relations with each other than with stakeholders from the other groups, explaining this fact by their more frequent meetings. This interrelation could be exactly represented in the simulation. Also, the simulated discussions in the regional assembly showed a high consistency with the empirical data of their real world counterparts.

Discrepancies were detected concerning the voting behavior: While according to the observation of the assembly meetings the stakeholders decided the projects predominantly concordantly positive, the interviews revealed that stakeholders were themselves skeptical about this procedure, because it resulted solely from a relatively low usage of the available budget in the first half of the period. They hence expected a need to prioritize and thus reject funding for at least some of the projects in the near future. This served as a justification to direct the simulation model towards addressing this prospective challenge for the stakeholders by formulating a dynamic decision process depending on the course of the discussions. This setup also served as a base model to experiment with the variability of the decisions in a number of different discussion settings.

However, some methodological challenges remain. While role playing games can monitor the overt behavior of stakeholders in specific situations and thus serve as a data source for a simulation, these situations will always be artificial in their setting. When using interviews, one has to be aware that the resulting narratives are nothing but linguistic representations of the interviewees' reflections rather than behavioral data. Therefore, it proved worthwhile to accompany the interview findings with techniques of observation and document analysis. Nevertheless, the process of data collection remains selective in the sense that not all involved stakeholders could be interviewed and not all meetings could be observed due to practicability reasons. It proved challenging to generate all information desired for the modeling process as well. From the methods used and data acquired it was not obvious which constellations of stakeholder goals triggered their cooperation. Although some basic mechanisms were identified, their situational occurrence varied to a larger extent.

A strong advantage of using the empirical methods used is however, that they are widely used and easily applicable. More interviews and observation can always be conducted to add to the information available and enhance the validity of the model.

References

1. Amdam, R.: Empowerment Planning in Regional Development. European Planning Studies 18, 1805–1819 (2010)
2. Healey, P.: The Communicative Turn in Planning Theory and Its Implications for Spatial Strategy Formation. Environment and Planning B 23, 217–234 (1996)
3. Janssen, M.A., Ostrom, E.: Empirically-Based, Agent-Based Models. Ecology and the Society 11, 37 (2006)
4. Polhill, J.G., Sutherland, L.-A., Gotts, N.M.: Using Qualitative Evidence to Enhance an Agent-Based Modelling System for Studying Land Use Change. Journal of Artificial Societies and Social Simulation 13(2), 10 (2010)
5. Robinson, D.T., Brown, D.G., Parker, D.C., Schreinemachers, P., Janssen, M.A., Huigen, M., Wittmer, H., Gotts, N.M., Promburom, P., Irwin, E., Berger, T., Gatzweiler, F., Barnaud, C.: Comparison of empirical methods for building agent-based models in land use science. Journal of Land Use Science 2(1), 31–55 (2007)
6. Barreteau, O.: Our Companion Modelling Approach. Journal of Artificial Societies and Social Simulation 6(2), 1 (2003)
7. Guyot, P., Honiden, S.: Agent-Based Participatory Simulations: Merging Multi-Agent Systems and Role-Playing Games. Journal of Artificial Societies and Social Simulation 9(4), 8 (2006)
8. Gotts, N.M., Polhill, J.G.: Narrative Scenarios, Mediating Formalisms, and the Agent-Based Simulation of Land Use Change. In: Squazzoni, F. (ed.) EPOS 2006. LNCS, vol. 5466, pp. 99–116. Springer, Heidelberg (2009)
9. Hollander, C.D., Wu, A.S.: The Current State of Normative Agent-Based Systems. Journal of Artificial Societies and Social Simulation 14(2), 6 (2011)
10. Schenk, T.A.: Using stakeholders' narratives to build an agent based simulation of a political process. Simulation Transactions (submitted)
11. Ormerod, P., Rosewell, B.: Validation and Verification of Agent-Based Models in the Social Sciences. In: Squazzoni, F. (ed.) EPOS 2006. LNCS, vol. 5466, pp. 130–140. Springer, Heidelberg (2009)

Defining Relevance and Finding Rules: An Agent-Based Model of Biomass Use in the Humber Area

Frank Schiller[1], Anne Skeldon[1], Tina Balke[1], Michelle Grant[1], Alexandra S. Penn[1], Lauren Basson[2], Paul Jensen[1], Nigel Gilbert[1], Ozge Dilaver Kalkan[1], and Amy Woodward[1]

[1] University of Surrey,
Guildford, Surrey, GU2 7HX, United Kingdom
{f.schiller,a.skeldon,t.balke,m.grant,a.penn,
p.jensen,n.gilbert,o.dilaverkalkan,a.woodward}@surrey.ac.uk
[2] The Green House,
Ubunye House, 70 Rosmead Avenue,
Kenilworth, 7708, South Africa
basson.lauren@gmail.com

Abstract. The field of industrial ecology applies ecosystem theory to industrial production, human consumption and societies. This article presents a case study of the development of the bio-based economy in the area surrounding the Humber estuary in the North-East of England. The study developed an agent-based model to simulate the evolution of the industrial system. We explain how the qualitative research process led to the development of a toy model that has successively been specified.

1 Introduction

The field of industrial ecology applies ecosystem theory to industrial production, human consumption and societies. As such, the models developed in this field are often conceptual, remote from actors' self-perception and sometimes on the brink of committing a natural fallacy. Yet the dynamics that can be observed in socio-metabolic systems are hugely interesting from an epistemic point of view and indeed for the future of industrialised societies with respect to their ontological dimensions.

Against this background, the Evolution and Resilience of Industrial Ecosystems project (ERIE) at the University of Surrey[1] is carrying out a case study of the area surrounding the Humber estuary in the North-East of England. The area is of interest for a variety of reasons: It hosts two deep water ports, existing chemical industry of national significance, and an agricultural hinterland. All of these factors are relevant in explaining the growth of a bio-based economy in the area. The industries utilising biomass include chemistry, the energy sector and agriculture. All are intricately linked although the policies that drive the development of the system are partly carried out independently. Our research was interested in the analysis of the development of a bio-based economy in the area. In particular, we wanted to explore the evolution of the bio-based economy in the Humber area with the help of an agent-based model.

[1] http//erie.surrey.ac.uk

B. Kamiński and G.Koloch (eds.), *Advances in Social Simulation*,
Advances in Intelligent Systems and Computing 229,
DOI: 10.1007/978-3-642-39829-2_33,

Our methodology consisted of two methods that we applied in three phases. In the first phase we used qualitative semi-structured interviews to elicit relations and interactions between actors that we intended to translate into behavioural agent rules. From this data we created a toy model, which, helped us to specify relevant problems and come up with a refined conceptual model. In the second phase of our research we collected more specific data by means of a participatory workshop. In this workshop we brought together the key stakeholders that we had identified in the first phase. The exercise also gave us an opportunity to receive feedback on our conceptual model. Finally, it established a basis for future collaborative modelling with the stakeholders in phase 3 as we plan to put successive versions of the model online during its development.

Fig. 1. The research stages

2 Information and Data Gathering

Since we had to approach the field with no particular (socio-economic) theory in mind the ERIE team opted for an altered grounded-theory approach [13]. As would be expected it took time and resources to establish working social relationships with relevant people working within the area's industries, and to obtain data about the system.

Our data was mainly collected through semi-structured interviews with 21 stakeholders. We employed two independent coding strategies to analyse the interviews leading to two different outputs: one for an agent-based model and the other for a network model. In what follows we will only be concerned with the former. The first coding strategy utilised established notions of cognition, social networks and institutions as foci [5] to distill relevant information from the interviews. We employed these frames to cover all relevant social aspects that might contribute to the economic-metabolic network but also to establish common ground between the different disciplinary orientations within the team.

Beckert's theory overcomes some of the difficulties associated with social network analysis or institutionalism as stand-alone social theories. It owes to Fligstein's work on fields, which states that a population of actors constitute a social arena by orienting their actions towards each other. Beckert argues that all three structural features, cognition, social networks and institutions, are important to explain the dynamics of markets.

1. Network structures position organisations and individual actors in a structural space. Fields can comprise specific structures of social networks which create power differences between firms and status hierarchies.

2. The relative performance of actors is anchored in regulative institutional rules which allow and support certain types of behaviour while discouraging others.
3. Cognitive frames provide the mental organisation of social environments and thereby contribute to order market fields.

Coding and analyzing the interviews provided us with an understanding of the networks in the area. We learned for instance that two local organisations became particularly important for facilitating the transfer of ecological capability in the area after the abolition of the Regional Development Agency Yorkshire Forward. These were the Humber Chemical Forum (HCF) and the Humber Industry Nature Conservation Association (HINCA). While HCF is highly relevant in developing the skill base for the industries in the area, HINCA has become highly important in overcoming obstacles resulting from the deficits of the planning process. Both organisations facilitate cooperation amongst the companies and the local authorities in the area. HINCA assumed its central role not least due to national and international regulation for protecting bird habitats. Nearly all changes to companies' operations need to go through the planning system which creates uncertainties companies are keen to avoid. Furthermore, dyadic relationships between companies exist that represent industrial symbiosis or the exchange of by-products, e.g. the supply of organic waste. These were promoted by another stakeholder, the National Industrial Symbiosis Programme represented by Linked2Energy in this area. In coding the interviews we defined actors cognition as including environmental management knowledge and skills, eco-innovations and others. Process innovations that we learnt of were all related to technical and social innovations. However, we discovered interdependencies between technical and social aspects of process innovations, e.g. when companies choose a particular form of financing to develop their eco-innovation. Our interviews were less conclusive with respect to institutions impacting on the area, which we ascribed to the change in government that took place immediately after the project started. However, we identified key policies that exerted downward pressures on the industry in the area including financial incentives. Several policy instruments provide various financial incentives to promote the uptake of eco-innovations.

This direct empirical information was supplemented by a literature review of the relevant policy documents. There are a variety of policies from national government and the EU that impact on firms in the area, e.g. both, national and European bodies provide relevant funding for the area that has some effect on a bio-based economy. For the purpose of the model we have only focused on policies supporting the development of a bio-based economy, which are: Landfill Tax, Renewable Obligation (RO), Feed-in Tariffs (FIT), Renewable Transport Fuel Obligation (RTFO), and Renewable Heat Incentive (RHI). These policies influence the incentive landscape to which economic actors adapt[2]. They can be rather easily translated into incentives on which agents in the ABM act.

The resulting information might have lent itself to setting up a SKIN-type of agent-based model [8]. However, two reasons spoke against the implementation of a SKIN model. Firstly, it had become clear from the interviews that the development of a regional bio-based industry was compromised by strong network externalities. Although

[2] We distinguish between actors in the real-world and agents in the model throughout this paper.

these were partly internalised for some biomass flows (e.g. used cooking oil contributed significantly to biofuel production in the UK) they were not internalised for others (e.g. large quantities of wood are imported from Scandinavia). Yet SKIN is not designed to cover negative network externalities. Secondly, the team had a strong interest in policy implementation and the effect of different policy instruments on the different process innovations. This resulted in the conceptual idea of a network (of nodes and ties) that is differentially influenced by the implementation of government policies. Two policy instruments were selected: facilitation of recyclate use (that is source-segregated organic waste that has been recycled and ceased to be considered waste by the regulator) and renewable energy policy.

3 The Toy Model

In specifying the agent-based model we followed ideas presented in [7]. In the toy model we considered only a limited number of actors relevant for the growth of the bio-based economy. We implemented food processors, composters, aerobic digesters, and biofuel refineries. In each time step, the following happens: (i) the waste suppliers, (ii) food processors supply any existing waste contracts that they have, (iii) new waste contracts are made. The exchange of biomass by necessity conforms to a constant material and energy balance. Material flow analyses are the stronghold of industrial ecology and they were integrated in the model by specifying in-flows and out-flows for each type of bio-processor, which allowed tracking the material energetic footprint of the biomass traded.

The maintenance of constant material and energy balance allows tracking of certain ecological effects of network evolution, including carbon emissions. Furthermore, a decision was made that proved to be crucial for the further development of the ABM: agents would exchange materials according to prices. While information on current prices for waste and primary resources was easy to obtain, prices are influenced by factors external to the system. The prices, which translate into costs or revenues for companies are matched by two cost components constituting the firm agent; on-going operational costs and start-up costs, modelled as a loan repayable with interest over time. Using this setup, the financial position of each company is updated each period by the income generated from energy production and any gate fees receivable, less the cost of materials, operating costs and realisation of start-up costs. This position must remain positive, if not the company goes bankrupt and is eliminated from the model. The rules of the agent-based model thus read:

1. The food processors produce waste.
2. Food processors supply any existing waste contracts that they have. Finances are updated accordingly.
3. New waste contracts are made and finances updated accordingly.
4. Food processors update the price that they pay to get rid of waste: if they have managed to get rid of all their waste, then they decrease the price they pay, if they have waste left, then they increase the price they pay.
5. Finances of food processors are updated - they receive money for the products that they develop and pay money to get rid of all remaining waste to landfill.

6. Companies that process waste, process any waste that they have received.
7. Finances of waste producing companies are updated.
8. Any company that has no money goes bust.
9. New contacts are made - how often depends on the contacts-rate slider.
10. New waste processors are created - how often depends on the slider called company-creation-rate. Biodiesel plants are created providing there is enough waste oil available to operate. Either composters or AD plants are started (50/50) depending on whether there is solid waste and, for AD plants, only if a "profitability" measure is satisfied.
11. Update any contract details - end any contracts that have reached the end of their contract period.

The above resulted in a toy ABM, which already reproduces some of the properties that we observed in the real world at the macro-level. For example, we can see from the model the impact that the RO incentives have in reducing organic waste going to landfill, or the volatility in waste prices, and the effect of this in the spread of company profitability. The macro data used for this hands-on validation exists in the form of reports and other qualitative data as well as quantitative data of particular material flows. Additionally, we can probe the relationship between the level of RO incentive and the proportion of anaerobic digesters to composters, and more generally the extent to which policy driven financial incentives affect the financial viability of different agents within the model.

In developing an ABM one problem is the difficulty in distinguishing between endogenous and exogenous drivers because they are commonly entangled in practice. Waste policy had clearly initiated the drive towards a bio-based economy but the introduction of a variety of climate change policy measures overlaid the initial patterns. HCF and HINCA were clearly facilitating endogenous growth of the knowledge network that might be considered one layer of the multiplex regional network, but both stand outside the production network. Furthermore, despite the empirical evidence we found for the coordinating role of HCF and HINCA we are unaware of a defendable method to determine the importance of these actions (in particular not in comparison to the exogenous influence on the network through, for instance, climate change policies). Since the ERIE model was intended to have the primary objective of representing the dynamic flows of biomass in the industrial production system, regulation not directly aimed at these flows has not been considered in the initial toy model. Even if such regulation may have indirectly induced additional and sometimes qualitatively important incentives.

The toy model does not allow discrimination between policy instruments. This does not allow for the intended comparison between a weak (facilitation) and strong policy instrument (economic incentives). Yet, interdependences also existed in practice, for example where facilitation was added to the existing UK waste policy, whereas the economic incentives were introduced as part of climate change policy. This shows the difficulty to empirically distinguish between and analytically model the co-evolution of policies and also the area network and national and international policies. This is partly because political actors (Local Authorities) are only represented in the model as suppliers of organic waste. They do not appear in their regulatory function, where they are left with the obligation to implement waste policy. By contrast the framework for waste and climate change policy is set at the national, European level, and indeed globally (Kyoto protocol).

Another reason for revising the model is that while policy-makers do consider interdependencies between the various instruments to some degree in the process of policy design, policies are u sually analysed using general equilibrium models. There are multiple financial incentives (including landfill tax escalator, ROCs, FITs, RTOCs and RHI) that overlap although the instruments are designed to act differentially on different actors. Yet for all practical purposes the micro-economic policy instruments are simply simulated with a macro-economic equilibrium model (MARKAL model). These considerations led to a more basic revision of the concept behind the toy model. We tried to separate more clearly endogenous and exogenous drivers and indeed, we came to conceptualise the network as an emerging market.

4 Refining the Concept

In order to distinguish more clearly between endogenous and exogenous drivers we saw the need to revisit the emergence of a network from the behaviour of individual firms. Parts of the emerging market for organic waste have a specific trait: agreed sales prices for bio waste are unknown to competitors. This situation very much resembles a fish market, where individual bids are made secretly (no auction). Such a market has been modelled empirically and it is considered valid economic theory [16]. This context however, is only true with regard to a fraction of processed biomass, namely the food waste arising from food processing. The situation is different for all other forms of biomass since these are traded nationally and globally (even if some supply is provided locally). In our revised concept of the agent-based model there are four providers of biomass: importers, forestry/agriculture, food processors, and municipalities (see Figure 2). The biomass they produce can be specified further. Food processors produce biodegradable waste and oil. The former requires a quick turnover while the latter can be stored for some time. Forestry and agriculture provide input such as corn, straw, and wood, which can all be stored but which have some seasonality attached to them in the case of the first two. By contrast, wood can be supplied according to demand and left standing as a living forest or stored to wait for a better price. Similarly, soy, rape and palm oil that importers provide can be sold according to world market prices. Thus, there are distinct differences between bilateral price negotiations and contracts that are informed by a global or regional price (a price known by all in an area).

In the interviews we learnt of three types of facilitators: consultants who can be divided into consultants who simply arrange contracts and oil brokers who also own the biomass. Both types of facilitators observe market price development of the different biomass primary resources for producers (mostly from the energy sector). Then we also have Link2Energy/NISP that used to offer a free service (mostly between composters and anaerobic digestion plants and their suppliers). Additionally, processors may decide to set up contracts by themselves. Drawing on external facilitators or setting up supply on their own corresponds to different transaction costs for all processors. These costs will differ for each company depending largely on the amount of biomass they require and the scale of the supply.

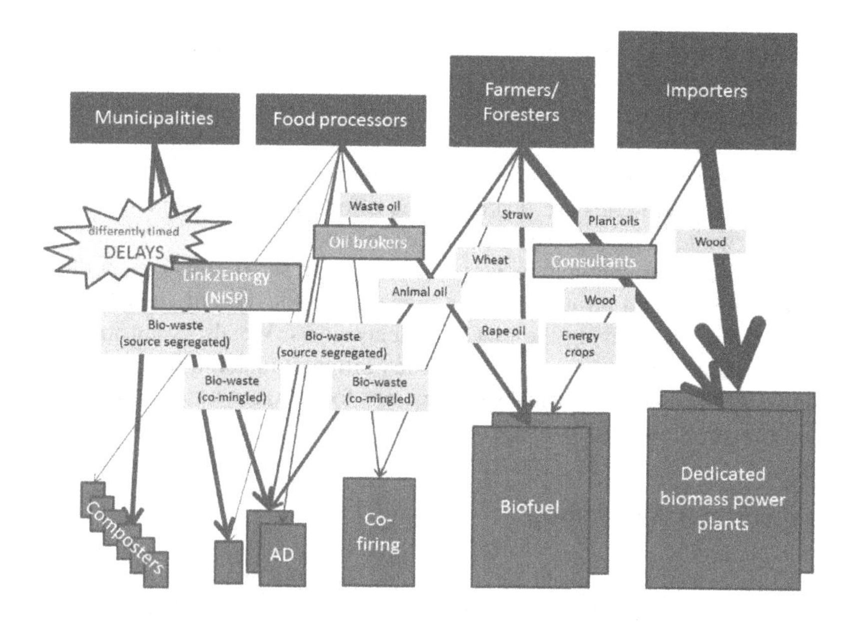

Fig. 2. Different interactions between companies and biomass flows

The different processors have sector specific ways of buying in supply. Dedicated biomass power plants secure their supply chain with long-term contracts of several years (as long as possible without freezing the market price). The same is true for biofuel producers. The situation is slightly more relaxed for anaerobic digesters since their capital costs are lower. However, they process considerable quantities of organic waste that are usually governed by long-term contracts (of up to 20 years). Composters and co-firing plants can take in biodegradable waste quite spontaneously (and potentially sporadically in the case of co-firing). They don't require upfront investment but simply have to calculate their fees against operational costs. As this already indicates, the duration of contracts with suppliers very much depends on the size of the processor and the specifics of the biomass resource.

Furthermore, the revised agent-based model will distinguish between material/energy and information flows. The use of biomass is path-dependent (cascading) or may entail trails for recycling organic waste whereas the diffusion of information is not. Following [6] we can discriminate between walks, trails, and paths. Walks are sequences of flows where nodes as well as ties are passed time and again e.g. money handed on in payments. Trails are sequences where nodes, but not ties, are passed repeatedly. Paths are sequences in which neither nodes nor ties are passed more than once. The flows are transferred in their entirety. Material/energy flows generally belong to this last category representing the irreversible dissipative economic structures of the real world. Although recycling reintroduces trail characteristics these may be neglected since we don't simulate regional energy supply or nutrient cycles.

With regard to information, [6] distinguishes between serial and parallel processing. The diffusion of information (price, quality and quantity) of agents operating in the waste context can be represented as a serial diffusion process (perhaps involving some kind of learning). By contrast the diffusion of information of consultants might be represented as a parallel process in which agents basically have access to the same information (prices in a global market). This leaves us with the oil brokers. Unlike facilitators and consultants, oil brokers could behave strategically and hold the flow (subject to long-term supply contracts) since they own the resource. (From the interviews we learnt that biofuel refineries are making some efforts for vertical integration but the high price for waste oil appears to hamper these attempts.)

To summarise, information is seen to spread through serial processes in the case of biodegradable waste and through parallel processes in the case of consultants and oil brokers with the latter having the power to delay flows. We also want to distinguish between: suppliers, facilitators, and processors, different material inputs and time delays. Yet we needed more empirical specifications for an agent-based model, which we obtained from the stakeholders in a workshop.

5 Stakeholder Workshop

It is widely accepted that stakeholders can bring valuable first-hand knowledge to the research process, can meaningfully contribute to selecting a model and developing it, can help in collecting data and integrating it, to develop scenarios, interpret results, and formulate collective strategies or policy alternatives [3,12]. On the other hand engaging stakeholders is time-consuming, may bring plural perceptions to the research process rather than unambiguous data, can be difficult to manage if it is not increasing the chances of project failure altogether, and might be difficult to carry out in monodisciplinary research projects [9,17]. Despite these potential pitfalls "participatory modelling, with its various types and clones, has emerged as a powerful tool that can (a) enhance the stakeholders knowledge and understanding of a system and its dynamics under various conditions, as in collaborative learning, and (b) identify and clarify the impacts of solutions to a given problem, usually related to supporting decision making, policy, regulation or management." [14].

This constitutes the context for the participatory workshop that we carried out in January 2013. We sought direct feedback on the existing conceptualisation of the agent-based model by presenting the concept to the stakeholders and prompting responses. The presentation of the concept as well as the introduction to the existing toy model passed the plausibility test as both were approved by the stakeholders. We also used the workshop to gather more information and data needed to empirically specify the agent-based model. Initially we had planned to divide the stakeholders into four different groups each group working through a particular set of questions to provide information and in some cases basic data. The topics of each group were:

1. Biomass Specification Organic Waste and Virgin Input
2. Firm Specification Bio-Processors
3. Specification of Service Providers Facilitators and Brokers
4. Policy Questions & Measures of Success

Each of these headlines involved several questions. Some of these were more qualitative in nature whereas others were more quantitative. We ask for the specification of biomass (1) in order to model delays that translate into dynamics in the system (e.g. supply chain fluctuations; price volatility). The specificities of the biomass may amplify or reduce these dynamics. The firm specification (2) related to the size and capital structure of companies and the agents properties in the model. We were also interested in obtaining information about the service providers (3). However due to an unexpectedly low turnout of stakeholders we decided to cancel this group and obtain this information in future via other means. Finally, we asked questions on policy in order to validate our concept according to these criteria later in the project (4).

The first two topics proved challenging because they involved a long list of sub questions. We collected data on the capital costs of the different technologies and on specific resource prices. Furthermore, we learnt that biomass processors can set up different production lines for biomass and oil to derive their intermediate inputs from different sources and to differentiate their products according to market price for either product category. This seemed to provide qualitative evidence for the hypothesis that price volatility is indeed a concern for producers. We also obtained some confirmation for properties demonstrated by the toy model. We heard that biofuel producers were indeed operating at a very tight margin as suggested by the model and that the path-dependencies we had observed appeared reasonable to stakeholders.

Overall we considered the workshop a success despite the fact that the turnout was lower than expected from advance confirmations. This was not least because stakeholders expressed interest in providing input into successive versions of the agent-based model, which will be made available to them online in the near future.

6 Revising the Rules of the ABM

Some of the information gathered at the workshop can directly be used to revise agents and their decision rules, e.g. we learnt about different processors' cost structures, durations of supply contracts according to processor type, and processors' flexibility in using different inputs. By contrast, other aspects can only be implemented together with the above conceptual approach, e.g. the effects of seasonality of (primary) biomass or the possibility to store biomass (thus mitigating possible supply volatility). (The toy model only covered organic waste without delays.) The revised ABM will be able to model the different streams and markets. These are partly interdependent and the interdependency between markets for primary biomass and secondary (recycled) biomass is crucial to keep the development of renewable energy from biomass on a sustainable trajectory. (It is also of considerable practical interest for instance for the stakeholders.) Interdependence is generally typical for markets emerging from the internalisation of environmental externalities. Another benefit of the above conceptual approach of differentiated flows is that it could also take the spread of pathogens into account. By contrast this is outside the realm of economic models.

- Food processors seek contracts (with introduction of waste tax) and individual offers are made secretly (no auction) Municipalities seek contracts and suitable (scale) processors.

- All processors seek suppliers. Depending on the scale of operation they preferably use the intermediaries – true for all processors equal or bigger than anaerobic digestors.
- In network terms the emerging process might be seen as a diffusion of information (price, quality and quantity). Agents operating in the waste context would be represented as a serial diffusion process (perhaps with some kind of learning). By contrast the diffusion of information of consultants might be represented as a parallel process in which agents basically have access to the same information (prices in a global market). This leaves us with the oil brokers. Unlike facilitators and brokers they could behave strategically and hold the flow (subject to long-term supply contracts) since they own the resource.

7 Discussion

The ERIE project progressed in two phases. Both relied on qualitative methods to define rules for the agents of an agent-based model. Despite this similarity we consider the two phases as distinct. The first phase consisting of semi-structured interviews effectively enabled us to understand relevant aspects of the developing bio-based economy in the Humber area. Whereas we had a clearly defined theoretical framework and collected sufficient data from the standpoint of qualitative social research we did not obtain enough data to create a defensible and relevant agent-based model. Instead we had to acknowledge that networks and institutions are empirically more complex than cognition resulting in an under-defined toy model. This led to a second phase in which we redesigned our conceptual model according to empirical problem observation and economic literature. We entered a third phase where we engage in a participatory modelling process, which is on-going. One of the reasons for choosing a participatory method over conventional qualitative methods such as interviews is the higher information density achievable with participatory exercises.

Once implemented in an agent-based model the empirically established rules should already result in complex dynamics such as market price fluctuations. The aim is to detect price volatility emerging from these interactions and mitigating effects resulting from e.g. long-term contracts or government policies. Price volatility is a widely recognised problem for companies and may become more pertinent in future as some studies suggest a strong relationship between bio energy and oil prices [10]. Furthermore, price volatility is a particular problem of secondary material (recyclate) markets more generally because these markets often lack market clearance of waste outputs according to supply and demand [1,2,4]. For this reason the UK government has endorsed the creation of recyclate markets from early on [15] by creating for instance one of the facilitators (NISP) we saw in the area network. This might be of less concern in respect to organic waste as it can either be recycled to land (source-segregated organic waste) or brown field sites (co-mingled organic waste). There is thus a host of questions around the evolution of the bio-based economic system we can now explore with the revised agent-based model.

We want to pursue three research questions in particular: firstly, we want to understand path dependency in the evolution of the bio-based economy in the Humber area.

We conceive of path dependency as emerging from the adaptation of companies to different incentives that are introduced sequentially. The concept of a bio-based economy involves biomass processing industries such as composters, anaerobic digesters, biofuel plants, bio plastics producers, and others [11]. While the processing options imply irreversible cascades of biomass use, any real-world implementation may fail to fully exploit the available technological potential if lock-ins occur early in the development. In addition to that, we would like to know whether the sequence in which the policies were introduced had a positive (dampening) effect on price volatility or what else might dampen price volatility (e.g. contract duration). Obviously, this is important for future policy implementation as it may allow or disallow the exploitation of the full technological and economic potential of biomass use.

Secondly, we are interested in explaining how resilient the system is to external shocks in the short term, e.g. seasonal loss of agricultural land due to flood, and how robust to long term stress e.g. of rising oil prices or permanent loss of agricultural land? We are thus interested in endogenous and exogenous shocks.

Finally, we want to know whether there are negative network externalities that might reintroduce price volatility (e.g. of biofuel)? While we believe that there is evidence for this in regard to the use of used cooking oil for biofuel production (e.g. sharp price increase in the last four years) we need to understand whether negative network externalities also applied to other biomass flows and indeed the bio-based economy at the Humber as a whole.

8 Conclusions

As this work has progressed we have become interested in the emerging interdependencies between the different biomass flows in the development of a bio-based economy, in particular in the sequence of the earlier waste policy and the later renewable energy policy. This problem, which resulted in a restated and better specified research question, was not obvious from the outset of the research project. In this respect the first phase of the research project represents a search for relevance. The second phase is now quickly leading to an agent-based model capable of answering very specific research questions. The re-conceptualisation of the agent-based model will allow a relevant simulation of the bio-based economic system. It will reflect endogenous development of the network more effectively, as much as complexity is now emerging from the interaction of economic actors. This should allow us to arrive at more reliable predictions of effects from different policies.

References

1. Ackerman, F., Gallagher, K.: Mixed signals: market incentives, recycling, and the price spike of 1995. Resources, Conservation and Recycling 35(4), 275–295 (2002)
2. Angus, A., Casado, M.R., Fitzsimons, D.: Exploring the usefulness of a simple linear regression model for understanding price movements of selected recycled materials in the uk. Resources, Conservation and Recycling 60, 10–19 (2012)

3. Antona, M., d'Aquino, P., Aubert, S., Barreteau, O., Boissau, S., Bousquet, F., Daré, W., Etienne, M., Le Page, C., Mathevet, R., Trébuil, G., Weber, J.: Our companion modelling approach. Journal of Artificial Societies and Social Simulation 6(1) (2003), `http://jasss.soc.surrey.ac.uk/6/2/1.html`
4. Baumgärtner, S., Winkler, R.: Markets, technology and environmental regulation: price ambivalence of waste paper in germany. Ecological Economics 47(2-3), 183–195 (2003)
5. Beckert, J.: How do fields change? the interrelations of institutions, networks, and cognition in the dynamics of markets. Organization Studies 31(5), 605–627 (2010)
6. Borgatti, S.P.: Centrality and network flow. Social Networks 27(1), 55–71 (2005)
7. Gilbert, N.: Agent-Based Models (Quantitative Applications in the Social Sciences). Sage Publications, Inc., Thousand Oakes (2008)
8. Gilbert, N., Pyka, A., Ahrweiler, P.: Innovation networks – a simulation approach. Journal of Artificial Societies and Social Simulation 4(3) (2001), `http://jasss.soc.surrey.ac.uk/4/3/8.html`
9. Hirsch Hadorn, G., Bradley, D., Pohl, C., Rist, S., Wiesmann, U.: Implications of transdisciplinarity for sustainability research. Ecological Economics 60(1), 119–128 (2006)
10. Kranzl, L., Kalt, G., Diesenreiter, F., Schmid, E., Stürmer, B.: Does bioenergy contribute to more stable energy prices? In: 10th IAEE European Conference (2009)
11. Nuss, P., Bringezu, S., Gardner, K.H.: Waste-to-materials: the longterm option. In: Karagiannidis, A. (ed.) Waste to Energy: Opportunities and Challenges for Developing and Transition Economies, pp. 1–26. Springer (2012)
12. Ramanath, A.M., Gilbert, N.: The design of participatory agent-based social simulations. Journal of Artificial Societies and Social Simulation 7(4) (2004)
13. Strübing, J.: Grounded theory. VS Verlag für Sozialwissenschaften, Wiesbaden (2004)
14. Voinov, A., Bousquet, F.: Modelling with stakeholders. Environmental Modelling & Software 25, 1268–1281 (2010)
15. Watts, B.M., Probert, J., Bentley, S.P.: Developing markets for recyclate: perspectives from south wales. Resources, Conservation and Recycling 32(3-4), 293–304 (2001)
16. Weisbuch, G.G., Kirman, A., Herreiner, D.: Market organisation and trading relationships. The Economic Journal 110(463), 411–436 (2000)
17. Zierhofer, W., Burger, P.: Disentangling transdisciplinarity: An analysis of knowledge integration in problem-oriented research. Science Studies 20(1), 51–72 (2007)

Author Index